Approaches to Global Sustainability, Markets, and Governance

Scopus indexed *Approaches to Global Sustainability, Markets, and Governance* book series takes a fresh and global approach to issues of corporate social responsibility, regulation, governance, and sustainability. It encompasses such issues as: environmental sustainability and managing the resources of the world; geopolitics and sustainability; global markets and their regulation; governance and the role of supranational bodies; sustainable production and resource acquisition; society and sustainability.

Although primarily a business and management series, it is interdisciplinary and includes contributions from the social sciences, technology, engineering, politics, philosophy, and other disciplines. It focuses on the issues at a meta-level, and investigates the ideas, organisation, and infrastructure required to address them.

The series is grounded in the belief that any global consideration of sustainability must include such issues as governance, regulation, geopolitics, the environment, and economic activity in combination to recognise the issues and develop solutions for the planet. At present such global meta-analysis is rare as current research assumes that the identification of local best practice will lead to solutions, and individual disciplines act in isolation rather than being combined to identify truly global issues and solutions.

This book series is part of the Social Responsibility Research Network (srrnet.org), a worldwide organisation with several thousand members. The network organises annual conferences, held at various locations around the world and many books come from these conferences. Submissions to the current conference are invited and details can be found at srrnet.org. Additionally submissions of chapters, or book proposals are always welcome at info@srrnet.org. If you are interested in contributing a monograph or curated volume to the series, please contact the Springer publishing editor.

Rubee Singh · David Crowther
Editors

Climate Neutrality and Sustainable Ecosystems

Practical Approaches to Addressing Global Warming

Springer

Editors
Rubee Singh
Institute of Business Management
GLA University
Mathura, Uttar Pradesh, India

David Crowther
Social Responsibility Research Network
UK

ISSN 2520-8772 ISSN 2520-8780 (electronic)
Approaches to Global Sustainability, Markets, and Governance
ISBN 978-981-96-7180-9 ISBN 978-981-96-7181-6 (eBook)
https://doi.org/10.1007/978-981-96-7181-6

This Springer imprint is published by the registered company Springer Nature Singapore Pte Ltd.
The registered company address is: 152 Beach Road, #21-01/04 Gateway East, Singapore 189721, Singapore

Preface

Climate neutrality and sustainable ecosystems are two of the most critical goals in the global fight against climate change. Climate neutrality refers to achieving net-zero carbon emissions, balancing the amount of carbon released into the atmosphere with an equivalent amount removed or offset. It requires radical shifts in how societies produce and consume energy, transport goods, and use natural resources. Sustainable ecosystems, on the other hand, emphasize the protection and preservation of natural habitats and biodiversity while promoting practices that allow human development to coexist with environmental stewardship. Together, these concepts form the foundation for a global strategy to address the mounting threats of global warming. Climate neutrality demands collaboration, innovation, and leadership at every level of society. The chapters within this book are designed to offer a comprehensive and interdisciplinary approach to addressing the multifaceted nature of global warming and sustainable ecosystem management.

This book, *"Climate Neutrality and Sustainable Ecosystems: Practical Approaches to Addressing Global Warming"*, is divided into three parts, each addressing distinct yet interconnected areas of this monumental challenge. The first part assesses the effects of climate neutrality on a global scale, emphasizing the role of governments, international organizations, and policies in driving climate action. It explores the challenges faced by developing nations, the importance of governance and resilience in vulnerable regions, and how global leadership can facilitate meaningful progress. This section provides an in-depth examination of how coordinated efforts across borders can lead to effective climate mitigation. The second part of the book focuses on the implementation of sustainability strategies in response to global warming. By highlighting the role of human capital, emerging technologies, and sectoral transformations in industries like energy, tourism, and transportation, this part illustrates how practical measures can be taken to meet climate neutrality goals. It also emphasizes the need for collaboration between public and private sectors, grassroots movements, and individuals in achieving global sustainability targets.

The third and final part examines the role of specific sectors in advancing climate neutrality and protecting ecosystems. This includes the integration of climate goals into national development strategies, the mobilization of climate finance, and the critical importance of green space, biodiversity, and natural resource conservation. The discussions provide practical insights into how various sectors, from finance to agriculture, can contribute to a climate-neutral future while ensuring the health and sustainability of ecosystems. This book draws on a wide range of global case studies and research from diverse regions, including India, Indonesia, Romania, Turkey, Uganda, Nigeria, Sri Lanka, UK, Maldives, among others. These examples highlight how different countries, each with unique challenges and opportunities, are addressing climate neutrality and sustainable ecosystem goals. By examining climate policies, sustainable development practices, and sectoral transformations in regions such as Sub-Saharan Africa, Southeast Asia, and Eastern Europe, the book provides a rich, comparative analysis of the strategies being employed around the world to combat global warming.

The importance of this book lies in its practical, solution-driven approach. While the challenge of climate change is vast, this book offers hope by presenting actionable strategies, backed by real-world examples, that can be adopted by governments, industries, and individuals alike. The interdisciplinary nature of the book provides readers with a comprehensive understanding of how different sectors and regions can come together to combat global warming. Policymakers, industry leaders, researchers, environmental advocates, and students will all find value in this book. It offers insights into the complex dynamics of achieving climate neutrality while protecting ecosystems, and it provides a clear roadmap for action. By addressing both the systemic changes required and the localized, sector-specific solutions, this book equips readers with the knowledge and tools necessary to make meaningful contributions to the global effort against climate change.

In a time when the impacts of global warming are being felt more acutely across the world, this book stands as a vital resource for those looking to be part of the solution. It emphasizes that the journey towards a climate-neutral future is not just possible but essential, and that with innovation, collaboration, and commitment, a sustainable and resilient future can be achieved. Let this be not only a call to action but a roadmap to the solutions we need, recognizing that the fight against global warming is one we must win—together.

Mathura, India — Rubee Singh
London, UK — David Crowther

Acknowledgements

We extend our heartfelt gratitude to all those who have contributed to the creation of this book, "*Climate Neutrality and Sustainable Ecosystems: Practical Approaches to Addressing Global Warming*". Our utmost appreciation goes to the climate scientists, researchers, and sustainability experts whose unwavering dedication to advancing climate solutions forms the backbone of this work. Their invaluable insights, rigorous research, and practical experiences have greatly enriched the content of this book, ensuring its relevance and applicability in tackling real-world environmental challenges.

We are especially grateful to the policymakers and leaders who have been at the forefront of integrating climate neutrality principles into strategic governance. Their dedication to developing policy frameworks that encourage global action on climate change has inspired much of the work presented here, offering a hopeful glimpse into the transformative potential of global cooperation and policy integration. It is our sincere hope that the insights within this book will spark meaningful dialogue, inspire innovative solutions, and catalyze tangible progress towards achieving climate neutrality and creating a more sustainable future for all.

Finally, we would like to express our immense gratitude to the entire team at Springer Nature, especially series editors Prof. David Crowther, Dr. Shahla Seifi, and Executive/Publishing Editor Ms. Mamun Mishra, whose professionalism and dedication have ensured the timely and successful publication of this work. Without their guidance and support, this project would not have been possible. Thank you for your continued belief in the importance of addressing global climate challenges through collaboration, research, and innovation.

Rubee Singh
David Crowther

About This Book

Climate Neutrality and Sustainable Ecosystems: Practical Approaches to Addressing Global Warming offers an in-depth exploration of the pathways to achieving climate neutrality and creating sustainable ecosystems in response to the escalating global climate crisis. This book examines the interdisciplinary approaches required to reduce carbon emissions, improve resilience, and promote sustainable development across various sectors. Divided into three parts, the book begins by assessing the broader impacts of climate neutrality initiatives. It explores the role of government leadership, strategic policy integration, and the challenges faced by developing countries, particularly in regions like Sub-Saharan Africa and the Indo-Pacific. Additionally, it addresses critical areas such as ocean health, migration, and green growth strategies.

It also describes practical implementation strategies for sustainability, highlighting the significance of human capital, emerging technologies, and sectoral transformations in industries such as tourism, energy, and transportation. By presenting global case studies and solution-based frameworks, it emphasizes how international cooperation, grassroots movements, and innovative policies can help balance economic growth with environmental sustainability. In the final part, the book focuses on specific sectors that play pivotal roles in achieving climate neutrality, including finance, agriculture, infrastructure, and biodiversity conservation. It provides actionable insights into integrating climate goals within national development plans, leveraging green finance, and protecting natural ecosystems.

With a focus on solutions, collaboration, and innovation, this book serves as a comprehensive guide for policymakers, industry leaders, researchers, and environmental advocates. It provides the tools and knowledge necessary to build a sustainable, climate-neutral future, addressing both the global and regional challenges posed by global warming.

Contents

About the Editors

Rubee Singh is an Assistant Professor of Human Resource at Institute of Business Management, GLA University India. She is a Post-Doctoral (D. Lit) Researcher at Department of Management Studies, Kumaun University, Nainital, India. She holds a Ph.D. in Management from Noida International University India, an MBA in HR from Dr A.P. J Abdul Kalam Technical University India. She has more than 11 years of experience in research and teaching. Apart from teaching and research, she has handled many administrative positions in premier institutes affiliated with A.K.T.U. She has conducted a number of FDPs, Conferences, Workshops, and HR Conclaves in the area of Sustainability, CSR, AI, HR, Human Factors, Women Empowerment and Human rights. She is a United Nations Volunteer, and a contributing author for Voices of Youth (UNICEF). She is an honorary life member of Women Economic Forum and National Secretary of IATM (Youth Wing) International Anti-Terrorism Movement India. Dr. Singh is a Managing Editor of IJARSH India. She is serving on the Editorial Board of more than 30 international journals. She serves as the Advisory Board Member of the *Heliyon Journal*, Elsevier, *Research in Globalisation*, Elsevier, *Circular Economy and Sustainability*, Springer and reviewer of many reputed journals like *European Journal of Innovation Management (EJIM), British food journal, Sustainability* etc. She is life member of International Association of Engineers, Thailand, IMRTC USA, ISRD UK etc.

She is a bestselling author with significant experience as an excellent Editor, Eminent researcher, great academician, Global Ambassador (Children's) and Social Activist. She has presented many papers at national and international conferences including IIT Bombay, IIM Calcutta, published 25 academic books, and more than 100+ research papers in Scopus, SCI, ABDC-A, & UGC index journals. She has an excellent research track record with 1500+ citations on Google Scholar and H-index-25. Her areas of research are sustainability, Digital HRM, Sustainable HRM, Circular Economy, Digital CE, Net Zero, Sustainable Innovation, Digital Innovation, Eco-Innovation, Industry 4.0, Artificial Intelligence, Blockchain and Corporate Social Responsibility. Being a thought leader, her book titled "Training and Development" was added to the MBA syllabus of "The Institute of Aeronautical Engineering" (Jawaharlal Nehru Technological University, Hyderabad). Dr. Singh has been honoured with many prestigious awards, which include Best Researchers consecutive award of the Year 2022–23, and 2023–24 by GLA University, Child Ambassador Award of The Year 2022, Iconic Faculty Award 2019, International Young Researchers Award-2021, and Young Scientist Award 2020. The list of awards and appreciation is long, but Dr. Singh has always prioritized her work over recognition, making her unique and commendable in her field.

David Crowther is expert in various fields: he is a specialist in business administration, a games theoretician and a psychologist while also a qualified accountant who had worked as an accountant, systems specialist and general manager in local government, industry and commerce for over 20 years. After a number of years in the financial services sector, including a spell as divisional managing director during which he set up and ran a credit card scheme, he decided to leave the business world and become an academic. In 1994, he joined Aston University and there obtained a Ph.D. in 1999 for research into corporate social performance. In 2001, he became the first Professor of Corporate Social Responsibility in the world. He now holds various positions as Professor as well as being visiting professor in various institutions. He gives advice as Chief Research Fellow at the Lithuanian Institute of Agrarian Economics and has consulted extensively around the world. At Ansted University he is also a member of the Board and President of the Sustainability Research Institute. He has consulted widely including to US Presidential candidates, UN, EU and governments. Prof. David has published more than 70 books and has also contributed more than 500+ research papers to academic, business and professional journals and to edited book collections. He has also spoken widely at conferences and seminars worldwide and acted as a consultant to a wide range of government, professional and commercial organizations. In 2003 he organized the first ever CSR conference and as a result of this he established the Social Responsibility Research Network and also launched *Social Responsibility Journal*, the premier journal in its field. In 2010, he established the Organisation Governance Network and conference series. He is listed in Who's Who in the World and various other directories. His research is into sustainability, governance and corporate social responsibility with a particular emphasis on the relationship between social, environmental and financial performance. His latest project has been to edit and produce the Palgrave major reference work on CSR.

Part I
Assessing Climate Neutrality Change Effects

Chapter 1
Government Partners in Climate Neutrality: Leadership, HRM, and Strategic Policy Integration

Cumali Kılıç and Gazi Kurnaz

Abstract The pursuit of climate neutrality, as a critical component of sustainable development, necessitates a strategic and comprehensive approach that integrates management practices, organizational strategies, and human resource management (HRM). It is imperative that governments and businesses collaborate to establish frameworks that facilitate the transition toward a climate-neutral economy. This transition is contingent upon robust leadership within organizations, wherein a dedication to sustainability is evidenced by the formulation of explicit environmental objectives, the allocation of suitable resources, and the integration of climate-related goals into overarching business strategies. The role of HRM is pivotal in this transformation, as it fosters a culture that aligns employee motivation and engagement with sustainability efforts, thus enhancing organizational performance in environmental initiatives. Moreover, the strategic integration of environmental management into core business operations has shifted from being a peripheral concern to a central organizational priority. This shift necessitates the development of innovative business models that not only prioritize sustainability but also achieve profitability. Thus, collaboration between government entities and businesses plays a crucial role in fostering an environment conducive to climate action. Businesses, through their operational insights and innovation capabilities, can offer critical feedback to policymakers, ensuring that climate-related regulations are both practical and effective. Therefore, the integration of climate neutrality into strategic government policy is not only an organizational imperative but also a societal one, requiring a multifaceted approach involving leadership, HRM, and government-business partnerships to achieve lasting climate-neutral outcomes.

C. Kılıç (✉) · G. Kurnaz
Faculty of Economics and Administrative Sciences, Department of Business Administration, Artuklu University, Mardin, Turkey
e-mail: cumalikilic@artuklu.edu.tr

G. Kurnaz
e-mail: gazi@artuklu.edu.tr

R. Singh and D. Crowther (eds.), *Climate Neutrality and Sustainable Ecosystems*, Approaches to Global Sustainability, Markets, and Governance,
https://doi.org/10.1007/978-981-96-7181-6_1

Keywords Climate neutrality · Public-private partnerships · Sustainability leadership · Human resource management · Strategic policy integration

1.1 Introduction

The incorporation of climate neutrality into strategic government policy represents a crucial stride in the direction of confronting the global challenge of climate change. Attaining this objective necessitates a comprehensive strategy that encompasses a multitude of domains, including managerial practices, organizational strategies, and human resource management (HRM). This multifaceted approach is of critical importance for the establishment of effective partnerships between government institutions and businesses, which are pivotal actors in the transition to a climate-neutral economy. Organizations that demonstrate robust leadership are more likely to successfully implement climate initiatives, as leadership commitment serves as the foundation for clear sustainability objectives and the requisite resource allocation to achieve these goals (Duong, 2023; Zhang et al., 2022). In addition to establishing sustainability targets, effective leadership ensures that these objectives are fully integrated into the organization's overarching business strategy, thereby creating a coherent and aligned approach to addressing climate challenges.

An organization's culture plays a significant role in reinforcing its commitment to climate goals. When the organizational culture is aligned with sustainability objectives, employee engagement and motivation increase, leading to enhanced performance in achieving environmental outcomes (Underhill et al., 2015). Organizations that foster a positive human resources (HR) climate—characterized by supportive HRM practices—tend to exhibit stronger environmental performance. Employees in such environments feel empowered to contribute to sustainability efforts, knowing that their input is valued and aligned with the organization's broader goals (Varnali, 2015). This synergy between leadership, culture, and employee motivation underscores the importance of embedding sustainability into the organizational ethos, as it not only drives performance but also fosters a sense of purpose and engagement among employees.

In addition to the importance of leadership, the strategic integration of environmental management into the core business operations of an organization has become increasingly significant. An increasing number of companies are acknowledging the necessity of integrating environmental management into the core of their business strategies rather than treating it as a mere peripheral issue (Chang & Yoo, 2023). This shift is driving the development of innovative business models that prioritize sustainability while maintaining profitability. For example, organizations that adopt green supply chain management practices are better positioned to improve resource efficiency, reduce waste, and enhance their competitive advantage in an environmentally conscious market (Nikseresht et al., 2023). Nevertheless, the implementation of these practices necessitates a comprehensive grasp of supply chain dynamics and the capacity to engage in efficacious collaboration with a diverse array of stakeholders, including suppliers and customers, in order to establish sustainable value chains (Chen et al., 2017; Manavalan & Jayakrishna, 2019).

HRM also plays a pivotal role in the transition toward climate neutrality. HRM practices that emphasize sustainability, such as training programs focused on environmental awareness and skills development, can significantly enhance an organization's ability to meet its climate goals (Akar & Bedük, 2023). Moreover, fostering a sustainability-focused culture within an organization not only improves employee retention and job satisfaction but also attracts individuals who value environmental stewardship, thereby contributing to the organization's long-term sustainability efforts (Underhill et al., 2015). Integrating sustainability into HRM practices is not just beneficial for organizational performance; it is also a key component in supporting broader societal goals related to climate change mitigation (Ravesangar, Ping, & Pachar, 2024).

Furthermore, strategic collaboration between businesses and government entities is indispensable for creating an environment that supports climate action. Government policies that provide incentives for sustainable practices—such as tax breaks for green investments or grants for renewable energy projects—can motivate businesses to adopt climate-neutral strategies (Andre & Valenciano-Salazar, 2022; Nwokolo et al., 2023a). Simultaneously, businesses can contribute to the policy-making process by offering insights based on their practical experiences, ensuring that regulations are designed to be both effective and feasible (Kumar et al., 2020). Such collaborative endeavors can result in the formulation of comprehensive climate policies that address the specific needs of both the public and private sectors.

Ultimately, the integration of climate neutrality into strategic policy requires a cohesive approach that encompasses leadership, management practices, and HRM. By promoting robust leadership, aligning business strategies with sustainability objectives, and fostering collaboration between government and business entities, organizations can play a pivotal role in facilitating the transition to a climate-neutral economy. This comprehensive and coordinated approach has the dual benefit of enhancing organizational performance and contributing to the broader societal imperative of mitigating climate change and achieving long-term sustainability.

1.2 Theoretical Approach

The theoretical foundation of climate neutrality policy implementation underscores the vital necessity of multistakeholder collaboration to effectively address the multifaceted challenges posed by climate change. The integration of climate neutrality into strategic policy frameworks necessitates a unified effort involving government agencies, businesses, non-governmental organizations, and local communities. It is imperative that these entities collaborate to develop and implement solutions that address the pressing need for sustainability. The formation of strategic partnerships, the allocation of human resources, and the provision of leadership are indispensable elements of this process. Their harmonization is of paramount importance for guaranteeing that climate policies are not only viable but also efficacious. This section examines the theoretical foundations of these three elements, emphasizing their importance in advancing climate-neutral initiatives and sustainable organizational practices.

1.2.1 Strategic Partnerships for Climate-Neutral Policy Implementation

The establishment of strategic partnerships is a fundamental aspect of the successful implementation of climate-neutral policies. These partnerships facilitate collaboration among a diverse range of stakeholders, including government agencies, private sector entities, non-profit organizations, and local communities (Van Langen et al., 2021). These partnerships are of paramount importance in facilitating the mobilization of resources, the dissemination of knowledge, and the enhancement of capacity, all of which are vital for the effective response to the complex challenges posed by climate change (Golhasany & Harvey, 2023).

A key advantage of strategic partnerships lies in their ability to leverage the strengths of different actors. For instance, public-private partnerships (PPPs) combine the innovation and efficiency of the private sector with the regulatory and policy-making expertise of government entities (Dolla & Laishram, 2020). Research indicates that such collaborations can lead to improved planning and development processes, particularly in urban settings where climate risks are prevalent (Broto & Bulkeley, 2013; Taylor & Harman, 2016). By engaging in consensus-building, these partnerships can create a neutral ground for stakeholders to share information and build authentic relationships, ultimately leading to more effective climate action (Clarke & MacDonald, 2016).

Moreover, partnerships can play a significant role in overcoming the regulatory and implementation deficits often associated with state-led climate policies. Cross-sector partnerships have been identified as a mechanism to accelerate climate change mitigation by addressing gaps in participation and resource allocation (Forsyth, 2010). For example, partnerships that focus on nature-based solutions can foster innovative governance approaches that integrate environmental considerations into urban planning and development (Ham & Klimmek, 2017; Kabisch et al., 2016). These partnerships are particularly important in cities, which have emerged as key actors in climate governance due to their substantial greenhouse gas emissions and their vulnerability to climate-related impacts (Streck, 2021).

By fostering collaboration between diverse stakeholders, including grassroots organizations, strategic partnerships can enhance the inclusivity and effectiveness of climate policies. In particular, grassroots organizations offer invaluable insights into the specific needs of local communities, thereby ensuring that climate policies are responsive to the unique challenges faced by those most vulnerable to climate change

(Nam & Lee, 2020; Prestwood et al., 2018). This participatory approach serves to both empower communities and ensure greater responsiveness of climate policies to local needs and conditions.

The structure and diversity of strategic partnerships significantly influence their effectiveness. Partnerships that feature a diverse range of stakeholders and strong governance frameworks tend to yield better climate change mitigation outcomes (Pattberg, 2010). Effective governance mechanisms enable the coordination of stakeholder actions, ensuring that resources are allocated efficiently and that decision-making processes are transparent (Alonso & Andrews, 2018). Furthermore, partnerships that are flexible and adaptive are better equipped to respond to the evolving challenges of climate change (Sun et al., 2020).

In conclusion, it can be stated that strategic partnerships are of vital importance for the successful implementation of climate-neutral policies. By fostering collaboration among various stakeholders, these partnerships can facilitate the mobilization of resources, the sharing of knowledge, and the promotion of inclusive decision-making processes. As governments and organizations endeavor to fulfill their climate commitments, the formation of efficacious partnerships will be indispensable for propelling substantial advancement toward the realization of climate neutrality.

1.2.2 Human Resources and Leadership in Climate Neutrality Efforts

The role of human resources and leadership in advancing climate neutrality efforts is increasingly recognized as a critical factor in achieving organizational sustainability goals (Singh, Ravesangar, et al., 2024). Effective leadership and a supportive HRM framework are essential for fostering a culture that prioritizes climate action and encourages employee engagement in sustainability initiatives (Adegoke et al., 2024).

Leadership plays a pivotal role in shaping organizational climate and influencing employee behavior toward climate neutrality. Top management support is a critical success factor in achieving climate neutrality, as leaders set the tone for organizational priorities and resource allocation (Bhushan & Singh, 2024; Duong, 2023). Leaders who demonstrate a commitment to sustainability can inspire employees to adopt environmentally friendly practices and contribute to the organization's climate goals (Unsworth et al., 2021). Transformational leadership, characterized by the ability to motivate and engage employees, has been shown to positively influence organizational climate and employee satisfaction, which are essential for fostering a proactive approach to climate initiatives (Zhang et al., 2022).

Moreover, the concept of "leadership climate" is significant in understanding how leadership behaviors impact employee well-being and engagement. A positive leadership climate can prevent emotional exhaustion among employees, thereby enhancing their capacity to engage in climate-related activities (Engel et al., 2018). This is particularly important in high-stress environments, where leaders can act as role models by promoting a culture of health and well-being that aligns with sustainability objectives (Schulz et al., 2017). By prioritizing employee mental health and creating a supportive work environment, leaders can enhance overall organizational performance in climate neutrality efforts.

Human resource management practices are equally crucial in supporting climate neutrality initiatives. A positive HR development climate, characterized by opportunities for training, development, and employee involvement, can significantly enhance employee engagement and commitment to sustainability goals (Obuobisa-Darko & Tsedzah, 2019). Organizations that invest in HRM practices that promote sustainability—such as training programs focused on environmental awareness and skills development—are more likely to achieve their climate objectives (Purohit & Verma, 2013). Furthermore, fostering a culture of collaboration and trust within teams can enhance collective efforts toward climate action, as employees feel empowered to contribute their ideas and initiatives (Chaudhary et al., 2012).

The integration of climate considerations into HRM practices can also facilitate the development of a workforce that is equipped to address the challenges of climate change. For instance, organizations can implement training programs that focus on sustainable practices and the importance of climate neutrality, thereby enhancing employees' understanding and commitment to these goals (Kılıç & Atilla, 2024; Nwokolo et al., 2023b; Schmidt et al., 2022). Additionally, organizations can leverage employee feedback to continuously improve their sustainability initiatives, ensuring that they are responsive to the needs and concerns of their workforce (Underhill et al., 2015).

Consequently, human resources and leadership are integral to the success of climate neutrality efforts within organizations. By fostering a supportive leadership climate and implementing effective HRM practices, organizations can enhance employee engagement, promote a culture of sustainability, and ultimately achieve their climate goals. The alignment of leadership strategies with HRM initiatives is essential for creating a cohesive approach to climate action that empowers employees and drives organizational performance.

1.2.3 Public-Private Collaborations for Climate-Neutral Innovation

Public-private collaborations (PPCs) have emerged as a pivotal mechanism for fostering climate-neutral innovation, particularly in the context of sustainable development (Mallick et al., 2024). These partnerships leverage the strengths of both public and private sectors to address complex environmental challenges and promote innovative solutions (Riege & Lindsay, 2006). The integration of diverse resources, expertise, and stakeholder interests is crucial for achieving sustainability objectives, as evidenced by various studies highlighting the effectiveness of PPCs in different sectors (Zarte et al., 2019).

One significant aspect of PPCs is their ability to facilitate knowledge diffusion and resource sharing, which are essential for driving innovation. For instance, public-private partnerships can enhance the innovation process by promoting knowledge exchange and minimizing technology transfer costs, thereby contributing to sustainable development outcomes (Khan et al., 2024; Wang & Ma, 2020). Similarly, sustainable smart city initiatives can benefit from the collaborative efforts of public and private entities, leading to improved quality of life and environmental sustainability (Mills et al., 2021). This synergy is particularly important in urban settings, where the complexities of sustainability require tailored solutions that address social equity, resource conservation, and economic vitality (Leshinka et al., 2023).

Moreover, the role of PPCs in specific sectors, such as water management and peatland conservation, illustrates their potential for fostering innovative practices. That partnerships in peatland management can combine resources and expertise to create sustainable practices that benefit both the environment and local communities (Zulkamaini et al., 2024). In the context of water infrastructure, the urgency of implementing sustainable desalination technologies through PPCs to address the growing demand for freshwater driven by climate change and urbanization (Kumar et al., 2024; Nwokediegwu, 2024). These underscore the versatility of PPCs in addressing diverse environmental challenges while promoting climate-neutral innovations.

The effectiveness of PPCs in driving innovation is also contingent upon the governance structures and collaborative mechanisms established within these partnerships. Successful PPCs must incorporate flexibility and adaptability to foster continuous innovation throughout the partnership's lifecycle (Singh, Khan, et al., 2024; Spraul & Thaler, 2019) and also key conditions for successful public-private collaboration includes the alignment of goals and the establishment of trust between partners (Callens & Verhoest, 2024). By creating an environment conducive to innovation, PPCs can effectively respond to evolving sustainability challenges.

Furthermore, the alignment of public and private sector goals is crucial for the success of PPCs. The tensions that may arise between profitability and sustainability within green public-private partnerships needs a balanced approach that considers stakeholder involvement and regulatory frameworks (Nwokolo et al., 2023c; Vassileva, 2022). This balance is essential for ensuring that the innovations developed through PPCs contribute positively to environmental sustainability while also being economically viable.

Subsequently, PPCs represent a promising approach for fostering climate-neutral innovation across various sectors. By leveraging the strengths of both sectors, these partnerships can facilitate knowledge sharing, resource optimization, and the development of innovative solutions to pressing environmental challenges. The success of PPCs hinges on effective governance, alignment of objectives, and the ability to adapt to changing circumstances, ultimately contributing to sustainable development goals and climate resilience.

1.2.4 Challenges and Opportunities in Building a Climate-Neutral Workforce

Building a climate-neutral workforce presents both significant challenges and opportunities across various sectors. As the urgency of addressing climate change intensifies, organizations are compelled to rethink their workforce strategies to align with sustainability goals (Newell, 2020). This transformation necessitates a multifaceted approach that encompasses education, training, and the integration of climate considerations into organizational practices (Wang et al., 2023). The following discussion synthesizes insights from various studies to elucidate the challenges and opportunities inherent in this transition.

One of the most significant obstacles to developing a climate-neutral workforce is the necessity for comprehensive educational and training programs that provide employees with the requisite skills to adapt to emerging technologies and practices (Pendrey et al., 2022). It is imperative that educational curricula integrate carbon neutrality concepts in order to cultivate a workforce capable of contributing to sustainability efforts in sectors such as agriculture (Shi, 2023). However, the current educational frameworks often lack the requisite focus on climate-related competencies, which can impede the development of a skilled workforce (Lemery et al., 2020). Furthermore, the accelerated rate of climate change requires ongoing skill development, as conventional training approaches may not fully equip workers with the necessary competencies to meet the evolving demands of their roles (Porter, 2024).

In addition to the educational disparities that have been identified, the psychological impacts of climate change on the workforce represent a significant challenge. The direct and indirect effects of climate change, such as extreme weather events and shifting health needs, have resulted in increased stress and anxiety among workers in various sectors, particularly healthcare (Souza, 2023; Zurynski, 2024). This psychological burden can result in decreased job satisfaction and productivity, which further complicates the task of building a resilient workforce. Despite their strong commitment to addressing climate change, health professionals frequently encounter obstacles that impede their ability to engage effectively (Kotcher et al., 2021). It is therefore incumbent upon organizations to prioritize mental health support and resilience training in order to mitigate these impacts and foster a more adaptable workforce.

Moreover, the risk of "carbon leakage," where production shifts to regions with less stringent climate regulations, presents a significant challenge for organizations striving for climate neutrality (Duffy et al., 2022). This phenomenon can undermine local efforts to reduce emissions and create a fair competitive landscape. To counteract this, companies must adopt comprehensive sustainability strategies that not only focus on internal practices but also engage with supply chain partners to ensure collective progress toward climate goals (DeLoach et al., 2023; Joshi et al., 2024; Sharma, 2023). This collaborative approach can enhance the overall effectiveness of climate initiatives and create a more robust workforce committed to sustainability.

Conversely, the transition to a climate-neutral workforce also presents a multitude of opportunities. For example, the growing demand for environmentally conscious employment opportunities presents a substantial opportunity for job creation across a range of sectors, including renewable energy, sustainable agriculture, and environmental management (Nwokolo et al., 2023d; Stilwell, 2021). As organizations transition toward sustainability, they can leverage new markets and innovations that emerge from this shift, thereby enhancing their competitiveness and resilience in a changing economic landscape (Firoiu et al., 2022). Moreover, the incorporation of digital technologies in the pursuit of climate neutrality can facilitate operational efficiency and enhance organizational responsiveness to emerging challenges (Maksymova & Kuryliak, 2022; Singh, Gupta, & Kandpal, 2024).

Furthermore, the cultivation of a culture of sustainability within organizations has the potential to enhance employee engagement and satisfaction. When employees perceive their organizations as committed to sustainability, they are more likely to exhibit higher levels of job satisfaction and organizational commitment (Doghan et al., 2019). Such a positive organizational climate may result in increased retention rates and the attraction of top talent, who are motivated by environmental stewardship (Nwokolo et al., 2024; Ravesangar, Halid, & Singh, 2024; Wolf et al., 2021). By integrating sustainability into the fundamental values and operational procedures of an organization, leaders can foster a workforce that is not only proficient but also enthusiastic about contributing to climate objectives.

It is imperative to recognize the pivotal role that leadership plays in enabling this transition. Effective leadership is of paramount importance in guiding organizations through the intricacies of developing a climate-neutral workforce. It is incumbent upon leaders to espouse sustainability initiatives, disseminate information regarding their significance, and furnish the requisite resources for training and development (Mallon, 2023). Moreover, interdisciplinary collaboration across departments can enhance the effectiveness of sustainability efforts, as diverse perspectives can lead to more innovative solutions (Zurynski, 2024). By prioritizing climate education and fostering a supportive organizational culture, leaders can empower their workforce to effectively address the challenges and seize the opportunities presented by the transition to climate neutrality.

Despite the difficulties inherent in the process of developing a climate-neutral workforce, the journey also presents considerable opportunities for innovation, engagement, and growth. It is incumbent upon organizations to proactively address educational deficiencies, psychological impacts, and the risk of carbon leakage while simultaneously leveraging the potential for job creation and enhanced employee satisfaction. By cultivating a culture of sustainability and emphasizing effective leadership, organizations can effectively navigate the complexities of this transition and contribute significantly to global climate goals.

1.3 Discussion

The integration of climate neutrality into government policies, business strategies, and organizational practices presents a multifaceted challenge, requiring collaboration across sectors and a commitment to sustainability from all involved stakeholders. The theoretical approaches outlined in this chapter underscore the pivotal role of strategic partnerships, leadership, and human resource management in propelling the transition to a climate-neutral economy. It is evident that a holistic and coordinated approach is necessary, as no single actor can achieve climate neutrality in isolation. It is evident that government entities occupy a pivotal position in the formulation of policy frameworks and the encouragement of sustainable practices. Concurrently, businesses are obliged to adopt innovative strategies that align environmental sustainability with profitability. The role of human resources is equally pivotal, as organizations must ensure that their workforce is equipped with the necessary skills and motivation to support climate initiatives.

A principal theme that has emerged from this discussion is the significance of leadership in cultivating a culture of sustainability within organizations. Effective leadership is instrumental in establishing a culture of climate action, ensuring that sustainability is not merely a peripheral concern but a core organizational priority. Those in leadership positions who demonstrate a commitment to climate goals can inspire their teams to engage in environmentally responsible practices, thereby creating a positive organizational climate that enhances overall performance. It is imperative that leadership strategies are aligned with human resource management practices in order to create a unified approach to climate action. Human resource management (HRM) practices, including the implementation of sustainability training programs and the development of green skills, are of paramount importance in enabling employees to contribute to the organization's climate goals.

The formation of strategic alliances between the public and private sectors serves to reinforce the necessity of collective action in the pursuit of climate neutrality. PPCs facilitate the sharing of knowledge, the mobilization of resources, and innovation, all of which are essential for addressing complex environmental challenges. By capitalizing on the unique capabilities of both sectors, PPCs can propel the advancement of climate-neutral technologies and solutions, thereby promoting sustainable growth and resilience in the context of climate change. Nevertheless, the efficacy of these partnerships hinges on the presence of robust governance structures, the convergence of stakeholder objectives, and the capacity to adapt to evolving environmental and economic conditions.

The transition to a climate-neutral workforce presents a duality of challenges and opportunities. Consequently, organizations are obliged to address deficiencies in the provision of education and training, ensuring that employees are equipped with the requisite skills to support sustainability initiatives. Furthermore, the psychological impacts of climate change, including increased stress and anxiety, present a significant challenge to workforce resilience. Conversely, the increasing demand for environmentally friendly employment opportunities and the potential for innovation in sustainable industries present significant opportunities for organizations that prioritize climate neutrality. The cultivation of a culture of sustainability within the workforce has the potential to enhance employee engagement and job satisfaction, which may in turn lead to higher retention rates and the attraction of top talent.

This discussion demonstrates that achieving climate neutrality necessitates a comprehensive and integrated approach, with leadership, human resources, and strategic partnerships assuming pivotal roles. By fostering collaboration, aligning organizational strategies with sustainability goals, and investing in workforce development, organizations can contribute to global climate goals while simultaneously enhancing their competitive advantage and resilience.

1.4 Concluding Remarks

The journey toward climate neutrality presents a complex and multifaceted challenge that necessitates a synthesis of theoretical and practical insights. The theoretical frameworks presented in this chapter underscore the pivotal role of leadership, HRM, and strategic partnerships in propelling the transition to a climate-neutral economy. From a theoretical standpoint, this study elucidates the interrelationship between organizational leadership, HRM practices, and public-private collaborations, offering a comprehensive perspective on how organizations can align their internal strategies with the overarching objectives of environmental sustainability. This integrated perspective contributes to the expanding body of literature that examines the dynamics of sustainability within organizational structures and governmental policy frameworks.

Theoretically, the chapter expands on the understanding of how leadership behaviors influence not only organizational culture but also the practical outcomes of sustainability initiatives. By framing leadership as a key driver of employee engagement in climate-related actions, this study contributes to the discourse on transformational leadership and its role in fostering environmentally conscious organizational climates. Moreover, the exploration of HRM practices provides a vital theoretical link between employee development and climate neutrality goals, showing how training, development programs, and a supportive organizational climate can empower employees to contribute to sustainability efforts. In terms of strategic partnerships, the chapter underscores the theoretical significance of PPCs as mechanisms for facilitating innovation, resource sharing, and policy implementation.

The practical implications of these findings are significant for both policymakers and organizational leaders. It is imperative that policymakers acknowledge the significance of cultivating cross-sector collaborations, particularly public-private partnerships that can bridge the divide between regulatory frameworks and the innovative capabilities of the private sector. The implementation of government policies that provide incentives for green investments, offer tax breaks for sustainable practices, and support renewable energy initiatives can serve to significantly accelerate the adoption of climate-neutral strategies across a range of industries. Conversely, businesses may bolster their competitive advantage and long-term resilience by integrating sustainability into their core strategies. It is incumbent upon leaders to ensure that climate action is not only a strategic priority but also integrated into the fabric of day-to-day operations and the organizational culture. The role of HRM in this process is of great consequence, as it is responsible for developing a workforce that is adequately equipped with the requisite skills and mindset to engage in sustainability initiatives.

The practical implications also extend to organizational development and operational efficiency. Companies that adopt green supply chain management and environmental management systems will find themselves better positioned to reduce waste, improve resource efficiency, and respond to the growing demands of environmentally conscious consumers. The integration of digital transformation and data-driven sustainability initiatives also presents an opportunity for businesses to streamline their climate-neutral efforts, enhancing both operational effectiveness and environmental outcomes. Additionally, the creation of a climate-neutral workforce, supported by targeted HRM practices such as sustainability training and employee engagement in green initiatives, can foster a culture of environmental responsibility, which is increasingly becoming a critical component of organizational success.

While the study employs a comprehensive approach, it is not without limitations. A significant limitation is the variation in governmental frameworks and industry-specific challenges, which may impact the generalizability of the findings across different sectors and regions. The efficacy of climate-neutral initiatives is contingent upon a multitude of contextual factors, including the regulatory environment, market conditions, and the availability of resources. Furthermore, although the chapter concentrates on the organizational and policy-related aspects, it does not sufficiently examine the behavioral and psychological factors that affect individual employee engagement in sustainability initiatives. Further insight may be gained into the ways of encouraging climate-related behaviors within the workforce by considering the role of intrinsic motivation, personal values, and behavioral nudges.

A further limitation is the restricted scope with regard to the scalability of successful public-private collaborations in different contexts. The chapter effectively demonstrates the efficacy of PPCs; however, it does not sufficiently examine the ways in which these partnerships can be adapted or scaled in diverse geographical or economic contexts. Moreover, the influence of emerging technologies, such as artificial intelligence and blockchain, on the advancement of climate-neutral initiatives is an area that requires further examination. Although the role of digital transformation is acknowledged, future research should undertake a more comprehensive assessment of how these technologies can support sustainability efforts across sectors.

In light of these constraints, future research should prioritize addressing several pivotal areas. First and foremost, further empirical studies are required to assess the efficacy of leadership styles, human resource management practices, and strategic partnerships across a range of industry contexts, with particular focus on high-carbon industries such as manufacturing, transportation, and energy. It is imperative to comprehend how these sectors can effectively implement climate-neutral strategies in order to make significant strides toward global sustainability objectives. Furthermore, future studies should examine the role of individual employee behaviors, such as the influence of intrinsic motivation and personal values on engagement in sustainability initiatives, and investigate potential interventions to enhance this engagement.

Further research is also needed to investigate how digital transformation can drive climate-neutral efforts, particularly in enhancing the efficiency of public-private collaborations and sustainability practices within organizations. Technologies such as artificial intelligence, the Internet of Things (IoT), and blockchain have the potential to revolutionize how organizations track, measure, and improve their environmental impact. Research into the practical applications of these technologies in facilitating real-time data sharing, transparency in sustainability practices, and more efficient resource management would provide valuable insights for both academics and practitioners.

Lastly, the scalability and adaptability of successful public-private collaborations across different geographical regions and economic contexts remain a critical area for future exploration. Research that investigates how these partnerships can be modified to fit the unique needs of developing economies, or regions with limited resources, could offer valuable strategies for extending the benefits of PPCs to a broader range of industries and regions. Understanding the conditions that facilitate the success of these collaborations globally would provide policymakers and organizations with more practical insights into implementing scalable and sustainable climate-neutral strategies.

In conclusion, the pursuit of climate neutrality necessitates a long-term commitment from all sectors, driven by robust leadership, strategic partnerships, and a comprehensive HRM framework. The theoretical contributions of this study offer a multidimensional perspective on the intersections between leadership, HRM, and public-private collaborations. In addition, the practical implications highlight the necessity for collaborative efforts at all levels. Further research is required to address the limitations of this study and to explore the suggested areas for future research. This will enhance the understanding of how organizations and governments can jointly contribute to global climate goals. The route toward climate neutrality is, while intricate, replete with significant possibilities for innovation, expansion and global sustainability.

References

Adegoke, A. A., Oyindamola, K. A., & Offonabo, N. (2024). The role of HR in sustainability initiatives: A strategic review. *International Journal of Research in Engineering, Science and Management, 7*(5), 31–37.

Akar, A., & Bedük, A. (2023). The effects of human resources management on organizational performance and organizational climate: A study in Konya industry. *Journal of Business Research.* https://doi.org/10.20491/isarder.2023.1689

Alonso, J., & Andrews, R. (2018). Governance by targets and the performance of cross-sector partnerships: Do partner diversity and partnership capabilities matter? *Strategic Management Journal, 40*(4), 556–579. https://doi.org/10.1002/smj.2959

Andre, F. J., & Valenciano-Salazar, J. A. (2022). Voluntary carbon neutral programs. Adoption and firms' strategies. *Journal of Cleaner Production, 381*, 135191. https://doi.org/10.1016/j.jclepro.2022.135191

Bhushan, M., & Singh, R. (2024). Talent management facilitates net-zero transition through employee green behavior. In R. Singh, S. Khan, A. Kumar, S. Luthra, & H. Chokshi (Eds.), *Net zero economy, corporate social responsibility and sustainable value creation. CSR, sustainability, ethics & governance*. Springer. https://doi.org/10.1007/978-3-031-55779-8_7

Broto, V., & Bulkeley, H. (2013). A survey of urban climate change experiments in 100 cities. *Global Environmental Change, 23*(1), 92–102. https://doi.org/10.1016/j.gloenvcha.2012.07.005

Callens, C., & Verhoest, K. (2024). Conditions for successful public-private collaboration for public service innovation. In *Collaborating for digital transformation* (pp. 52–79). Edward Elgar Publishing. https://doi.org/10.4337/9781803923895.00012

Chang, Y., & Yoo, J. (2023). How does the degree of competition in an industry affect a company's environmental management and performance? *Sustainability, 15*(9), 7675. https://doi.org/10.3390/su15097675

Chaudhary, R., Rangnekar, S., & Barua, M. (2012). Relationships between occupational self efficacy, human resource development climate, and work engagement. *Team Performance Management, 18*(7/8), 370–383. https://doi.org/10.1108/13527591211281110

Chen, L., Zhao, X., Tang, O., Price, L., Zhang, S., & Zhu, W. (2017). Supply chain collaboration for sustainability: A literature review and future research agenda. *International Journal of Production Economics, 194*, 73–87. https://doi.org/10.1016/j.ijpe.2017.04.005

Clarke, A., & MacDonald, A. (2016). Outcomes to partners in multi-stakeholder cross-sector partnerships: A resource-based view. *Business & Society, 58*(2), 298–332. https://doi.org/10.1177/0007650316660534

DeLoach, C., Soliz, A., & Mesa, H. (2023). Building community and technical college capacity through state funding. *Community College Review, 51*(2), 246–265. https://doi.org/10.1177/00915521221145295

Doghan, M., Bhatti, M., & Juhari, A. (2019). Do psychological diversity climate, HRM practices, and personality traits (big five) influence multicultural workforce job satisfaction and performance? Current scenario, literature gap, and future research directions. *SAGE Open, 9*(2), 21582440198. https://doi.org/10.1177/2158244019851578

Dolla, T., & Laishram, B. (2020). Bundling in public-private partnership projects-a conceptual framework. *International Journal of Productivity and Performance Management, 69*(6), 1177–1203. https://doi.org/10.1108/IJPPM-02-2019-0086

Duffy, C., Prudhomme, R., Duffy, B., Iannetta, P., O'Donoghue, C., Ryan, M., & Styles, D. (2022). Randomized national land management strategies for net-zero emissions. *Nature Sustainability, 5*(11), 973–980. https://doi.org/10.1038/s41893-022-00946-0

Duong, A. (2023). Critical success factors of food safety management for achieving climate neutrality: A multilevel moderated approach with industry revolution 4.0. *International Journal of Logistics Management, 35*(3), 865–891. https://doi.org/10.1108/ijlm-03-2023-0112

Engel, S., Wörfel, F., Maria, A., Wolter, C., Kleiber, D., & Renneberg, B. (2018). Leadership climate prevents emotional exhaustion in German Police Officers. *International Journal of Police Science and Management, 20*(3), 217–224. https://doi.org/10.1177/1461355718786295

Firoiu, D., Pirvu, R., Jianu, E., Cismaş, L., Tudor, S., & Latea, G. (2022). Digital performance in EU member states in the context of the transition to a climate neutral economy. *Sustainability, 14*(6), 3343. https://doi.org/10.3390/su14063343

Forsyth, T. (2010). Panacea or paradox? Cross-sector partnerships, climate change, and development. *Wiley Interdisciplinary Reviews: Climate Change, 1*(5), 683–696. https://doi.org/10.1002/wcc.68

Golhasany, H., & Harvey, B. (2023). Capacity development for knowledge mobilization: A scoping review of the concepts and practices. *Humanities and Social Sciences Communications, 10*(1), 1–12. https://doi.org/10.1057/s41599-023-01733-8

Ham, C., & Klimmek, H. (2017). Partnerships for nature-based solutions in urban areas—Showcasing successful examples. In *Nature-based solutions to climate change adaptation in urban areas: Linkages between science, policy and practice* (pp. 275–289). Springer. https://doi.org/10.1007/978-3-319-56091-5_16

Joshi, A., Singh, R., Kumar, V., & Rani, S. (2024). Net zero strategy: Exploring opportunities, confronting challenges for developing countries. In *Transition towards a sustainable future* (pp. 233–252). Springer. https://doi.org/10.1007/978-981-97-5756-5_11

Kabisch, N., Frantzeskaki, N., Pauleit, S., Naumann, S., Davis, M., Artmann, M., & Bonn, A. (2016). Nature-based solutions to climate change mitigation and adaptation in urban areas: Perspectives on indicators, knowledge gaps, barriers, and opportunities for action. *Ecology and Society, 21*(2), 39. https://doi.org/10.5751/es-08373-210239

Khan, S., Singh, R., Alnahas, J., Abbate, S., & Centobelli, P. (2024). Navigating the smart circular economy: A framework for manufacturing firms. *Journal of Cleaner Production, 2024*, 144007. https://doi.org/10.1016/j.jclepro.2024.144007

Kılıç, C., & Atilla, G. (2024). Industry 4.0 and sustainable business models: An intercontinental sample. *Business Strategy and the Environment, 33*(4), 3142–3166. https://doi.org/10.1002/bse.3634

Kotcher, J., Maibach, E., Miller, J., Campbell, E., Alqodmani, L., Maiero, M., & Wyns, A. (2021). Views of health professionals on climate change and health: A multinational survey study. *The Lancet Planetary Health, 5*(5), e316–e323. https://doi.org/10.1016/s2542-5196(21)00053-x

Kumar, K., Singh, R., Sharma, A., & Lehri, B. P. S. (2024). Climate change adaptation strategies for achieving net-zero economy. In R. Singh & D. Crowther (Eds.), *Transition towards a sustainable future. Approaches to global sustainability, markets, and governance*. Springer. https://doi.org/10.1007/978-981-97-5756-5_8

Kumar, S., Banerjee, B., & Duarte, F. (2020). The business—Government nexus: Impact of government actions and legislation on business responses to climate change. *Journal of Management & Organization, 26*(6), 952–974. https://doi.org/10.1017/jmo.2020.21

Lemery, J., Balbus, J., Sorensen, C., Rublee, C., Dresser, C., Balsari, S., & Hynes, E. (2020). Training clinical and public health leaders in climate and health. *Health Affairs, 39*(12), 2189–2196. https://doi.org/10.1377/hlthaff.2020.01186

Leshinka, P., Tembo, M., & Mwanaumoi, E. M. (2023). Role of public-private partnerships in developing sustainable cities: Framework, opportunities, mechanisms. *International Journal of Innovations and Interdisciplinary Research, 1*(1), 1–15. https://doi.org/10.61108/ijiir.v1i1.1

Maksymova, I., & Kuryliak, V. (2022). World industry digitization in the context of ensuring climate neutrality. *Journal of European Economy, 21(3*, 343–360. https://doi.org/10.35774/jee2022.03.343

Mallick, P. K., Salling, K. B., Pigosso, D. C., & McAloone, T. C. (2024). Designing and operationalising extended producer responsibility under the EU green Deal. *Environmental Challenges, 2024*, 100977. https://doi.org/10.1016/j.envc.2024.100977

Mallon, W. (2023). Reasons for optimism about academic medicine's actions against climate change. *Academic Medicine, 98*(11), 1243–1246. https://doi.org/10.1097/acm.0000000000005331

Manavalan, E., & Jayakrishna, K. (2019). A review of internet of things (IoT) embedded sustainable supply chain for industry 4.0 requirements. *Computers & Industrial Engineering, 127*, 925–953. https://doi.org/10.1016/j.cie.2018.11.030

Mills, D. E., Izadgoshasb, I., & Pudney, S. G. (2021). Smart city collaboration: A review and an agenda for establishing sustainable collaboration. *Sustainability, 13*(16), 9189. https://doi.org/10.3390/su13169189

Nam, A., & Lee, S. (2020). Students as partners. Implementation of climate change education within the Harvard graduate school of education. In *Education and climate change: The role of universities* (pp. 153–180). Springer. https://doi.org/10.1007/978-3-030-57927-2_6

Newell, P. (2020). The business of rapid transition. *Wiley Interdisciplinary Reviews: Climate Change, 11*(6), e670. https://doi.org/10.1002/wcc.670
Nikseresht, A., Golmohammadi, D., & Zandieh, M. (2023). Sustainable green logistics and remanufacturing: A bibliometric analysis and future research directions. *The International Journal of Logistics Management, 35*(3), 755–803. https://doi.org/10.1108/ijlm-03-2023-0085
Nwokediegwu, Z. (2024). Review of innovative approaches in water infrastructure: Sustainable desalination and public-private partnerships. *International Journal of Science and Research Archive, 11*(1), 324–332. https://doi.org/10.30574/ijsra.2024.11.1.0072
Nwokolo, S. C., Singh, R., Khan, S., & Kumar, A. (2024). Threats to the rapidity of sustainability transitions posed by technological changes. In *Technological pathways for Africa's net-zero economy*. Elsevier. https://doi.org/10.1016/B978-0-443-31486-5.00002-1
Nwokolo, S. C., Singh, R., Khan, S., Kumar, A., & Luthra, S. (2023a). Global investment and development in Africa. In *Africa's path to net-zero. CSR, sustainability, ethics & governance*. Springer. https://doi.org/10.1007/978-3-031-44514-9_2
Nwokolo, S. C., Singh, R., Khan, S., Kumar, A., & Luthra, S. (2023b). Impacts of climate change in Africa. In *Africa's path to net-zero. CSR, sustainability, ethics & governance*. Springer. https://doi.org/10.1007/978-3-031-44514-9_7
Nwokolo, S. C., Singh, R., Khan, S., Kumar, A., & Luthra, S. (2023c). Africa's awakening to climate action. In *Africa's path to net-zero: Exploring scenarios for a sustainable energy transition* (pp. 299–310). Springer. https://doi.org/10.1007/978-3-031-44514-9_9
Nwokolo, S. C., Singh, R., Khan, S., Kumar, A., & Luthra, S. (2023d). Remedies to the challenges of renewable energy deployment in Africa. In *Africa's path to net-zero. CSR, sustainability, ethics & governance*. Springer. https://doi.org/10.1007/978-3-031-44514-9_3
Obuobisa-Darko, T., & Tsedzah, V. (2019). Human resource development climate and employee engagement in developing countries. *International Journal of Business and Management, 14*(6), 150. https://doi.org/10.5539/ijbm.v14n6p150
Pattberg, P. (2010). Public-private partnerships in global climate governance. *Wiley Interdisciplinary Reviews: Climate Change, 1*(2), 279–287. https://doi.org/10.1002/wcc.38
Pendrey, C., Quilty, S., & Lucas, R. (2022). Surveying the changing climate of northern territory medical workforce retention. *Australian Journal of Rural Health, 30*(3), 402–409. https://doi.org/10.1111/ajr.12858
Porter, D. (2024). Designing agile pathways for climate adaptation skill development. *The Open/Technology in Education Society and Scholarship Association Journal, 3*(2), 1–17. https://doi.org/10.18357/otessaj.2023.3.2.52
Prestwood, E., Longhurst, J., Townsend, I., Haines, T., & Tsiarapa, E. (2018). Facilitating stakeholder dialogues on a carbon neutral city: We need to talk about carbon (and air quality). *WIT Transactions on Ecology and the Environment, 230*, 501–510. https://doi.org/10.2495/air180461
Purohit, B., & Verma, R. (2013). A study of human resource development climate in government health centres in India. *Journal of Health Management, 15*(3), 431–443. https://doi.org/10.1177/0972063413491878
Ravesangar, K., Halid, H., & Singh, R. (2024). Unveiling the influence of employees' intrinsic motivational factors on work performance in the Malaysian banking industry: Investigating the role of psychological ownership as a mediator. *Academic Journal of Business and Social Sciences, 8*(1), 1–31.
Ravesangar, K., Ping, L. L., & Pachar, S. (2024). A review on the sustainable HRM practices in building net-zero transformation: An emerging trend in the workplace. In *Net zero economy, corporate social responsibility and sustainable value creation: Exploring strategies, drivers, and challenges* (pp. 131–144). Springer. https://doi.org/10.1007/978-3-031-55779-8_8
Riege, A., & Lindsay, N. (2006). Knowledge management in the public sector: Stakeholder partnerships in the public policy development. *Journal of Knowledge Management, 10*(3), 24–39. https://doi.org/10.1108/13673270610670830
Schmidt, A., Köster, D., & Strube, J. (2022). Climate neutrality concepts for the German chemical–pharmaceutical industry. *PRO, 10*(3), 467. https://doi.org/10.3390/pr10030467

Schulz, H., Zacher, H., & Lippke, S. (2017). The importance of team health climate for health-related outcomes of white-collar workers. *Frontiers in Psychology, 8*, 74. https://doi.org/10.3389/fpsyg.2017.00074

Sharma, S. (2023). Developing climate neutrality among supply chain members in metal and mining industry: Natural resource-based view perspective. *The International Journal of Logistics Management, 35*(3), 804–832. https://doi.org/10.1108/ijlm-03-2023-0108

Shi, J. (2023). Pathways for integrating the concept of carbon neutrality into the talent cultivation process: A case study of animal production programs in Chinese agricultural colleges and universities. *Sustainability, 15*(23), 16317. https://doi.org/10.3390/su152316317

Singh, R., Gupta, A., & Kandpal, V. (2024). ESG excellence in the digital world: Circular economy E-business solutions. In *Fintech for ESG and the circular economy* (pp. 193–204). Wiley. https://doi.org/10.1002/9781394234035.ch10

Singh, R., Khan, S., Dsilva, J., Akram, U., & Haleem, A. (2024). Modelling the organisational factors for implementation of corporate social responsibility: A modified TISM approach. *Global Journal of Flexible Systems Management, 25*(2), 283–301. https://doi.org/10.1007/s40171-024-00388-x

Singh, R., Ravesangar, K., Narayanan, S., & Singh, K. K. H. (2024). HRM practices in creating net zero: Supporting organizational sustainability. In *Net zero economy, corporate social responsibility and sustainable value creation: Exploring strategies, drivers, and challenges* (pp. 1–18). Springer. https://doi.org/10.1007/978-3-031-55779-8_1

Souza, M. (2023). Healthy patients, workforce and environment: Coupling climate adaptation and mitigation to wellbeing in healthcare. *International Journal of Environmental Research and Public Health, 20*(22), 7059. https://doi.org/10.3390/ijerph20227059

Spraul, K., & Thaler, J. (2019). Partnering for good? An analysis of how to achieve sustainability-related outcomes in public–private partnerships. *Business Research, 13*(2), 485–511. https://doi.org/10.1007/s40685-019-0097-3

Stilwell, F. (2021). From green jobs to green new deal: What are the questions? *The Economic and Labour Relations Review, 32*(2), 155–169. https://doi.org/10.1177/10353046211009774

Streck, C. (2021). Strengthening the Paris agreement by holding non-state actors accountable: Establishing normative links between transnational partnerships and treaty implementation. *Transnational Environmental Law, 10*(3), 493–515. https://doi.org/10.1017/s2047102521000091

Sun, X., Clarke, A., & MacDonald, A. (2020). Implementing community sustainability plans through partnership: Examining the relationship between partnership structural features and climate change mitigation outcomes. *Sustainability, 12*(15), 6172. https://doi.org/10.3390/su12156172

Taylor, B., & Harman, B. (2016). Governing urban development for climate risk: What role for public–private partnerships? *Environment and Planning. C, Government & Policy, 34*(5), 927–944. https://doi.org/10.1177/0263774x15614692

Underhill, E., Blunsdon, B., & McNeil, N. (2015). Guest editors' note: Human resource climates. *Asia Pacific Journal of Human Resources, 53*(4), 395–396. https://doi.org/10.1111/1744-7941.12089

Unsworth, K. L., Davis, M. C., Russell, S. V., & Bretter, C. (2021). Employee green behaviour: How organizations can help the environment. *Current Opinion in Psychology, 42*, 1–6. https://doi.org/10.1016/j.copsyc.2020.12.006

van Langen, S. K., Vassillo, C., Ghisellini, P., Restaino, D., Passaro, R., & Ulgiati, S. (2021). Promoting circular economy transition: A study about perceptions and awareness by different stakeholders groups. *Journal of Cleaner Production, 316*, 128166. https://doi.org/10.1016/j.jclepro.2021.128166

Varnali, R. (2015). An exploratory study of the cultural context of organisational climate and human resource practices. *Asia Pacific Journal of Human Resources, 53*(4), 432–447. https://doi.org/10.1111/1744-7941.12080

Vassileva, A. (2022). Green public-private partnerships (ppps) as an instrument for sustainable development. *Journal of World Economy Transformations & Transitions*. https://doi.org/10.52459/jowett25221122

Wang, N., & Ma, M. (2020). Public–private partnership as a tool for sustainable development—What literatures say? *Sustainable Development, 29*(1), 243–258. https://doi.org/10.1002/sd.2127

Wang, T., Olivier, D. F., & Chen, P. (2023). Creating individual and organizational readiness for change: Conceptualization of system readiness for change in school education. *International Journal of Leadership in Education, 26*(6), 1037–1061. https://doi.org/10.1080/13603124.2020.1818131

Wolf, S., Teitge, J., Mielke, J., Schütze, F., & Jaeger, C. (2021). The European green deal—More than climate neutrality. *Intereconomics, 56*(2), 99–107. https://doi.org/10.1007/s10272-021-0963-z

Zarte, M., Pechmann, A., & Nunes, I. L. (2019). Decision support systems for sustainable manufacturing surrounding the product and production life cycle - A literature review. *Journal of Cleaner Production, 219*, 336–349. https://doi.org/10.1016/j.jclepro.2019.02.092

Zhang, A., Tay, H., Alvi, M., Wang, J., & Gong, Y. (2022). Carbon neutrality drivers and implications for firm performance and supply chain management. *Business Strategy and the Environment, 32*(4), 1966–1980. https://doi.org/10.1002/bse.3230

Zulkamaini, Z., Nasution, M. S., Rinto, R., Meiwanda, G., & Bedasari, H. (2024). Public private partnerships in peatland management: A design for sustainable practices. *E3S Web of Conferences, 506*, 08001. https://doi.org/10.1051/e3sconf/202450608001

Zurynski, Y. (2024). Bolstering health systems to cope with the impacts of climate change events: A review of the evidence on workforce planning, upskilling, and capacity building. *The International Journal of Health Planning and Management, 39*(3), 781–805. https://doi.org/10.1002/hpm.3769

Chapter 2
The Impact of Climate Change on Migration: A Historical and Scenario Based Approach

Rubee Singh, David Crowther, and Seema Rani

Abstract Concerns on the climate change (CC) have resulted in forecasts about the reactions of populations residing in regions susceptible to drought, flooding, and temperature fluctuations to these occurrences. Preliminary research indicated that due to CC there are significant surges in human migration (HM) as households became unable of adapting to its effects. Recent empirical research examining the CC and its impacts have recorded the fluctuations in migration patterns fluctuate as a result of the extreme events as well as the traveling capacity of the households. This chapter presents a comprehensive review of the existing research on the interconnections between CC and migration. The chapter synthesizes a broad range of publications to demonstrate that how environmental changes are related to migration dynamics. Key focus concepts include the differentiation between climate-induced migration resulting from fast-onset events and slow-onset events. Additionally, it is also described that how migration impacts the origin and destination regions. The review offers implications to effectively manage and navigate the challenges resulting from climate-induced migration, emphasizing the requirement for informed policy responses and adaptive frameworks. This chapter is aimed at contributing to the understanding of how CC influences human mobility, offering guidelines for areas that can be explored in future research and action in this area.

Keywords Climate change · Environmental challenges · Human mobility · Migration · Sustainability

---

R. Singh (✉)
Institute of Business Management, GLA University, Mathura, India

D. Crowther
Social Responsibility Research Network, Derby, UK

S. Rani
Chandigarh School of Business, Chandigarh Group of Colleges, Jhanjeri, India

R. Singh and D. Crowther (eds.), *Climate Neutrality and Sustainable Ecosystems*, Approaches to Global Sustainability, Markets, and Governance,
https://doi.org/10.1007/978-981-96-7181-6_2

2.1 Introduction

Climate change (CC) is intricately linked to human actions related to adaptation for mitigating environmental challenges. Recently, the impact of CC on human migration is increasingly garnering the interest of academics and politicians. CC is acknowledged as a principal catalyst of movement in the agenda for humanity under sustainability goals for the 2030 year which also encompass a target to "facilitate orderly, safe, and responsible migration and mobility of people, including through the implementation of planned and well-managed migration policies," alongside the 13th SDG is dedicated to climate action, while it was not explicitly addressed in millennium goals previously (Bhushan & Singh, 2024). Prominent media outlets have extensively addressed the subject, with migration occasionally characterized as the "human face" of CC (Klepp, 2017). In recent decades, developing nations have numerous climate migrants including Bangladesh, China, and India as a result of extreme weather conditions including floods, droughts, storms in desert areas, melting of glaciers, and volcanic eruptions (Nwokolo et al., 2024; Smith & Smith, 2013). As per the estimates from the United Nations Organization, till the year of 2020 there were around 40 million climate migrants and by 2025 this number is estimated to surpass the count to 200 million (Wu & Wang, 2018). By the year 2050, it is anticipated that 1.2 billion individuals globally will move as a result of environmental anomalies and CC.

Migration behavior of individuals serves as a coping measure confronting the pressures of CC, specifically in instances of involuntary permanent relocation triggered by significant natural disasters. This phenomenon is exemplified by residents of the Tonga, Haiti, Maldives, and many smaller island nations, who frequently endure such calamities and may ultimately turned out to be "people who lost their country" (Cao & Chen, 2013). Furthermore, there are various reasons from both origin as well as the destination regions playing the push and pull dynamics in case of human migration (HM) (Khan et al., 2023), leading to significant migratory behaviors. The most discussed prediction on the people residing in regions most vulnerable to CC is by Myers (2002) as the author described about the potential movement of around 200 million individuals. Similarly, it has been estimated by Biermann and Boas (2010) that there could be approximately 200 million or more refugees due to CC by 20,250 by aggregating specific regional migration projections resulting from various sorts of climate shocks.

Further, Rigaud et al. (2018) made a forecasted for Latin America, South Asia, and Sub-Saharan Africa that these regions could experience a high rate of migration of nearly 3% as exceeding 143 million individuals, will be compelled to relocate internally by 2050 due to climate change. Various studies have sought to quantify the current environmental migratory flux, estimating it to be between 10 and 30 million individuals annually (Myers, 2002). In 2016, it was projected that 24 million individuals will be displaced forcibly on account of CC due to rapid-onset events (IDMC, 2017). On the other hand, in general mobility cases, the global number of migrants is estimated at over 750 million internally, which includes nearly 70 million of those individuals who are displaced forcibly while approximately 250 million individuals are migrants from cross-border (UNDESA, 2017; UNHCR, 2018).

The migration process, the significant ability of humans to adapt in diverse situations, is a more intricate phenomenon than previously depicted in earlier literature, which frequently relied on estimates concerning the probability of natural disasters and the potential number of individuals impacted (Stapleton et al., 2017). One approach characterized as a "maximalist" perspective, positing that environment related forces are the principal determinants of HM. The opposing viewpoint, termed a "minimalist" approach, contends that environmental influences serve just as contextual elements for which insufficient knowledge exists to establish causality. Both represent simplifications that, when they influence policy, can restrict alternatives for enhancing household mobility or for fostering resilience in different situations, so constraining the process of adjusting to a changed environment (Stapleton et al., 2017). Neither extreme perspective accurately reflects several scenarios. Since the onset of the twenty-first century, the majority of experts have expressed worry regarding the effects of CC on HM. In 2019, the migration of individuals was abruptly halted due to the outbreak COVID-19 pandemic, leading to a decrease in carbon emissions and significantly affecting CC (Sta Maria, 2023).

This growing significance of climate neutrality has led the researchers to enhance their understanding of the mechanics underlying climate-induced migration, alongside prominent research calculating possible migrant populations. Nevertheless, there is little focus on the examination of the interplay between CC and HM systematically, as well as the subsequent impacts of this relationship on the environment of the origin as well as destinations regions. Present study attempts to fill this gap by incorporating the findings arising from the literature, highlighting the linkage between CC and HM. The chapter will enhance our understanding on different scenarios related to migration and ways to navigate through various challenges arising out of climate-induced migration. The chapter has various sections. First, literature review section containing different sub-sections related to the topic is provided followed by research methodology section. Then, there is implications guided by literature, detailing the adaptation strategies in light of increasing migration. Lastly, conclusion section presents the study's limitations and a few suggestions on aspects where more research is required.

2.2 Literature Review

2.2.1 Climate Change

Over the past few decades, CC has emerged among the most contested issues and urgent challenges facing humanity. Consequently, comprehending societal perspectives on CC and its adverse effects is essential for the effective execution of strategies designed to mitigate and adapt to its effects (Nwokolo et al., 2023a). The literature has numerous research examining climate-related disasters at a regional level to provide management strategies for effectively addressing climatic uncertainty (Chophel, 2022; Joshi et al., 2024; Singh et al., 2024). Simultaneously, the research can help in designing policies to enhance the level of awareness among local communities regarding environmental disasters, highlighting the importance of human influence on global climate change (Millock, 2015; Nwokolo et al., 2023d). Climate mitigation pertains to specific human interventions aimed at diminishing sources emitting greenhouse gases in the atmosphere, whereas anticipated or actual stimuli regarding climate prompt the human systems towards adaptation to alleviate detrimental impacts or capitalize on advantageous opportunities (Khan et al., 2024).

Numerous studies investigate the views and attitudes of principal stakeholders concerned with or accountable for adaptation, intending to enhance their involvement in climate change (Kumar et al., 2024). Fadeyi and Maresova (2020) investigated the important concerns of concerned individuals for CC, encompassing advocates, experts, activists, government representatives, non-governmental organizations, and various other stakeholders, in West African developing nations. They concluded that ongoing efforts by local authorities and governments are essential to leverage these perspectives for climate action effectively. The perceptions of stakeholders in Sweden regarding CC and constraints in adapting with such changes has been evaluated, concluding that the principal impediment to such adjustment is the ability to respond (André et al., 2012).

The extant research on CC has considered both governmental and non-governmental sectors regarding stakeholders' engagement and role (Sebos et al., 2016). In the process of decision-making the involvement of stakeholders is really important. The use of stakeholder engagement mitigates potential future risks by allowing relevant groups to voice their opinions and enhance the effectiveness of policies and activities. Authors have evaluated the awareness level regarding CC impacts on coastal regions among institutional stakeholders (Piwowarczyk et al., 2012) as well as viewpoints on the repercussions of CC on tourism activities in destinations heavily reliant on climatic resources (Dincă et al., 2014).

2.2.2 A Historical Context of Human Migration

The climate has influenced life on Earth. Species have evolved to endure and flourish in their environments; if conditions alter, they must either adapt to the new circumstances or relocate to seek more favorable conditions. The history of human evolution illustrates this distinctly. The most significant migration waves occurred around every 20,000 years, coinciding with the cold and arid conditions of the early Pleistocene, which were intermittently interrupted by warm summers in the northern hemisphere due to the axial precession of the Earth (Praetorius et al., 2023). The warm intervals established verdant "corridors" linking Africa and Eurasia, facilitating human migration in both directions across the continents.

Likewise, fluctuations in the intensity of the Atlantic Meridional Overturning Circulation—a significant oceanic circulation pattern—resulted in "wet phases" that rendered the typically "inhospitable" Sahara and Sahel "amenable" to human transit during specific periods (Castañeda et al., 2009). Alterations in ice sheets may have impacted human migration patterns, since "early populations may have utilized winter sea ice that linked islands and coastal refugia" for movement, according to a study (Kelman, 2015). Approximately 12,000 years ago, the Earth transitioned into a novel geographical era known as the Holocene, distinguished by "exceptionally stable" climatic conditions. According to a feature article in the journal *Nature*, the consistent temperatures and availability of freshwater "sustained the conditions that facilitated human development" (Rockström et al., 2009).

In the newly stable climate, nomadic groups established themselves in areas with favorable conditions to create the initial communities. Agricultural practices rapidly evolved, and presently, only a minor fraction of humanity continues to engage in nomadic lives. Consistent seasons have enabled individuals to establish patterns of seasonal travel. Many communities engage in transhumance, which involves the seasonal relocation of animals between designated summer and winter pastures (Xu et al., 2020). Experts caution that the Earth is currently exiting its secure climatic zone. Numerous experts contended that the earth has entered a new another period of geological transformation termed the Anthropocene, distinguished by human impact on the climate. In this new climatic period, regions of the Earth that have been inhabited for millennia may become inhospitable.

2.2.3 The Interconnection Between Environment and Migration

Since the 1980s, various conceptual ideas of mobility across human beings in CC context have emerged, including the "refugees and migrants" in the context of environment as well as climate. Firstly, the term "environmental refugee" was originally documented in a 1985 report by the "United Nations Environment Program (UNEP)" and referenced in political and academic discourse (Han et al., 2024). It typically denotes significant harm to the habitat and the pressing necessity to obtain refuge due to a distinct pressure and urgency. "Climate refugees" has remained a buzzword on news channels as well as social websites frequently which is referred as a novel category of individuals compelled to migrate across borders due to CC (Li, 2012). "Refugees" term is associated with both the environmental and climate-related aspects however, because this term has specific political implications and broadening the definition of refugees could result in significant adverse consequences (de Siqueira & da Luz, 2018). Consequently, during past few years, other impartial terms such as "environmental or ecological migration" and "climatic migration" have gained widespread usage. These are both subcategories of environmental migration (Cao & Chen, 2013).

In contrast to the term "refugees," "migrants" exhibit a greater degree of subjective human agency, while being influenced by climatic and environmental factors. The terms "migrants" and "refugees" are secondary to HM induced by CC. The scholarly community has developed a more detailed classification of HM types in the relation to the CC. Current concept may be categorized into many forms, including temporal measurement, degree of impact, scale of migration scale, inclination to migrate, and distance of migration (Han et al., 2024). On the basis of the extent of impact of CC on migrants, the categories can be (a) severe environmental refugees, (b) environmentally compelled migrants, and (c) environmentally induced migrants. On the other hand, under a temporal perspective, the migrants can be grouped into and phased migrated individuals and permanent migrated individuals. The spectrum of HM encompasses individual labor movement, family migration, and extensive regional concentrated migration.

Undoubtedly, within the framework of the global community, the substantial human migration of both "migrants" and "refugees" would significantly influence the sustainable development of both the destination and origin environments (Hartmann, 2010). People may adopt alternative tactics, including liquidating assets, enhancing source of revenue, capitalizing in resilience measures, curtailing unnecessary costs, and utilizing communal systems or public programs for support. Households are inclined to select some approach predicated on anticipated yields and the associated risks, in addition to their contextual comprehension, morals, social connections, and traditional bonds (Adger, 2009). Migration is a multifaceted phenomenon, with droughts, floods, and other climate-related issues being merely a subset of the numerous elements influencing the decisions of families or individual members to relocate (Stapleton et al., 2017). Consequently, migration persists as a significant and prevalent response.

2.2.4 Scenarios of Environmental Migration

The tendency to migrate in relation to environmental change is different as per situation. First is migration during the early phases of progressive environmental alteration. In the initial phases of environmental deterioration, farmers harvest diminished crops, fisherman capture reduced quantities of fish, and pastoralists encounter increasingly limited grazing areas for their livestock (Millock, 2015). As earning potential diminishes, household members may resort to internal or cross-border migration for employment and to produce additional income through remittances. Labor mobility may positively influence environmental restoration initiatives by reducing demographic strain on limited natural resources. Upon their return, migrants might utilize the funds and skills they have learned to enhance infrastructure and land-use practices, ultimately mitigating environmental damage (IPCC, 2014).

Migration at advanced phases of progressive environmental alteration is the next. Ongoing environmental degradation, potentially intensified by climate change, may entirely jeopardize local ecosystems and livelihoods (Gallopín, 2006). Previously transient phases of natural resource constraint may evolve into prolonged or potentially permanent conditions. Complete industries may disintegrate, resulting in elevated unemployment and a total depletion of household income. Increasing sea levels pose an impending risk of displacement. Consequently, households that once participated in temporary or circular labor mobility may arrive at a tipping point when the expenses of return migration surpass the benefits of sustaining increasingly diminishing local livelihoods (Gray & Mueller, 2012). This scenario may lead to a transition from circular to permanent migration, potentially harming social connections, living standard, social status, and right to use the services offered to public.

Then there could be migration resulting from severe environmental occurrences. The most recognized situation involves extensive human displacement following natural and industrial calamities (Han et al., 2024). Although not exclusively caused by environmental factors, catastrophic tsunamis, earthquakes, and floods have rendered millions homeless and deprived of essential services. Migration resulting from extensive development and land preservation is also prevails. The development of significant infrastructure projects, such as dams, may necessitate the temporary or permanent displacement of communities (Millock, 2015). By redirecting conventional resource flows, such initiatives can transform patterns of human settlement, mobility, and health outcomes.

2.2.5 Causes of Climate-Induced Migration

In the chapter, for understanding migration concept, the household is regarded as the decision-making entity that perceives migration as an approach to avoid risks. It is to be noted that migration decisions are impacted by weather conditions as they alter vulnerability and capability of households. Household's vulnerability is a manifestation of susceptibility to climatic disturbances and sensitivity to these disturbances (Adger, 2006). Extreme weather situations heighten vulnerability by compromising the plans of a household for earning in the future. These weather anomalies could adversely impact the production efficiency of a household's in relation to natural resources in the long run, including a livelihood approach, such as rain-fed agriculture or fertility of land (Gray & Mueller, 2012). Capability on the other hand refers to ability to select and lead a life deemed valuable in the eyes of such households (Sen, 1999). It encompasses both tangible and intangible resources together enabling the households to opt for relocation in reaction to harsh climatic disturbances.

Alternatively, due to structural connection, climatic disturbances could compromise general assurance systems. Vulnerability encompasses is not only a current condition, in fact it is a prospective scenario, particularly a threat that a household may enter or persist in scarcity (Calvo & Dercon, 2013). Contrastingly, capability is recognized as a factor of vulnerability (McLeman & Hunter, 2011). A household having substantial capacity has power to mitigate the vulnerability towards prospective shocks and reduce overall sensitivity about these conditions. Nonetheless, vulnerability being a more expansive situation which is merely the lack of competence and, therefore, although interconnected, the former is not an antithesis to the latter. A household may possess significant wealth; nevertheless, due to situational variables or methods of earning, it might concurrently exhibit considerable vulnerability. Due to the potential for variations in capabilities to influence the decisions to migrate in a manner that may contradict alterations in responsiveness as well as exposure, we acknowledge its intrinsic role in a household's vulnerability (Adger, 2006; Gallopín, 2006).

Furthermore, certain non-climatic factors may influence both competence and vulnerability as previously outlined (Adger, 2009; Black et al., 2011). Wealth along with the existence of social connections positively influence migration capacity and concurrently diminish vulnerability. Land tenure exerts an equivocal influence: as in a case households with enhanced tenure security are anticipated to possess a stronger capacity for migration, owing to the association of land ownership with wealth generation. Nonetheless, those possessing larger landholdings reflect increased demands of workforce and heightened vulnerability to fluctuations in agricultural productivity, such as climatic disturbances. We see a contrast in two migration categories based on their risk profiles: "risk-reducing" migration and other is "discretionary." Migration could elevate or diminish risks to household income and, hence, wellbeing.

According to NELM, individuals from more vulnerable households are more likely to become migrants for diversifying their means of earning and mitigate risk in reaction to CC. Vulnerable households may engage in strategically "risk-reducing" movement, for example pursuing short term employment in adjacent metropolitan regions in response to a poor agricultural harvest or its anticipation (Hunter & Nawrotzki, 2016). Conversely, less vulnerable households with enhanced skills may opt for "discretionary" migration, including on account of education, nuptial, and moving abroad, that might not inherently mitigate risks and could even elevate risks in the near term. This type of "discretionary" migration seeks to enhance anticipated income in the future or for improving other aspects, encompassing utility derived from marriage. This type of migration does not inherently mitigate risk but it is typically associated with higher costs, whereas risk-reducing migration tends to incur lower costs. This differentiation among migration types elucidates the variations in migratory patterns.

The research regarding the effects of CC on HM frequently differentiates between "fast onset events (FOE) and slow onset events (SOE)" occurrences (Bohra-Mishra et al., 2014). In addition to this, it differentiates between "direct and indirect" connections.

2.2.5.1 Fast Onset: Landslides, Intense Rainfall, Flooding, Hurricanes

The effects of storms and floods on displacing the residents are readily apparent, as they present themselves in a harsh and direct fashion. Displacements are generally transient and occur across brief distances (McLeman & Gemenne, 2018). Those who suffer are predominantly residing in impoverished nations, possess limited means for traveling to long-distance (Zickgraf & Perrin, 2016), and most displaced individuals return promptly to rebuild their houses in the disaster-affected area. The findings from various global study programs consistently validate this assertion. Globally, the prevailing opinion is that the storms and severe rains possess only a limited capacity of inducing migration in extensive way or for long time period, particularly concerning international migration that necessitates crossing the border of a country.

Recent studies on CC acknowledges the anticipated rise in the occurrences of extreme weather events significantly affecting the livelihoods of populations (Kim & Marcouiller, 2017). The effects of such series of events can significantly differ from a singular disaster, irrespective of the severity of each event. The perceptions of impacted communities regarding their future significantly influence their decision to relocate. Several studies have noted a correlation between the frequency of hazards and migration, albeit without comprehensive empirical investigation. Owing to insufficient data, a limited number of research can accurately measure the precise effects of hazard repetitions (Bohra-Mishra et al., 2014). The comprehension of these effects on mobility and the corresponding thresholds thus represents a pivotal inquiry for forthcoming research and data gathering initiatives.

2.2.5.2 Slow Onset: Drought, Desertification, and Rising Temperatures

The impact of desertification and droughts on HM is typically slow as compared to the FOE discussed earlier, as droughts are usually linked to progressively gradual departures. Diverse case studies have illustrated a divergent perspective on the impacts of CC on patterns of migration. Numerous prominent instances of huge population departures, primarily internal displacement, are evident in Southern and Central Asia, the Middle East, South America, and Africa. Hammer (2004) provides a comprehensive chart detailing forced migration resulting from droughts (and floods) in the Sahel from 1973 to 1999, highlighting a specific instance in 1985 where one million individuals were evacuated, either temporarily or permanently, due to drought in Niger. Conversely, other writers contend that these figures remain quite minor relative to the population impacted by drought, as the push factors being merely one of numerous elements affecting decisions to migrate (Black et al., 2011).

Certain experts share perspectives like to those of Nobel laureate Amartya Sen, asserting that mobility, akin to famines, is predominantly only weakly linked to environmental variables, with migration primarily driven by political reasons. A study analyzed migration among different provenances of Burkina Faso and indicated that CC account for just a minor portion of human mobility (Henry et al., 2003). A comparable outcome is proposed in Syria also (Selby et al., 2017). In specific instances, the effect was found to be inverted as in the case of Mali in mid-1980s on account of drought because a decline in HM cross boarders was noted because of insufficient financial resources to support the travel (Findley, 1994). As per Cattaneo and Peri (2016), a steady rise in temperatures diminishes global migration from impoverished nations, due to the existence of financial limitations.

The portrayal of drought-affected individuals as powerless sufferers compelled to escape is misleading, since it disregards the adaptation efforts made at individual, community, and national levels (Kaczan & Orgill-Meyer, 2020). Thus, quantifying the extent of individuals at threat and the possibilities of HM as a consequence of droughts driven or, more broadly, due to increased temperature would be challenging. In summary, SOE such as land degradation and droughts typically induce HM that is regarded as intentional and primarily economically driven, whereas FOE like hurricanes, storms, and floods causes abrupt, spontaneous, and immediate displacements.

2.2.6 *The Effect of Migration on the Environment*

Environmental variables influence human mobility, and conversely, movement can affect the environment. Instances of abrupt mass migration can significantly impact the environmental conditions of the host region (Ionesco et al., 2016). Historically, extensive areas of land have been cleared of forests to establish camps or villages. Displaced individuals may engage in unsustainable resource management to survive in exceptional and prolonged situations (Hunter & Nawrotzki, 2016). Certain development and land conservation projects have produced environmental impacts akin to those of resettled populations that lack viable alternative livelihoods or are situated in uninhabitable regions, leading to the overexploitation of natural resources (Black et al., 2011). This situation hampers preparations for SDGs for regional and municipal administrations, as it is expected that, for the first time, a majority of the global population will inhabit urban areas instead of rural ones (Peri & Sasahara, 2019).

Further, in the absence of inexpensive housing and adequate sanitation, migrants may resort to deforesting and acquiring coastal sand for construction materials, as well as eating contaminated water and food supplies (Bohra-Mishra et al., 2014). These coping strategies not only expedite deforestation and soil erosion but also restrict the water drainage capacity of increasingly impervious surfaces and the population's access to safe, clean water, leading to further public health decline (Cattaneo & Peri, 2016). Consequently, previously benign rainfalls can result in severe flooding and mudslides, widespread contamination of essential water and food resources, and a swift decline in public health. Even when urban planning has accounted for migrant inflows, irregular migration may nonetheless exert pressure on infrastructure and services (Kibreab, 2009).

The impact of HM on environment in origin areas also warrants attention. The impact of migration on the resilience of ecosystems in places of origin varies based on the characteristics and magnitude of migratory flows, as well as the initial condition of environmental deterioration (Gedikli et al., 2022). Mobility can alleviate population density concerning available resources and resource-management strategies; nevertheless, extensive permanent migrations may deplete the communities of origin, hindering their capacity to address subsequent environmental degradation (McLeman & Hunter, 2011). Following disaster-induced displacement, large-scale returns may pose challenges to environmental management in the places of origin. The challenge intensifies when reconstruction and rehabilitation efforts lag behind the rate of returns.

2.3 Research Methodology

This chapter is a review-based research. A comprehensive range of publications have been studied for writing this chapter. To get a clear understanding of the basic terms, first of all, the literature have been reviewed on the topic of CC as well as HM individually. History of human migration is offered to build the foundation of the topic. Then the research published on the topic of integration and linkage between migration and environment is reviewed including different scenarios of environment related migration. Further, climate-induced migration is discussed in detail along with fast onset and slow onset events. Finally, effect on migration on destination and origin regions are presented. At the end, implications to navigate the climate-induced migration are provided.

2.4 Implications

Recent research indicated that migrants use the migration as a strategy for adapting with CC, conserving the environment, and reducing the risk their home countries. Initially, the evidence must be enhanced to more effectively guide policies and actions, ensuring that responses align with actual needs (Nwokolo et al., 2023b). Increased efforts are necessary to enhance capacity to conduct research at local level and knowledge to elevate data quality. Enhancing the comprehension of HM's effects on migrants, origin, and destination areas is essential for accurately evaluating whether migration alleviates vulnerability or exacerbates risks and precariousness. Policies must be formulated properly. Due to climate change, demographic shifts, economic developments, and geopolitical dynamics, various migratory causes are expected to become more interconnected, challenging conventional explanations and classifications of migration and rendering current frameworks of governing migration obsolete. Effectively addressing such difficulties requires the formulation of complete remedies grounded in substantial and credible information and data.

Additionally, migration and social issues ought to be more effectively incorporated into policies regarding the climate, sustainability, environment, natural disaster coping methods, conversely, environmental and climatic considerations should have more recognition within migration policies. It may entail acknowledging the capacity of migrants to enhance adaptation with CC and management of sustainability in their origin nations, fostering a conducive atmosphere for the movement's "green" investments, or meeting the distinct requirements and susceptibilities of migrants within local or regional strategies of development (Nwokolo et al., 2023c). There are other instances, which may entail enhancing management of resource, restoring those ecosystems which are degraded, generating substitute "green" options for employment, and upgrading essential public services to deter out-migration, so enabling societies to remain and facilitating the return of migrants and displaced populations.

Addressing environmental migration at the institutional level necessitates enhanced dialogue, coordination, and collaboration among entities and stakeholders engaged in environmental, sustainability, and climate initiatives, as well as those focused on migration and humanitarian efforts, across intergovernmental, public, academic, and private sectors. Simultaneously, enhanced communication between research and policy sectors will be crucial to guarantee that actions address actual need. The efficacy of our response to current and impending difficulties related to migration and environmental change will hinge totally on our capacity for effective communication, the integration of diverse skills, and collaborative efforts.

Ultimately, initiatives to foster a more equitable dialogue on migration must persist. The advantageous characteristics and benefits of migration are frequently disregarded, while its adverse impacts are overstated, leading to the perception of migrants as a danger. In conjunction with enhanced integration strategies, a more robust acknowledgment of the beneficial facets of migration is essential to counteract misconceptions around migrants and movement, so fostering greater acceptance of migrants in host nations and regions. Simultaneously, it is crucial to assist those who wish to remain in their land, houses, and communities by implementing regulations that enable them to do so. Environmental migration is a multifaceted phenomenon. The present moment necessitates not only the acknowledgment but also the resolution of environmental migration through constructive, practical, and inventive strategies that enhance and transcend conventional methods and existing frameworks.

2.5 Conclusion

Migration can both exacerbate and mitigate the vulnerability of individuals and communities. An increasing number of professionals in academia and policy acknowledge migration as a viable adaptation and risk mitigation technique in response to environmental challenges. Relocation from high-risk places can preserve lives, alleviate environmental strain and protect delicate resources, while offering people access to alternative, more stable sources of income. But present climate change projections, a specific degree of compelled climatic migration is inevitable and the extent of impact is contingent upon the international community's current mitigation and adaptation strategies. The world community must confront the likelihood of extensive displacement resulting from climate change. A substantial amount of research is required to comprehend the causes and effects of climate migration and to track the statistics. Practitioners should enhance communication and collaborative ties among the various human rights, population, environmental, and migratory agencies tasked with addressing population displacement.

Despite the significant implications offered by this chapter, we acknowledge a few limitations. First, the chapter is entirely based on literature and implications derived from within, hence empirical evaluations should be made to ensure their practical applicability. The review may be limited by the availability of studies on specific regions or populations. Certain areas affected by climate change may not be well-researched, leading to potential gaps in understanding various patterns of migration. The interplay between CC, HM, and socioeconomic factors is not be explored in this chapter which should be examined in future. Researchers can also examine how communities adapt to CC and the role of HM as a coping approach.

References

Adger, W. N. (2006). Vulnerability. *Global Environmental Change, 16*(3), 268–281. https://doi.org/10.1016/j.gloenvcha.2006.02.006

Adger, W. N. (2009). Social capital, collective action, and adaptation to climate change. *Economic Geography, 79*(4), 387–404. https://doi.org/10.1111/j.1944-8287.2003.tb00220.x

André, K., Simonsson, L., Swartling, G., & Linnér, B.-O. (2012). Method development for identifying and analysing stakeholders in climate change adaptation processes. *Journal of Environmental Policy & Planning, 14*(3), 243–261.

Bhushan, M., & Singh, R. (2024). Talent management facilitates net-zero transition through employee green behavior. In R. Singh, S. Khan, A. Kumar, S. Luthra, & H. Chokshi (Eds.), *Net zero economy, corporate social responsibility and sustainable value creation. CSR, sustainability, ethics & governance* (pp. 117–129). Springer. https://doi.org/10.1007/978-3-031-55779-8_7

Biermann, F., & Boas, I. (2010). Preparing for a warmer world: Towards a global governance system to protect climate refugees. *Global Environmental Politics, 10*(1), 60–88. https://doi.org/10.1162/glep.2010.10.1.60

Black, R., Bennett, S. R. G., Thomas, S. M., & Beddington, J. R. (2011). Climate change: Migration as adaptation. *Nature, 478*, 447–449. https://doi.org/10.1038/478477a

Bohra-Mishra, P., Oppenheimer, M., & Hsiang, S. M. (2014). Nonlinear permanent migration response to climatic variations but minimal response to disasters. *Proceedings of the National Academy of Sciences, 111*(27), 9780–9785. https://doi.org/10.1073/pnas.1315945111

Calvo, C., & Dercon, S. (2013). Measuring individual vulnerability. *Social Choice and Welfare, 41*(4), 721–740.

Cao, Z. J., & Chen, S. J. (2013). Resources and environment in the Yangtze Basin. *National Science Library, Chinese Academy of Sciences, 22*(4), 527–534.

Castañeda, I. S., Mulitza, S., Schefuß, E., Lopes dos Santos, R. A., Sinninghe Damsté, J. S., & Schouten, S. (2009). Wet phases in the Sahara/Sahel region and human migration patterns in North Africa. *Proceedings of the National Academy of Sciences, 106*(48), 20159–20163.

Cattaneo, C., & Peri, G. (2016). The migration response to increasing temperatures. *Journal of Development Economics, 122*, 127–146. https://doi.org/10.1016/j.jdeveco.2016.06.002

Chophel, Y. (2022). Global warming and climate change (GWCC) realities. In *The nature, causes, effects and mitigation of climate change on the environment* (Vol. 3). IntechOpen.

de Siqueira, E. C. V., & da Luz, C. K. (2018). Contextualizing environmental migration: The gap between the legal nature of refuge and environment during the age of global warming and natural catastrophes. *Revista de Direito Econômico e Socioambiental, 9*(1), 125–141.

Dincă, A. I., Surugiu, C., Surugiu, M., & Frenţ, C. (2014). Stakeholder perspectives on climate change effects on tourism activities in the northern Romanian Carpathians: Vatra Dornei resort case study. *Human Geographies, 8*(1), 27–41.

Fadeyi, O., & Maresova, P. (2020). Stakeholders' perception of climate actions in some developing economies. *Climate, 8*(66), 1–21.

Findley, S. E. (1994). Does drought increase migration? A study of migration from rural Mali during the 1983-85 drought. *International Migration Review, 28*(3), 539–553. https://doi.org/10.2307/2546950

Gallopín, G. C. (2006). Linkages between vulnerability, resilience, and adaptive capacity. *Global Environmental Change, 16*(3), 293–303. https://doi.org/10.1016/j.gloenvcha.2006.02.004

Gedikli, A., Taş, N. B., & Yalçın, A. K. (2022). Breaking the vicious cycle between migration and environmental degradation: The role of government. In A. Gedikli, N. B. Taş, & A. K. Yalçın (Eds.), *Analyzing sustainability in peripheral, ultra-peripheral, and low-density regions* (pp. 156–178). IGI Global.

Gray, C., & Mueller, V. (2012). Drought and population mobility in rural Ethiopia. *World Development, 40*(1), 134–145. https://doi.org/10.1016/j.worlddev.2011.05.023

Hammer, T. (2004). Desertification and migration: A political ecology of environmental migration in West Africa. In J. D. Unruh, M. S. Krol, & N. Kliot (Eds.), *Environmental change and its implications for population migration*. Kluwer.

Han, Q., Kumar, R., & Kumar, A. (2024). Climate change and human migration: Perspectives for environmentally sustainable societies. *Journal of Geochemical Exploration, 256*, 107352.

Hartmann, B. (2010). Rethinking climate refugees and climate conflict: Rhetoric, reality, and the politics of policy discourse. *Journal of International Development, 24*(6), 233–246. https://doi.org/10.1002/jid.1676

Henry, S., Boyle, P., & Lambin, E. F. (2003). Modelling inter-provincial migration in Burkina Faso: The role of socio-demographic and environmental factors. *Applied Geography, 23*(2–3), 115–136. https://doi.org/10.1016/S0143-6228(03)00019-5

Hunter, L. M., & Nawrotzki, R. (2016). Migration and the environment. In M. White (Ed.), *International handbook of migration and population distribution* (pp. 465–484). Springer.

IDMC. (2017). *Global internal displacement database.* Internal Displacement Monitoring Centre. Retrieved http://www.internal-displacement.org/database/displacement-data

Ionesco, D., Mokhnacheva, D., & Gemenne, F. (2016). *The atlas of environmental migration.* Routledge.

IPCC. (2014). Summary for policymakers. In O. Edenhofer, R. Pichs-Madruga, & Y. Sokona (Eds.), *Climate change 2014: Mitigation of climate change* (pp. 1–33). Cambridge University Press. https://doi.org/10.1017/CBO9781107415324

Joshi, A., Singh, R., Kumar, V., & Rani, S. (2024). Net zero strategy: Exploring opportunities, confronting challenges for developing countries. In R. Singh & D. Crowther (Eds.), *Transition towards a sustainable future. Approaches to global sustainability, markets, and governance* (pp. 233–252). Springer. https://doi.org/10.1007/978-981-97-5756-5_11

Kaczan, D. J., & Orgill-Meyer, J. (2020). The impact of climate change on migration: A synthesis of recent empirical insights. *Climatic Change, 158*(3), 281–300.

Kelman, I. (2015). Difficult decisions: Migration from small Island developing states under climate change. *Earth's Future, 3*(4), 133–142.

Khan, I., Alharthi, M., Haque, A., & Illiyan, A. (2023). Statistical analysis of push and pull factors of migration: A case study of India. *Journal of King Saud University Science, 35*(8), 102859. https://doi.org/10.1016/j.jksus.2023.102859

Khan, S., Singh, R., Alnahas, J., Abbate, S., & Centobelli, P. (2024). Navigating the smart circular economy: A framework for manufacturing firms. *Journal of Cleaner Production, 480*, 144007. https://doi.org/10.1016/j.jclepro.2023.144007

Kibreab, G. (2009). Climate change and human migration: A tenuous relationship. *Fordham Environmental Law Review, 20*, 357.

Kim, H., & Marcouiller, D. W. (2017). Mitigating flood risk and enhancing community resilience to natural disasters: Plan quality matters. *Environmental Hazards*. https://doi.org/10.1080/17477891.2017.1400331

Klepp, S. (2017). Climate change and migration. In *Oxford research encyclopedias, climate science*. Oxford University Press. https://doi.org/10.1093/acrefore/9780190228620.013.42

Kumar, K., Singh, R., Sharma, A., & Lehri, B. P. S. (2024). Climate change adaptation strategies for achieving net-zero economy. In V. Kandpal, E. D. Santibanez-Gonzalez, P. Chatterjee, & M. K. Nallapaneni (Eds.), *Transition towards a sustainable future* (pp. 163–183). Springer.

Li, W. (2012). On the dilemma and solution of international legislative protection for climate refugees. *Journal of Hainan University, 30*(1), 67–72.

McLeman, R., & Gemenne, F. (Eds.). (2018). *Routledge handbook of environmental migration and displacement*. Routledge.

McLeman, R. A., & Hunter, L. M. (2011). Migration in the context of vulnerability and adaptation to climate change: Insights from analogues. *Wiley Interdisciplinary Reviews: Climate Change, 1*(3), 450–461. https://doi.org/10.1002/wcc.51

Millock, K. (2015). Migration and environment. *Annual Review of Resource Economics, 7*(1), 35–60.

Myers, N. (2002). Environmental refugees: A growing phenomenon of the 21st century. *Philosophical Transactions of the Royal Society B: Biological Sciences, 357*(1420), 609–613. https://doi.org/10.1098/rstb.2001.0953

Nwokolo, S. C., Singh, R., Khan, S., & Kumar, A. (2024). Threats to the rapidity of sustainability transitions posed by technological changes. In *Technological pathways for Africa's net-zero economy* (pp. 39–75). Elsevier. https://doi.org/10.1016/B978-0-443-31486-5.00002-1

Nwokolo, S. C., Singh, R., Khan, S., Kumar, A., & Luthra, S. (2023a). Remedies to the challenges of renewable energy deployment in Africa. In *Africa's path to net-zero. CSR, sustainability, ethics & governance* (pp. 59–74). Springer. https://doi.org/10.1007/978-3-031-44514-9_3

Nwokolo, S. C., Singh, R., Khan, S., Kumar, A., & Luthra, S. (2023b). Africa's awakening to climate action. In *Africa's path to net-zero. CSR, sustainability, ethics & governance* (pp. 299–310). Springer. https://doi.org/10.1007/978-3-031-44514-9_9

Nwokolo, S. C., Singh, R., Khan, S., Kumar, A., & Luthra, S. (2023c). Global investment and development in Africa. In *Africa's path to net-zero. CSR, sustainability, ethics & governance* (pp. 15–58). Springer. https://doi.org/10.1007/978-3-031-44514-9_2

Nwokolo, S. C., Singh, R., Khan, S., Kumar, A., & Luthra, S. (2023d). Impacts of climate change in Africa. In *Africa's path to net-zero. CSR, sustainability, ethics & governance* (pp. 237–262). Springer. https://doi.org/10.1007/978-3-031-44514-9_7

Peri, G., & Sasahara, A. (2019). *The impact of global warming on rural-urban migrations: Evidence from global big data*. National Bureau of Economic Research.

Piwowarczyk, J., Hansson, A., Hjerpe, M., Chubarenko, B., & Karmanov, K. (2012). Climate change in the Baltic Sea region: A cross country analysis of institutional stakeholder perceptions. *Ambio, 41*, 645–655.

Praetorius, S. K., Alder, J. R., Condron, A., Mix, A. C., Walczak, M. H., Caissie, B. E., & Erlandson, J. M. (2023). Ice and ocean constraints on early human migrations into North America along the Pacific coast. *Proceedings of the National Academy of Sciences, 120*(7), e2208738120.

Rigaud, K. K., DeSherbinin, A., Jones, B., Bergmann, J., Clement, V., Ober, K., et al. (2018). *Preparing for internal climate migration*. The World Bank. https://doi.org/10.7916/D8Z33FNS

Rockström, J., Steffen, W., Noone, K., Persson, Å., Chapin, F. S., Lambin, E. F., Lenton, T. M., et al. (2009). A safe operating space for humanity. *Nature, 461*(7263), 472–475.

Sebos, I., Progiou, A., Kallinikos, L., Eleni, P., Katsavou, I., Mangouta, K., & Ziomas, I. (2016). Mitigation and adaptation policies related to climate change in Greece. In P. Grammelis (Ed.), *Energy, transportation and global warming* (pp. 35–49). Springer. https://doi.org/10.1007/978-3-031-25748-3_8

Selby, J., Dahi, O., Fröhlich, C., & Hulme, M. (2017). Climate change and the Syrian civil war revisited. *Political Geography, 60*, 232–244. https://doi.org/10.1016/j.polgeo.2017.05.002

Sen, A. (1999). *Development as freedom*. Oxford University Press.

Singh, R., Gupta, A., & Kandpal, V. (2024). ESG excellence in the digital world: Circular economy E-business solutions. In E. D. R. Santibanez Gonzalez, V. Kandpal, P. K. Ozili, & P. Chatterjee (Eds.), *Fintech for ESG and the circular economy* (pp. 193–204). Wiley. https://doi.org/10.1002/9781394234035.ch10

Smith, K., & Smith, K. (2013). *Environmental hazards: Assessing risk and reducing disaster* (6th ed.). Routledge. https://doi.org/10.4324/9780203805305

Sta Maria, A. D. (2023). Labour migration and exclusive state amidst the global pandemic of COVID-19. In S. Petcharamesree & M. P. Capaldi (Eds.), *Migration in Southeast Asia, IMISCOE research series* (pp. 91–106). Springer. https://doi.org/10.1007/978-3-031-25748-3_8

Stapleton, S. O., Nadin, R., Watson, C., & Kellett, J. (2017). *Climate change, migration, and displacement: The need for a risk-informed and coherent approach*. Overseas Development Institute and United Nations Development Programme.

UNDESA. (2017). *International migration report 2017*. United Nations Department of Economic and Social Affairs.

UNHCR. (2018). *Global trends in forced displacement in 2017*. United Nations High Commissioner for Refugees. https://doi.org/10.1080/09584939308719709

Wu, X., & Wang, Q. (2018). Review on research progress of climate migration. *Yuejiang Journal, 10*(2), 61–70.

Xu, C., Kohler, T. A., Lenton, T. M., Svenning, J. C., & Scheffer, M. (2020). Future of the human climate niche. *Proceedings of the National Academy of Sciences, 117*(21), 11350–11355.

Zickgraf, C., & Perrin, N. (2016). Immobile and trapped populations. In F. Gemenne, D. Ionesco, & D. Mokhnacheva (Eds.), *The atlas of environmental migration*. Routledge.

Chapter 3
Ocean Health and Climate Neutrality: Interdisciplinary Sustainable Perspectives

Wasswa Shafik

Abstract Ocean health and climate neutrality were discussed prominently at the Sustainability Research Biennial, as well as a side event of a public discussion organized within the United Nations Climate Change Conference in Bonn, Germany, in 2017. This chapter touches critically on both topics and expands on areas relevant to innovative approaches to addressing the interconnected challenges of climate change and ocean health. Specifically, this interdisciplinary sustainable technique focuses on ocean health and ultimately integrates interconnected aspects such as acidification, overfishing, eutrophication, and marine plastic pollution in more detail. Several international organizations in marine sciences and governance advise a drastic reduction of CO_2 emissions, as it is believed this is required to preserve ocean and Earth system stability. Limited consensus, however, remains on how to proceed without further affecting marine ecosystems, industries, or communities in the relatively near future. This chapter further discusses the roles of different stakeholders and invites interested readers to follow up on specific interconnections as they prefer a holistic perspective. This perspective looks at "ocean health" as directly relevant to a sustainable future. Oceans constitute the habitat for the majority of living organisms, including single-celled plankton and complex life like fish, marine birds, and mammals. In turn, the sea provides more than 50% of atmospheric oxygen and is a major nutrient, and hence food, provider for the principal human food sources: land-based plants and animals. Thus, one task of an *Anthropocene* global society is to intentionally align itself with ecological systems to avoid detrimental effects resulting from the inevitable interactions of complex systems. One way forward, this essay suggests, is to argue for an integration of parallel, albeit conceptually discrete, mechanisms.

Keywords Climate neutrality · Ocean health · Sustainable perspectives · Sustainable development goals · Sustainable ecosystem

---

W. Shafik (✉)
Dig Connectivity Research Laboratory (DCRLab), Kampala, Uganda
e-mail: wasswashafik@dcrlab.org

R. Singh and D. Crowther (eds.), *Climate Neutrality and Sustainable Ecosystems*, Approaches to Global Sustainability, Markets, and Governance,
https://doi.org/10.1007/978-981-96-7181-6_3

3.1 Introduction

The earth is at war due to human activities that degrade it either directly or indirectly. Oceans cover almost three-quarters of the Earth's surface and harbor a rich and growing diversity of ecosystems, including the flora, fauna, and microorganisms that inhabit them. The term "ocean health" reflects the state of these systems in terms of their structure, functions, and resilience to human activities (Silva et al., 2024). It includes marine ecosystem components, such as water and seafloor habitats, as well as the flora and fauna of which they are composed. Definitions of ocean health are not only limited to marine ecosystem health but may also include physical, chemical, and biological water quality parameters affecting human health. Another term that has become increasingly popular is "climate neutrality." Climate neutrality refers to achieving a balance between the amount of greenhouse gas emissions produced and the amount removed from the atmosphere for net-zero greenhouse gas emissions. The two terms, ocean health and climate neutrality, are thus closely related and pivotal toward achieving sustainable development (Kelly et al., 2024).

Consensus is frequently attained on the direct general meanings of sustainability-associated terms like climate neutrality but faces obstacles at a strategic and operational level. Both the absence of internationally agreed methods and metrics for calculating, or even measuring, net-zero emissions and the notable discrepancy in measuring ocean health represent an open scientific question on how to assess and monitor ocean health at a global scale due to significant variations in ocean health indicators. The need for a unified framework underscores the complexity of the issue, as ecological and social components are interconnected and cannot be analyzed separately (Zhong et al., 2024). Ensuring the coordination of activities between researchers and policymakers, as well as the integration of local, regional, national, and international stakeholders, cannot be underestimated either.

The ocean plays a significant role in the provision and upholding of life support systems for all the organisms circulating in the biosphere. It regulates Earth's thermal balance, our climate system, and the generation of oxygen. The ocean is one of the central carbon sinks, providing a marine food source for more than half of the world's population and offering a number of essential but still underexplored valuable drugs. However, anthropogenic pressures, as well as climate change impacts, have undermined the ocean's role in life support systems, indirectly threatening human health and development (Araujo, 2024; Shafik, 2024a). Sustainable development requires a substantial effort on several fronts to re-establish sustainable relations with the ocean and recognize the ocean and climate system's importance for achieving the UN Sustainable Development Goals (SDGs). The transition toward a sustainable blue society provides multiple opportunities for economic growth, for new industries, and scientific and technological advances. The exploitation of these opportunities could lead to a net increase in job demand, with job benefits not being equally distributed in affected regions. Measures must be foreseen so that the transition path can foster all communities and avoid social dumping in sustainable development (Khan et al.,

2024). Creating a healthy ocean and reaching climate neutrality targets requires collaborating at different levels.

3.1.1 Defining Ocean Health and Climate Neutrality

The number of references to "ocean health" suggests that the phrase is of relatively recent provenance. However, the meaning of the term is far from clear, and it is regularly used with reference to circumstances as diverse as ocean acidification, biodiversity, climate regulation, sustainable resource use, pollution, and marine ecosystems. Consequently, the term is required to bear an enormous amount of conceptual weight. The term "ocean health" would, therefore, seem to be operating as an unmistakably speculative conceptual hodgepodge, encompassing many discrete difficulties related to the ocean (Leng et al., 2024). Direct analysis of ocean health, therefore, disentangles a great many distinct problems. By failing to identify and address these issues separately, the health of the oceans and the role of research and educational institutes in maintaining that health are endangered. Drawing on the precautionary principle in law, as it is commonly represented, it is argued that while we are uncertain of the consequences of various forms of human intervention in our oceans, it is better to opt for that interpretation of "ocean health," which presents less risk of irreversible damage (Lasch et al., 2024).

3.1.2 The Role of Oceans in Climate Regulation

Interdisciplinary approaches between various fields should tackle the effects of human activity with relevant pathways to investigate in order to promote ocean health and climate neutrality. Throughout the past few years, several frameworks and strategies have been developed focusing on increasing ocean health by observing multiple aspects for action. Given the increase in demand for alternative solutions, methodologies are also required that include different bodies of knowledge. Only recently, global ideas have underlined that system leadership addresses, decentralizes, and empowers sustainability science and practice, arguing for the ocean to become a sustainable biological system (Wang et al., 2024). The ocean exerts a critical effect on the global environment, but not all such effects are positive. This chapter primarily focuses on the roles that the ocean can fulfill to regulate the Earth's climate system. The roles of the ocean in regulating the Earth's climate system are as follows: control of Earth's water cycle, storage and redistribution of heat, regulation of atmospheric structure, and regulation of the Earth's average temperature (Shafik, 2024b; Vonk et al., 2024). To continue to maintain these roles, protecting and restoring marine ecosystems must become more of a priority.

In concluding this chapter, in order for the ocean to continue to function successfully as the main engine of Earth's climate system, more long-term interdisciplinary research is required. More action is needed to protect and restore marine ecosystems using integrated solutions based on the best scientific evidence available. It is necessary to support an increase in ocean literacy and consciousness globally. One of the most challenging issues facing humanity is how not only to achieve a reduction in greenhouse gases but to reduce them to net zero after accounting for emissions from the already heavily burdened atmosphere. The ultimate aim is climate neutrality, where the ocean has a vital role to play (DuPrey Diederichsen et al., 2024; Kumar et al., 2024). We must make the most of the ocean's potential to contribute to climate neutrality, reduce atmospheric CO_2, and improve the overall well-being of planet Earth.

3.2 Ocean Carbon Sequestration

According to current scientific understanding, most climate change is driven by the increased greenhouse gas concentration in the atmosphere caused by human activities. More than 60% of the increased CO_2 released into the atmosphere is absorbed by the sea, which results in a number of adverse climate changes. The sea stores a large amount of heat and transports it into the depths, reducing the greenhouse effect and the Earth's temperature, which will result in an alteration of the global equilibrium (Lasch et al., 2024; Silva et al., 2024). Changes in the properties of the deep sea, such as water temperature, can affect deep ocean circulation, and changes in deep ocean circulation will also affect the Earth's climate. The increase in temperature at the sea surface will lead to the decline of the solubility of atmospheric carbon dioxide, enlarge the degree of ocean carbon dioxide saturation, and affect the pH value in seawater, causing the shells of some marine creatures to dissolve (Araujo, 2024).

The impact of climate change on the ocean is mutual. On the one hand, the ocean can exacerbate climate change, while on the other hand, climate change will seriously endanger the ocean ecosystem on which the Earth depends. Therefore, research on ocean health and climate neutrality and the development of sustainable ocean management policies are urgently needed as measures to respond to climate change. The ocean is the most significant carbon dioxide sink in the world. The absorption of large amounts of carbon dioxide by the ocean contributes to climate stability (Khan et al., 2024). In today's economic and environmental transformation, the concept represented by the development of an economy with 80% renewable energy has become more widely recognized and calls for a system of zero net carbon emissions requirement. Under this background, there are many pioneering works associated with energy development, although they may have different focuses. The carbon dioxide in the atmosphere can promote the photosynthesis of planctomycetes in the ocean, a process referred to as Fischer's effect, and accumulate a large amount of biological ash in the ocean, changing the deep-sea average carbon-to-carbon ratio (Leng et al., 2024; Shafik, 2024c).

3.3 Impacts of Climate Change on Ocean Health

Scientists have documented dramatic and often unprecedented changes in marine ecosystems as a consequence of climate change. The overview of observed and predicted effects on the physical and chemical environment, species distributions, genetic diversity, survival and growth, nutrition and trophic interactions, shores and freshwater habitats, and socio-economic impacts of marine climate change, focusing on the Mediterranean Sea.

3.3.1 Planet's Climate and Dynamics

The discussion is supported by findings from projects in which these effects were assessed, for instance, by sea level, ocean circulation, and biogeochemical model simulations, regional high-resolution circulation modeling, the evaluation of salinity tolerance, and hatchery trials and experimental studies. The ocean is a principal part of the climate system: the exchange of heat and freshwater with the atmosphere controls the planet's climate, and dynamics in the ocean interior help to sequester anthropogenic CO_2, thereby mitigating the impacts of climate change. The ocean also provides food, energy, and other resources and services to a growing human population; participates in and modulates water, energy, and nutrient exchange across interfaces such as the air-sea and land-sea boundaries; and influences air quality and the severity of natural hazards such as hurricanes (Leng et al., 2024; Wang et al., 2024). The condition of the ocean can alter the established relationships between these systems and can lead to feedbacks that influence climatic conditions and the ecosystem health of all connected interfaces.

3.3.2 Ocean Acidification

Marine life is accustomed to relatively constant, naturally occurring pH levels. However, since the Industrial Revolution, rapid changes in ocean chemistry have been observed as a result of increasing atmospheric carbon dioxide concentrations. The uptake of CO_2 by the ocean stimulates a series of chemical reactions that result in increasing hydrogen ion and bicarbonate ion concentrations and a decrease in carbonate ion concentration, ultimately resulting in a decrease in seawater pH and a decrease in the availability of calcium carbonate minerals. This shift in chemistry is referred to as ocean acidification and lowers the pH of the ocean, giving rise to an impairment of calcification by marine organisms that rely on calcium carbonate, primarily invertebrates with shells and echinoderms (DuPrey Diederichsen et al., 2024; Vonk et al., 2024). Despite many unknowns and complexities, ocean acidification has been globally characterized based on substantial efforts to determine both the ongoing state of the oceans' carbonate balance and projected future changes therein using the concentration of CO_2 either dissolved in the surface ocean or equilibrium with current atmospheric or future projected atmospheric CO_2 concentration. Over thousands of years, ocean acidification will gravely and fundamentally alter the assemblages of marine organisms and their ecosystems that exist today. This will have significant implications for ecosystems worldwide and an enormous impact on the 2.2 billion people in the developing world who rely on seafood as a primary protein source (Samiullah & Khanum, 2024).

3.4 Interconnectedness of Ocean Health and Climate Neutrality

The ocean performs many critical services that sustain life as we know it on Earth. It has absorbed more than 90% of the excess heat in the climate system during the period from 1971 to date.

3.4.1 Earth's Greenhouse Gas Levels

The effects of this heat absorption are pervasive, and the ocean now stores more than 60% of the heat associated with ongoing global warming. The ocean has also absorbed 23% of the carbon dioxide from anthropogenic sources, mitigating the rise of Earth's greenhouse gas levels. The oxygen bill provided by marine organisms is crucial for oxygen-breathing organisms. Approximately 50–70% of Earth's oxygen is produced by marine photosynthetic organisms in the upper sunlit layer of the ocean. Marine biology plays a curious role in the Earth's duality as an integrated living system. The decomposition of organic materials from the land, as well as marine-produced organic matter, is performed by the hydrolytic and concomitant reactions caused by biodiversity bounded with marine sediments or muds (Laiolo et al., 2024). By doing these processes, marine geobiodiversity in accelerating ocean acidification and deoxygenated zones buffers the natural variety of atmospheric CO_2 and N_2O concentrations. Yet, as a living organization, the ocean is highly integrated from the shore to the open sea, demanding cooperative integration among the oceanographic sciences measurements, models, simulations, forecasts, projections, understanding, and ocean planning. This provides different interdisciplinary approaches to assessing the health of the ocean (Hua et al., 2024). These backend and frontend integrations as detailed below.

3.4.2 Natural Sciences Perspectives

Today, we know better than ever about the state of the earth's oceans and the damages caused by human activities, mainly through the lens of natural science. A wide range of methods is used in natural science to understand the oceans, monitor ocean health and marine ecosystems, and identify the needed measures to reach climate neutrality. This can range from methods of collecting quantitative or qualitative data on chemicals, organisms, or ecosystems on board research vessels at a local level to more specific field studies and experiments about single organisms or ecosystems. Research about scientific facts shows how complex marine ecosystems are organized globally, how they work, and what long-term changes they have undergone (Fauville et al., 2024). The impacts of climate change on oceans include an increase in temperature, rise in sea levels, increase in ocean acidification, changes in the distribution of nutrients, atmospheric oxygen content, alteration of storm and weather patterns, overfishing, pollution, marine litter, and microplastics, and other jellyfish blooms. There is also cooling of the Atlantic Ocean surface, changes to current patterns, significant uplift risk of marine and coastal life, and the economic loss of marine and coastal habitats and the services and indicators that such habitats hold. Evidence bases and existing gaps about the overall state of the earth's oceans and the cause-and-effect relationship between human activities and harmful impacts on the oceans in various fields of natural sciences are informed (Omedes et al., 2024).

3.4.3 Social Sciences Perspectives

In the literature about the sustainability of a society, energy is considered the key to social metabolism, and economics has provided valuable perspectives. This approach is equally valid in the case of the viability of an ocean settlement, where social and natural variables have been integrated. Social sciences contribute to a better understanding of the ocean and human societal needs as an intrinsic part of both ecological and human systems. In general, natural scientists are moving toward an understanding of social-ecological systems characterized by adaptive governance, considering the importance of the dynamics between human depredation and the capacity for ocean endurance (Kelly et al., 2024). Social sciences are interested in socio-economics, not in the "economy of the sea" or "the behavior of fishermen." It is important to understand under what circumstances fishing communities represent one subset of the process by which societies exploit the ocean and interact with ecological systems and fisheries. In this approach, the community cannot be encapsulated by the ecology of fishing; the stock-based, single-species, developed country, expensive labor offshore or coastal sectors dimension solely defines a community's interactions with resource stock and ecosystem (Araujo, 2024). Different forms of community interaction are, of course, described and analyzed, but the emphasis is on spatial and temporal diversity and variability in the way societies exploit different resources and systems.

3.4.4 Ecosystem-Based Approaches to Climate Neutrality

Marine species perform 4.2% of carbon sequestration. The great importance of oceans as carbon sinks justifies the development of approaches for the sustainable positive impact of the ocean on the global carbon balance against the background of global climatic challenges. The concept of Ocean Health is proposed as the basis of the Blue Economy development strategy. It is suggested to advance and develop systemic approaches applied to the multifunctional use of coastal zones through the development of green-blue infrastructure based on the principles of ecosystem-based adaptation and disaster risk reduction. The creation of green-blue infrastructure is an approach that incorporates the working of ecosystems into planning and designing specifically engineered solutions that stem from traditional ecological planning (Khan et al., 2024; Leng et al., 2024). Taking into account the ongoing research focused on the engineering of reefs and the development and improvement of various methods for increasing the quantity and variety of species attracted to the reefs and their habitats, it is possible to start lengthening the list of carbon sink functions of reefs. The implementation of strategies minimizing and preventing the breakage of coral reefs as carbon sinks will not only accomplish the goals of ocean health but also generate income from climate neutrality projects. It is only in recent years that marine ecosystems have been mentioned as a source of ecological engineering methods. The current paper puts marine ecosystems in the place they deserve and emphasizes not only the danger factor but also the adaptive capabilities of marine ecosystems, which remain largely unexplored. Oceans are attempting to compensate for our anthropogenic errors and mistakes (Lasch et al., 2024). Therefore, it is time to stop taking them for granted. Enhancing the biological value of marine ecosystems and their ecological services could be offered in return for the reduction of the present anthropogenic pressure on the sea.

3.5 Technological Innovations for Ocean Health and Climate Neutrality

The current technological advances in computing, artificial intelligence, machine learning, and big data are fundamentally reshaping fields as diverse as life sciences, energy, transportation, finance, and leisure services.

3.5.1 Sustainable Disruptive Technologies

The next wave of technological innovations and their potential transformative impacts on ocean health and climate neutrality point toward an emerging research area, "Ocean Analytics." The use of the term "analytics" aspires to go beyond merely harnessing big data; in addition, it draws across scientific disciplines, integrating different ways of knowing and engendering awareness as a means of both envisioning and achieving systemic and fundamental change. Oceans, playing a critical role in the production and mitigation of greenhouse gases, are not only heavily affected by the dramatic global rise in emissions caused over the last two centuries, but the resulting acidification and climate change also hamper productivity with the most damaging consequences for human society (Vonk et al., 2024; Wang et al., 2024).

3.5.2 Integrated Ocean Observing Systems

Like many other societal and academic rosters of issues, the ocean domain has undergone significant specialization and division of labor, with the emergence of new "subfields" focusing on such missions as managing lakes and rivers, biological conservation, clean water supply, coastal development, integrated coastal zone management, and eventually the recent emergence of terms such as the "blue economy" or "blue growth." Integrated Ocean Observing Systems have also taken shape, sowing the seeds for potential ocean "big data" in the form of high-frequency ocean observation networks (Samiullah & Khanum, 2024). Assuredly, the added field specialties and global observatory networks are beneficial for collective missions, but do they risk exerting a centrifugal influence by dissolving the collective vision? This is yet to be effectively answered.

3.5.3 Renewable Energy Solutions for Marine Environments

Energy derived from renewable sources is sustainable. The EU has committed to achieving climate neutrality and meeting the objectives of the Paris Climate Agreement; scientific knowledge is a crucial tool to achieve these objectives. The research for ocean health and climate neutrality has explored the potential for renewable energy derived from offshore power-generating installations for the EU, and the research also focused on the synergy between renewable energy production and nature conservation. The research has concluded that the installation of wind and wave installations in the Atlantic Sea basin has the most significant potential to produce energy, and an assessment of their reduced carbon emissions and energy supply would provide future climate-neutral energy needs (Hua et al., 2024). The research has also determined which areas of the maritime area could be assigned to be suitable for spatial planning of nature conservation areas with energy production. The existing research acknowledges an essential element when floating installations are used, which is how such installations would connect to the national grids. The research determined that electric eel-like connections, which are floatable, could be a potential solution for connecting floating installations to the grid since there would be no need for subsea cables (Fauville et al., 2024).

3.5.4 *Industry Standard Nacelles*

Attaining reliable data for operational floating offshore wind energy installations is complex to collect due to certifying organizations not having such data since almost all of the offshore installations developed and constructed so far are bottom-fixed. Towing of retired offshore facilities should be avoided; tow lines have cut a significant number of wire-injuring species, and the finding is that wire injury have significant implications on not only the injured animal but also the functionality of sentinel devices (Omedes et al., 2024). The data collected from interactions with wind installations has also found that some of the warning signals on the top of the industry-standard nacelles do not fully meet their objective. This is because the cables, saving lines, and fixing equipment cannot carry the necessary structural weight, and the weight and stress of such equipment can cause severe implications on the industry's power nanogrid design. The research concluded that lessons learned from the past should be considered to avoid similar failures in the future (Guimbeau et al., 2024). In essence, the Green Deal offers a common-sense approach to stimulate long-term sustainable growth through research, sharing of knowledge, and applied solutions.

3.6 Policy and Governance Frameworks for Ocean Health and Climate Neutrality

These policy decisions and governance frameworks, which manage or advise on the ocean's participation in climate change mitigation, will determine the ocean's long-term condition and the services it provides.

3.6.1 *Climate-Ocean Governance Reforms*

Policies and governance depend on perceptions and the knowledge held. This involves an understanding of ocean and climate functioning and dynamics and also incorporates political and public values. An example of diving straight into the existing challenges is the latter framework, which has addressed many global environmental problems, including oceans and climate, and encapsulates much of what we know about social and natural systems and environmental behavior. The shared values and norms that emerged from these societal experiences are mainly captured in treaties and conventions (Guimbeau et al., 2024). To offer stakeholders and experts a guide of input on types of governance recommendations and policy options, we observed the target audiences and coping strategies proposed in preceding literature about situation detection and both ocean health and climate neutrality regions. The potential climate-ocean governance reforms, for instance, collective arrangements, are essential if the magnitude of the ocean health and climate neutrality twin challenge is to be controlled, adaptation policies and actions developed, and the highest ocean health and climate neutrality regions found.

3.6.2 International Agreements and Treaties

The modern spirit of cooperation in preserving the world's oceans and the sustainable use of their resources began to form only at the beginning of the second half of the twentieth century. Protection of the oceans and their components and the challenges faced became the reasons for creating new international legislative instruments. Meanwhile, the need for cooperation with developing countries on marine issues generally began to grow. The 1960s and 1970s became the years of the most active work on the creation of the legal framework for the protection of the marine environment and its practical application (Watson-Wright et al., 2024). The 1980s and 1990s, in turn, set sail as the years of institutional and legal modernization of the world ocean governance system despite the complex socio-political conditions globally and regionally.

3.6.3 World Governance System

The ocean was gaining significance as an object of international governance; its agenda was getting complete and complex, and the objective grounds and potential for all humanity to take part in the implementation were increasingly actively discussed. The mentioned world governance system of the ocean includes the institutions and mechanisms established for cooperation and decision-making in the international ocean management sector on a global, regional, and subregional level, as well as the set of rules and regulations ensuring the implementation of its goals, tasks, and functions (Gerhardinger et al., 2024). The regulatory framework of world governance is formed by a number of international instruments created explicitly within the framework of international cooperation. These instruments are capable of ensuring the international order in the ocean in terms of collective actions, collective responsibility, and goals.

3.6.4 Indigenous Knowledge in Promoting Ocean Health and Climate Neutrality

A discourse analysis of interviews with ten Pacific Indigenous knowledge (IK) experts and governmental officials, reflecting the governance approaches to Pacific climate and oceans within two critical international negotiations in 2019, the Bonn climate change talks and the United Nations General Assembly Pre-action briefing. It is reported that an Indigenous governance approach is not consistently reflected in either the forum's output or the official statements, demonstrating a disjuncture within the Pacific community between Indigenous and Western approaches. Therefore, the key message is that the Pacific community's own IK experts actively promote an Indigenous governance approach that recognizes the complementary and equal relationship between Indigenous and Western knowledge systems (Laiolo et al., 2024; Samiullah & Khanum, 2024). On the Pacific Ocean are Pacific people whose livelihoods are integrally connected to it and who, throughout thousands of years, have developed deep knowledge about the ocean and its resources. Therefore, the importance and value of Traditional Ecological Knowledge (TEK) and IK for management purposes should be recognized and incorporated into ocean management processes. Moreover, this IK should be integrated into policy considerations. The purpose of this paper is to draw on the ocean cries of Pacific cultural protestors to examine the role of IK in promoting a relationship between people and the ocean that recognizes their ecology (Khan et al., 2024).

3.6.5 Traditional Ecological Knowledge

Indigenous communities often have intricate understandings of the seas, honed through millennia of experience. This traditional ecological knowledge can fill in scientific gaps about the ocean and its inhabitants, benefiting the scientific process as well as our collective understanding of the seas. TEK is the accumulated comprehensive understanding that the community has gained through long association with a habitat (Vonk et al., 2024). Traditional knowledge of oceans and seas is very strongly aligned with the people who live adjacent to the coastal habitats; therefore, we included the term community traditional knowledge (CTK) to more specifically refer to traditional knowledge obtained from communities living adjacent to the coastal habitats. In many instances, CTK is unique to a specific community, and in some cases, traditional knowledge might be widespread and universal to a region. For example, the northern Indigenous peoples of Canada know that an incredible amount of their culture comes from water in the Arctic and what waters mean to them. Understanding these waters and ecosystems and sharing that knowledge equitably allows the mutual respect and understanding needed for good relations in the Arctic (DuPrey Diederichsen et al., 2024).

3.7 Economic Valuation of Ocean Health and Climate Neutrality

The idea of natural capital and the ecosystem services it provides, in goods and non-material well-being, is vital to the modern understanding of the use of know-what and know-how for natural resources and the environment. These resources are both crucial and rare, both of which criteria play an essential role in the determination of goods trade and environmental quality. The system has to balance the good of present consumers and the welfare of future generations. When deciding what to use, for whom, and at what cost, the concept and the knowledge of place value and financial value will also determine incentives and penalties for conservation or degradation of the services provided (Samiullah & Khanum, 2024). The text presents methodological experiences of how to determine the economic value of ocean health indicators associated with the bounds of climate neutrality. The use of the Natural Capital Accounting a method elaborated at the international level is matched with the valuation of ocean-related services. The application involves Portuguese ocean-based activities at the national and regional scales. The method is of potential prospection for other activities at the ocean or land sector level. Initial benchmark valuations are communicated (Laiolo et al., 2024).

3.7.1 Blue Economy Concepts

Species from shallow-water coral reefs are the most diverse group in the oceans, and reef-hosting communities are prevalent recreational areas. They are some of the most biologically productive and diverse environments in the world, thus having a high commercial value in terms of fisheries, leisure, and tourism. It has been reported that their direct economic value is estimated at over $30 billion per year, but the benefits are estimated to continue with the help of several ecosystem functions to claim the revenues even higher, at $170 billion per year (Hua et al., 2024). However, it is essential to note that interest in these services is not permanent and that overextended utilization or even destruction of coral reefs would lead to the loss of beneficial contributions in terms of services to human society.

3.7.2 Sustainable Economic Approach

The concept of the "Blue Economy" is associated with the goal of a sustainable ocean economy that uses marine resources to promote economic growth and improve livelihoods and jobs while maintaining the health of marine and aquatic ecosystems. To achieve this goal, it is necessary to have due regard for the relationship between economic and social development, protection of the environment, and respect for equality and democratic values. Although the concept of the Blue Economy has been introduced only in the last two decades, it has quickly gained worldwide attention because it explores a new form of balance and economic model for the ocean that combines economic development with the conservation of marine ecosystems (Fauville et al., 2024). In this context, the Blue Economy is defined as "any activity that can economically sustain increasing numbers of marine creatures and leads to economic values pursued by humans." This sustainable economic approach takes into account the relationship between increasing economic potential, conservation of marine ecosystems, and social equity, defined as a "balance that allows society to spread economic value across ecosystems and human cultures while creating efficient market and non-market signals that shape human decision-making." A study aims to identify the main challenges that currently affect the development of the Blue Economy (Omedes et al., 2024).

3.8 Educational Initiatives for Promoting Ocean Health and Climate Neutrality

Oceans play a critical role in the Earth's climate system and represent a unique educational potential for understanding and feeling the need for ocean protection and climate neutrality. While a comprehensive ocean teaching plan does not form part of most national education systems, there are numerous informative, educational resources available about the ocean. The concentration is on a selection of educational programs that are dedicated to ocean learning. The classification of environmental education types, strategies, and principles will be discussed in this chapter as well. These might be interesting for policymakers managing school curricula in different countries and researchers interested in concepts applied and theoretical foundations of awareness-raising activities, including the field of applied environmental social sciences (Guimbeau et al., 2024).

3.8.1 Integration and Effective Partnerships

Understandably, the ocean is currently at the forefront of discussions both in governments and research communities interested in reaching the SDGs, and questions concerning the incorporation of new educational themes are particularly urgent today. The problem is not simple. We can already categorize: "Theorists who doubt almost everything except the multiplicity of their borrowings. Specialists who predict everything except the risk of infanticide." Indeed, awareness, information, resource mobilization, integration, and effective partnerships are critical aspects of almost all 17 goals of the SDGs, and the cultural and ethical components described thus far must be restructured to focus on the poorest (Watson-Wright et al., 2024). The SDGs are challenging, and the most enormous difficulties are connected with the application of knowledge to the areas where the worst problems are most urgent.

3.8.2 Ocean Literacy Programs

The widespread adoption of ocean literacy programs addresses this problem and forms relatively self-contained policy interventions. However, the question arises as to their effectiveness, particularly since public and private investment in these programs continues. This study combines interdisciplinary perspectives from marine scientists to investigate the current state of these programs and subsequently specify the ocean literacy research agenda. We apply bibliometric and text analysis over the ocean literacy corpus of academic sources with some sophistication at stage one, with only fundamental analysis at stage two. Our study concludes by revealing understandable concerns over whether these programs are successfully delivering education that meets these challenges (Gerhardinger et al., 2024). They found contrasting areas of empirical and descriptive knowledge for their importance and effectiveness. Based on a variety of scientific methods, this study argues that a coherent specification of the ocean literacy research agenda is still desirable. Furthermore, the multidisciplinary implementation of marine science concepts led to different perspectives on the extent to which such concerns are valid. For instance, geologists hold somewhat opposing views on the call for increased geological knowledge for ocean-literate citizens and imply that ocean literacy draws a distorted image of scientific knowledge. In their view, ocean literacy programs should be careful to associate citizenship primarily with an attitude of responsible use of oceans (Hua et al., 2024).

3.8.3 Community Engagement and Citizen Science

The 2015 Paris Climate Treaty, together with various recent global initiatives related to sustainable development, climate neutrality, and nature-based solutions, is a welcome step. Nevertheless, one critical fault lies in the lack of linkages to the biogeochemical cycles in the ocean. As a result, targets proposed to reduce atmospheric greenhouse gases and alleviate global warming could indirectly create serious environmental problems. The biogenic volatile organic compounds produced during photosynthesis are mainly responsible for the characteristic ultraviolet-visible spectra absorbed by the blue-dark color of Earth's Ocean. This absorption generates a more stratified ambiance where marine phytoplankton grow and emit dimethyl sulfide (Fauville et al., 2024). Through its oxidation, the products of dimethyl sulfide serve as cloud condensation nuclei, thus linking to the radiative properties of clouds and affecting ocean color and temperature of seawater, which are used to estimate the contribution of dimethyl sulfide to marine dimethyl sulfide concentration.

3.8.4 Community-Based Monitoring

The effective conduct of scientific research and the translation of fundamental discoveries into technologies that facilitate economic growth, protect national security, and improve healthcare, among other societal goals, require both the participation of diverse groups and knowledge exchange between science and the broader public. Such needs compel considerations of citizen science and science communication techniques. In recent decades, science and the scientific endeavor have become more integrated into society, education, culture, and the daily lives of laypeople worldwide, yet many feel that science has had adverse effects (Hua et al., 2024; Laiolo et al., 2024). Formally, citizens returning to science is summarized as educating the public and encouraging more young people, particularly those from groups that are underrepresented in science and students attending colleges and universities that have not traditionally produced a strong pipeline of future scientists and engineers, to pursue careers in scientific fields. Informally, the conceptualization of citizen science builds on the foundation of community-based monitoring, academic citizen science, and volunteer environmental monitoring and education, including nationwide federal, state, and local agencies' programs, as well as community-based, non-profit, and academic organization activities (Fauville et al., 2024). Despite its popularity, multiple challenges exist, many of which lie in the residents' weak understanding of science.

3.8.5 Volunteer Monitoring Programs

The importance of long-term and standardized monitoring to track changes in and respond to ocean health is increasingly recognized by the public, management agencies, and the research and philanthropic communities. The growing demand for monitoring ocean health has outstripped the ability of state and federal agencies to provide comprehensive environmental monitoring programs. In response to the shortage of government-run monitoring programs, as well as in response to public interest in having hands-on involvement in natural resource issues, conservation organizations often turn to a core group of professional volunteers who respond to environmental accidents, the marine environment, and monitor ocean health (Omedes et al., 2024). The most significant and longest-standing marine volunteer effort is the Beach Watch volunteer program in Puget Sound. The program's mission is to measure trends and better understand pollution-related effects on coastal ecosystems. To accomplish this goal, Beach Watch uses thousands of volunteers to collect marine debris information, gather information on nonpoint source water pollution in Puget Sound, and monitor trends in marine algae. Other programs such as Harbor Seal Fetal Corpse programs, shellfish disease monitoring initiatives, Kelp Watch, and Whale Scout all help to create a shared link of stewardship into political influence that generates change in allowing us to preserve and conserve our marine environment and ocean health (DuPrey Diederichsen et al., 2024; Samiullah & Khanum, 2024).

To undertake a "so-what" interdisciplinary analysis of some of the most successful efforts of communities in their ocean health and climate neutrality initiatives over the past years. These efforts require the transformation of society to enable more inclusive, sustainably vibrant, and resilient communities capable of successfully coexisting with the world's biodiversity and rapidly changing and evolving climatically and economically. The success stories, albeit in various stages of development and execution, demonstrate that it is possible to address global and intractable challenges in ways that are simultaneously local and personal (Fauville et al., 2024; Hua et al., 2024). In this complex interdisciplinary co-production of knowledge, the realization of successful initiatives is highly dependent on trust in others to fulfill promises and commitments. However, the use of market mechanisms to promote commerce and social behaviors can conceal the extensive array of "human capital" underpinning the initiative, likely the primary determinant of success and sustainability. Ultimately, the transformation of society is achieved one person at a time (Guimbeau et al., 2024).

To address the complex challenges of ocean health and climate neutrality, interdisciplinary solutions are often required rather than action from a single discipline. In the coming years, it is essential to advance such approaches across disciplinary boundaries. To advance interdisciplinary work, researchers need to make data, methods, findings, hypotheses, theories, and rigor more transparent across disciplinary boundaries. This chapter provides a perspective on an everyday ocean health and climate neutrality research community and outlines some research questions and priorities that cut across these themes (Fauville et al., 2024). The course is designed to allow law school students and other graduate students with extensive conservation or legal training to collaborate to advance theory and empirical research on global issues of ocean health and climate neutrality.

3.9 Conclusion

The call for urgent and proactive policy actions is based on principles of sustainable ocean health and climate neutrality. Transformative practices, such as polycentric governance, mobilizing finance and investment at scale, and creating cross-sectoral partnerships are discussed. Noting the value of technologies to address challenges, there is a call for accountable innovation that respects the principles discussed. However, there is no single path to success; indeed, it will be a mixture of many nuanced and context-appropriate solutions that ensure human well-being and a healthy ocean. The work has also highlighted apparent deficiencies in the research base. For this to inform adaptive governance, a concerted effort is required to capture the nuances of sectoral activities, to improve capacity to understand dynamic tipping points in both ocean health and climate neutrality and to apply a systems approach to analyze complex, trade-off laden trade-offs among SDG 14. Given the time-critical nature of many of these challenges, the time to act and to research is undoubtedly now. Such heroics are required. Indeed, quite the reverse; there is still time, and the people are fit for purpose. What is needed most of all is understanding both of the intertwined challenges and that there is agency and collective power to enable the transformations needed from momentous challenges facing us.

References

Araujo, V. A. (2024). Ocean literacy strategies using flag animal as environmental awareness model. *International Journal of Zoology and Animal Biology, 7*(1), 554. https://doi.org/10.23880/izab-16000554

DuPrey Diederichsen, S., Decker Sardinha, G., Eymael Garcia Scherer, M., & Nicolodi, J. L. (2024). Opportunities to overcome the ocean decade challenges in Brazil's ocean and coastal governance system. *Ocean and Coastal Management, 247*, 106907. https://doi.org/10.1016/j.ocecoaman.2023.106907

Fauville, G., Voşki, A., Mado, M., Bailenson, J. N., & Lantz-Andersson, A. (2024). Underwater virtual reality for marine education and ocean literacy: Technological and psychological potentials. *Environmental Education Research*. https://doi.org/10.1080/13504622.2024.2326446

Gerhardinger, L. C., Colonese, A. C., Martini, R. G., da Silveira, I., Zivian, A., Herbst, D. F., Glavovic, B., Calvo, S. T., & Christie, P. (2024). Networked media and information ocean literacy: A transformative approach for UN Ocean decade. *npj Ocean Sustainability, 3*(1), 2. https://doi.org/10.1038/s44183-023-00038-2

Guimbeau, A., Ji, X. J., Long, Z., & Menon, N. (2024). Ocean salinity, early-life health, and adaptation. *Journal of Environmental Economics and Management, 125*, 102954. https://doi.org/10.1016/j.jeem.2024.102954

Hua, T., He, L., Jiang, Q., Chou, L. M., Xu, Z., Yao, Y., & Ye, G. (2024). Spatio-temporal coupling analysis and tipping points detection of China's coastal integrated land-human activity-ocean system. *Science of the Total Environment, 914*, 169981. https://doi.org/10.1016/j.scitotenv.2024.169981

Kelly, R. P., Lodge, D. M., Lee, K. N., Theroux, S., Sepulveda, A. J., Scholin, C. A., Craine, J. M., Andruszkiewicz Allan, E., Nichols, K. M., Parsons, K. M., Goodwin, K. D., Gold, Z., Chavez, F. P., Noble, R. T., Abbott, C. L., Baerwald, M. R., Naaum, A. M., Thielen, P. M., Simons, A. L., & Weisberg, S. B. (2024). Toward a national eDNA strategy for the United States. *Environmental DNA, 6*(1), 432. https://doi.org/10.1002/edn3.432

Khan, M. L., Hassan, H. U., Khan, F. U., Ghaffar, R. A., Rafiq, N., Bilal, M., Khooharo, A. R., Ullah, S., Jafari, H., Nadeem, K., Siddique, M. A. M., & Arai, T. (2024). Effects of microplastics in freshwater fishes health and the implications for human health. *Brazilian Journal of Biology, 84*, 272524. https://doi.org/10.1590/1519-6984.272524

Kumar, K., Singh, R., Sharma, A., & Lehri, B. P. S. (2024). Climate change adaptation strategies for achieving net-zero economy. In R. Singh & D. Crowther (Eds.), *Transition towards a sustainable future. Approaches to global sustainability, markets, and governance*. Springer. https://doi.org/10.1007/978-981-97-5756-5_8

Laiolo, E., Alam, I., Uludag, M., Jamil, T., Agusti, S., Gojobori, T., Acinas, S. G., Gasol, J. M., & Duarte, C. M. (2024). The global ocean genome: A "catalog" of ocean life. *Frontiers for Young Minds, 11*, 1052361. https://doi.org/10.3389/frym.2023.1052361

Lasch, K., Gollner, S., Oude Elferink, A., Rühs, S., Sangiorgi, F., van Sebille, E., & Wang, J. (2024). Whose Ocean? Exploring multidisciplinary perspectives towards ocean sustainability and implications for the un(der)represented. *Research Ideas and Outcomes, 10*, e114485. https://doi.org/10.3897/rio.10.e114485

Leng, T., Liu, B., Xie, J., Zou, L., & Lu, D. (2024). Essential and non-essential elemental composition and health risks in pacific sardine in the Northwest Pacific Ocean. *Journal of Marine Science and Engineering, 12*(3), 368. https://doi.org/10.3390/jmse12030368

Omedes, S., Crespo-Picazo, J. L., Robinson, N. J., García-Párraga, D., & Sole, M. (2024). Identifying biomarkers of pollutant exposure in ocean sentinels: Characterisation and optimisation of B-esterases in plasma from loggerhead turtles undergoing rehabilitation. *Chemosphere, 348*, 140770. https://doi.org/10.1016/j.chemosphere.2023.140770

Samiullah, M., & Khanum, R. (2024). Climate change and ocean acidification: Unraveling the complex interactions and implications for marine ecosystems. *International Journal of Natural Sciences, 4*(1), 2397. https://doi.org/10.47604/ijns.2397

Shafik, W. (2024a). Community and artificial intelligence-enabled disaster management and preparedness. In G. Meraj, S. Hashimoto, & P. Kumar (Eds.), *Navigating natural hazards in mountainous topographies. Disaster risk reduction.* Springer. https://doi.org/10.1007/978-3-031-65862-4_13

Shafik, W. (2024b). Sustainable development goal 14: Explainable AI (XAI) for ocean health. In D. De, D. Sengupta, & T. A. Tran (Eds.), *Artificial intelligence and edge computing for sustainable ocean health. The Springer series in applied machine learning*. Springer. https://doi.org/10.1007/978-3-031-64642-3_8

Shafik, W. (2024c). *Sustainable development. Building community resiliency and sustainability with tourism development* (pp. 1–25). IGI Global. https://doi.org/10.4018/979-8-3693-5405-6.ch001

Silva, F. S. V. F., Carvalho, T. A. D. A., Mota, P. D. D. B., & Vasconcelos, V. N. D. S. A. (2024). Health promotion of the military police: Advances, challenges and trends in mental health. *Revista Brasileira de Seguranca Publica, 18*(1), 1764. https://doi.org/10.31060/rbsp.2024.v18.n1.1764

Vonk, A. N., Bos, M., Smeets, I., & Sebille, E. (2024). A comparative study of frames and narratives identified within scientific press releases on ocean climate change and ocean plastic. *Journal of Science Communication, 23*(1), 201. https://doi.org/10.22323/2.23010201

Wang, Z., Xue, C., & Ping, B. (2024). A reconstructing model based on time–space–depth partitioning for global ocean dissolved oxygen concentration. *Remote Sensing, 16*(2), 228. https://doi.org/10.3390/rs16020228

Watson-Wright, W. M., Wells, P. G., Duce, R. A., Gilardi, K. V., Girvan, A. S. T., Huber, M. E., Kershaw, P. J., Linders, J. B. H. J., Luit, R. J., Vivian, C. M. G., & Vousden, D. H. (2024). An ocean science-policy interface standing the test of time. *Marine Pollution Bulletin, 199*, 115917. https://doi.org/10.1016/j.marpolbul.2023.115917

Zhong, K. X., Chan, A. M., Collicutt, B., Daspe, M., Finke, J. F., Foss, M., Green, T. J., Harley, C. D. G., Hesketh, A. V., Miller, K. M., Otto, S. P., Rolheiser, K., Saunders, R., Sutherland, B. J. G., & Suttle, C. A. (2024). The prokaryotic and eukaryotic microbiome of Pacific oyster spat is shaped by ocean warming but not acidification. *Applied and Environmental Microbiology, 90*(4), e0005224. https://doi.org/10.1128/aem.00052-24

Chapter 4
Impossible Is Nothing: Achieving Climate Neutrality in Developing Countries

Michael Oyedele Oyenuga

Abstract Climate neutrality in developing countries is a key challenge while sustainable development is an essential objective, calling for all hands to be on deck to halt climate disruption. In this manuscript we explore a range of perspectives on how to enable the transition, focusing in particular on sustainable and resilient land management including agriculture; and increased uptake of what are now mainstream renewable energy options (though these will be only part of the solution); agency-led community engagement as well deployment by business. Sustainable land management (SLM) practices like agroforestry, or soil conservation can greatly enhance carbon sequestration and make the landscape climate-resilient. The sharp switch to renewable fuels is also required in part, as it helps lessen greenhouse gas emissions and encourage policies that "make investments [in green energy] more attractive". I also argue for improved international and technological support to help developing countries meet their climate goals. This paper illustrates how different mechanisms can be exploited to ensure that developing countries—enveloped in their diverse socio-economic patterns—can move towards a carbon neutral future. The manuscript highlights that in the long run, climate neutrality should be employed as a policy advocacy discourse peering from an environmental stewardship and sustainable development standpoint specifically for policymakers and stakeholders.

Keywords Carbon neutral · Climate-neutrality · Developing countries · Sustainable development · Sustainable land management

4.1 Introduction

Climate neutrality in developing countries is a major global challenge that calls for an approach which encompasses environmental sustainability and economic development. With the effects of climate change more prevalent, good strategies that

M. O. Oyenuga (✉)
School of Business, Woxsen University, Hyderabad, Telangana, India
e-mail: oyenuga.michael@woxsen.edu.in

R. Singh and D. Crowther (eds.), *Climate Neutrality and Sustainable Ecosystems*, Approaches to Global Sustainability, Markets, and Governance,
https://doi.org/10.1007/978-981-96-7181-6_4

reduce greenhouse gas emissions and ensure sustainable growth are in high demand. While most of the international focus is on G20 countries such as China, in reality, it's developing nations that suffer disproportionately from climate change—even though they are least responsible for global emissions. For instance, Africa accounts for only about 2–3% of global emissions but is the most vulnerable region in the world (UNEP, 2024). This vulnerability is underpinned by the dismal level of socio-economic development on the continent. Climate change is planetary, but its impacts are vast—even partly defined by being unequally—distributed to the poorest sections of society. This is the result of a resource scarcity leading to goods and services they require for buffering themselves and, recuperating from worst-hit consequences of climate change being out of their reach. This is the case in Africa. In this regard, the United Nations Environment Programme Africa office has targeted its climate change work on supporting countries to establish an architecture for the implementation of their NDCs following the highest priority social economic imperatives which are food security, provision of income, and entrepreneurial opportunities, particularly for youth employment as well as a path for sustainable growth.

Developing nations can hardly be expected to pursue their version of carbon neutrality for several reasons. This involves, among other things, the scarcity of financial resources or a technological infrastructure in place but rather for capacity building on good governance and communication (Digitemie, 2024). The notion of climate neutrality includes not just the need to reduce emissions but also that for increasing carbon sinks, through sustainable land-use approaches. The Sahel region's Great Green Wall and other relevant initiatives demonstrate how regenerative land management practices may restore degraded ecosystems, store carbon, as well as enhance food security (Mechiche-Alami et al., 2022). Developing countries can benefit by adapting practices for achieving land degradation neutrality to build resilience against climate impacts, at the same time contributing towards global climate goals (Rozanov, 2024). In addition, climate finance has a critical role to play in helping provide these nations the financial capital they require for their transitions into low-carbon economies. Financial support can also give way to the development of indispensable renewable energy, sustainable agriculture and infrastructural projects that are paramount in realising climate neutrality (Digitemie, 2024). The success of developing countries in achieving their climate targets has social equity and economic inequalities but renders international cooperation and significance to action agreements with interests (Chen, Msigwa, et al., 2022; Wang, Chen, et al., 2022). By 2050, the goal is to have a climate-neutral world, and many countries and organisations are already working towards this by setting net zero targets. As of November 2023, around 145 countries had announced or were considering net zero targets, covering close to 90% of global emissions.

4.1.1 What Is Climate Neutrality?

Climate neutrality is the point at which an entity such as a country, organisation, or individual emits no net greenhouse gases; only reduces and offsets greenhouse gas emissions. The idea of climate neutrality is closely associated with efforts to address global warming by reducing carbon dioxide and other greenhouse gas emissions into the atmosphere down to net zero. It also means a state where the net greenhouse gas (GHG) emissions are equivalent to zero, following an equal balance of GHGs being emitted and removed from the atmosphere. Recently understanding of this has developed including an awareness that it is needed for climate change mitigation and also to achieve global temperature stabilisation targets such as the 1.5 °C limit (Guntuka, 2024; Yangka et al., 2018). This well exceeds the planned emission reductions to carbon neutrality (Duffy, Prudhomme, Duffy, Iannetta, et al., 2022; Sharma, 2023), and it is commonly estimated as a goal by combining with carbon-offsetting strategies like reforestation or new capture methods of its kind for various sectors such as energy, transportation forestry, and agriculture.

While achieving net zero greenhouse gas emissions by balancing the amount of emitted greenhouse gases with an equivalent measure of the GHGs that are either taken from the atmosphere or offset elsewhere adds a criterion relative to climate neutrality, even more important is accounting for regional and local bio-geophysical effects whereby certain human activities can result in these opposite climatic differential impacts. Put simply, it is that we reduce our emissions through climate action and have no net impact on climate change. At the outset of this, UN Climate Change instituted Climate Neutral Now in 2015 so that individuals and organisations across the globe could act to reduce their emissions commodities, that is, go climate-neutral. Even though the target is a world aiming for "climate neutrality by 2050", Climate Neutral Now stresses that we must take action today to achieve it.

Climate Neutral is when you reduce your emissions down to zero, be it in an individual or a corporate context. Many organisations cannot become carbon neutral instantaneously though, and will require years to implement the technological and behavioural changes, necessary for going green. However, for those less patient—as well as both shortcomings implemented today and extra cuts possible now with investment in projects outside the organisation—they can also invest to avoid emissions taking place elsewhere (so-called offsets). This gives an immediate leg up for the world in reducing emissions and it puts a "carbon price" on their Corporate—ladder that wants them to offset those credits even faster. Compensation units can also be purchased by private individuals and households because—like businesses, they will not be able to achieve climate neutrality overnight. Offsets are not a permanent solution and they should not be seen as an alternative to an organisation's getting to zero as quickly as possible, but instead provide the mechanism by which this can be accelerated.

The way to neutrality varies, and each country or even region should set their targets as per the kind of context they are based in with what technique is feasible. For example, in many developed economies there are plans to achieve territorial climate neutrality by 2050 which requires an integrated approach across governance processes involving policy, technology, and engagement (Duffy, Prudhomme, Duffy, O'Donoghue, et al., 2022; Wolf et al., 2021). Transitioning to climate neutrality is not just a technical issue; it also requires society-wide acceptance and engagement, as well as the coordination of different actors—such as governments, businesses and civil society (Albrecht et al., 2022; Christidis, 2024). Moreover, this definition of climate neutrality is broader than carbon emissions and does not exclude other greenhouse gases; in context, it calls all attention to a systemic approach needed to tackle sustainability (Guntuka, 2024). So today to have more sustainable development and address some of the pressing challenges that are increasingly presented by climate change, industries as well as communities tend to adopt innovative practices and technologies contributing towards achieving a state closer if not equal with "climate neutrality" (Schmidt et al., 2022; Seyfried et al., 2023). Therefore, climate neutrality is a major goal in the worldwide fight against mounting climate change that demands orchestrated steps of multiple sectors and focus towards emission cuts as well as strengthening carbon sequestration. Some standards for carbon neutral certification permit high levels of carbon offsetting, as an example. However, net zero standards mandate that emissions reductions must reach 90%+ with an offset for the remainder to apply in a true sense of 1.5-degree aligned product or service delivery (The Net-Zero Standard, 2023).

The terms net zero emissions, carbon neutrality and climate neutrality are often used interchangeably and occasionally inappropriately. These terms, however, in some cases notional the same meaning (Jeudy-Hugo et al., 2021; Rogelj et al., 2021), they are some terms in the international language of global climate change. Having a net zero carbon footprint, or being "carbon neutral" is the point at which all carbon emissions released into the atmosphere are balanced by the removal of an equal amount of atmospheric CO_2 directly from some form of long-term storage site in one way or another. For all intents and purposes, carbon neutrality will reduce your impact on climate change. Not all greenhouse gases would be included in becoming carbon neutral—although, as the dominant one responsible for up to four-fifths of global warming CO_2 is prioritised—and sectors don't necessarily need to reach that point to limit their room for manoeuvre.

Climate neutrality covers all greenhouse gases, not only carbon. Similar to carbon neutrality, climate neutrality refers to producing greenhouse gases at the rate at which they are removed from the atmosphere. A climate-neutral source in not increasing/climbing the global rise of atmospheric temperature. Climate neutrality signals a broader transition to sustainability and environmental accountability that needs input from all sectors of society in every region to tackle emissions and address the threats posed by global warming. The biggest contribution of any organisation on the journey towards climate neutrality and a sustainable future for this planet is through energy efficiency, adapting renewable sources, and absorbing carbon.

4.2 Importance of Climate Neutrality to Developing Nations

Climate neutrality is of high importance not only to developing countries but serves numerous collaterals in achieving the goals and objectives set out by global agreements on combating climate change for sustainable development. The impacts of climate change (e.g. extreme weather events, food insecurity and economic fragility) are often most keenly felt by developing countries which are typically the least responsible for increasing CO_2 emissions per capita on a national level that drive global warming and associated effects (Shively et al., 2012). The aim of a climate neutral world is to mitigate these effects and reduce greenhouse gas emissions, thereby increasing resilience against them. Developing nations will gain the following from achieving climate neutrality:

(a) Economic growth: Through the adoption of low-carbon technologies and practices. Investments in renewable energy, energy efficiency, and sustainable agriculture can create jobs, stimulate local economies, and reduce dependence on fossil fuels (Colenbrander et al., 2015). For instance, the implementation of cost-effective low-carbon options can support social and economic development and build institutional capacities for climate mitigation (Colenbrander et al., 2015). This is particularly important in middle-income countries, where early investments in climate change mitigation can yield significant economic returns (Chen, Yang, et al., 2022).

(b) Food security and agricultural resilience: As climate change poses significant risks to agricultural productivity, particularly in low-income countries, adopting sustainable agricultural practices can help farmers adapt to changing climatic conditions while reducing emissions (Hertel & Lobell, 2014). By focusing on agricultural adaptation, developing nations can improve their food systems and ensure that they are better equipped to handle the challenges posed by climate change (Shively et al., 2012). Low-income countries are most vulnerable to climate change because the vast majority of their populations live in rural areas and depend on agriculture for sustenance, with limited capacity or land resources that might allow food production at higher elevations (Hertel & Lobell, 2014). Sustainable agricultural practices, such as conservation tillage that improves nutrient use efficiency while conserving water will additionality reduce emissions associated with adaptation—from changes in management activities for these farmers. Attention to agricultural adaptation would allow developing countries to increase their food systems and manage climate change as it arises (Shively et al., 2012).

(c) Addressing social inequalities: There is an uneven impact of climate on the already unequal distribution of resources and opportunities in developing countries (Diffenbaugh & Burke, 2019). Local Priority for Climate Neutrality Sustainable development is required to be equitable and socially inclusive (Cevik & Jalles, 2020), rather than dealing with novel adaptation or mitigation measures. Debt-for-nature swaps, for example, help to address both climate change and inequality by creating resources that go into conservation efforts as well as reducing the debt burden on many low-income countries (Cevik & Jalles, 2020).

(d) International support: The race to transition to be climate neutral also corresponds with multinational obligations for climate protection, as the Paris Agreement highlights collective efforts to restrain worldwide temperature rises above pre-industrial levels (Jabakhanji et al., 2022). However, when many nations are taking actions to being or maintaining neutrality in regard of their climate contributions, working actively toward this objective may increase the credibility and thus international support for receiving these means from other countries (Lopez et al., 2022). Support is also critical to help developing countries implement strong climate policies and meet their Sustainable Development Goals.

(e) Innovation and capacity building: Developing countries can build a sustainable economy that is less dependent on fossil fuels by investing in renewable energies, which will also be more resilient to overdependence shifting patterns (Shu et al., 2022). Not only does this help to meet climate goals, it also improves the general well-being of their populations by providing cleaner air, better health outcomes and future-proof economies (Borghi, 2024).

(f) Mitigate the risks of carbon leakage: Developing nations' commitments to climate neutrality propose an opportunity for leadership through sustainable practices that are more likely to appeal to or receive backing from donors and investors (Duffy, Prudhomme, Duffy, Iannetta, et al., 2022). This is important because many developing countries are already under duress due to factors relating to climate change; with heightened vulnerabilities in terms of natural disasters, economic instability and poverty looming large above their heads.

(g) Financial aid: Climate neutrality is consistent with the international climate agenda and calls for complementary contributions of all countries to address global warming according to the Paris Agreement. Climate financing mechanisms can facilitate access to climate finance and allow countries to attain their mitigation commitments by adopting neutrality goals and meeting other development needs (Digitemie, 2024). This funding is vital for enacting practices to build economic resilience in the long term and sustainable environmental outcomes. In addition to the above, by establishing national parks and other conservation areas, carbon sink capacities may be raised, further bringing about climate neutrality, biodiversity retention, and ecotourism advancement (Zhao et al., 2022). This twofold benefit highlights the need to combine environmental conservation with economic development strategies.

4.3 Challenges Faced by Developing Nations in Implementing Climate-Neutrality

Developing nations face numerous challenges in their pursuit of climate neutrality, which are compounded by their unique socio-economic contexts and vulnerabilities. These challenges can be categorised into several key areas:

1. Resource constraints: Most developing countries face a scarcity of financial resources, making it difficult for them to invest in required climate neutrality technologies and infrastructure. Renewable energy projects and energy efficiency improvements require high initial capitalisation which can dissuade investment (Fu et al., 2011) The situation is aggravated by the North-South technology gap as developing countries are often hampered by not having access to those superior and more advanced technologies which could expedite their transition toward low-carbon economies (Fu et al., 2011). For developing nations, one of the most serious challenges is to find funds for investments in climate-neutral technologies and infrastructure. The whole field is one of low budget in many nations, so competing claims afoot are on child poverty and healthcare or education (Xue et al., 2023). Consequently, climate funds are then overshadowed to direct more at developmental programmes which take precedence. One reason may be that there are high upfront costs from shifting to renewable energy, and an increase in efficiency due to the pollution created by non-energy-saving companies or investment-eating projects but across sustainable agricultural practices (Ma, 2023). In addition, developing nations may find it difficult to obtain international climate finance as a result of often convoluted application mechanisms with strict eligibility requirements (Xue et al., 2023).
2. Dependence on fossil fuels: One of the largest and most difficult challenges for emerging nations is their dependence on fossil fuels to drive energy access, as well as contribute to greater economic growth. However, this dependency furthers the GHG emissions and creates barriers to switching to renewable energy. The very structures of our economy which place greater value on fossil fuel extraction and use rather than sustainable alternatives can be one major hurdle to moving the transition along (Giri et al., 2023). The economies of developing countries depend on fuels from the fossil class which are coal and oil, the two biggest sources causing greenhouse gas emissions (Kaewunruen, 2024). Beyond the technical difficulties are economic ones—many of these sources provide jobs and contribute to national revenues. A case in point is China where coal has played a pivotal role as one of the main engines for economic growth, and moving toward carbon neutrality brings remarkable challenges to its job security, and regional economies reliant on mining (Ma, 2023). Over time, such overreliance has created an uncomfortable double bind where the need for growth runs into conflict with emission reductions that few of these countries have been willing to accept.
3. Vulnerability to climate impacts: Countries in the process of development are hit hardest by climate change: they endure violent weather situations, suffer from malnutrition and fall into health emergencies. Immediate needs frequently

overshadow longer-term mitigation strategies, diverting attention and resources from climate neutrality initiatives. For example, agricultural sectors in some of these nations are highly climate variable as well with substantial implications for balancing food security and emissions reduction efforts (Giri et al., 2023).

4. Institutional and governance issues: Large swaths of this require effective governance to put climate policies into action, in a context where too many developing countries struggle with poor institutional frameworks and bad-practice-style governance. Corruption, political will and the participatory approach of the public are some reasons which distort the development and implementation of such policy (Ojha et al., 2015). In addition, climate policy-making is itself complex as it typically deals with anticipatory risks (instead of current problems) that are difficult to determine and account for in an intervention (Ojha et al., 2015).
5. Technological barriers: The technology transfer process is the key to attain climate neutrality, but many developing countries are struggling in terms of technology access and commercialization. Some of these barriers raised are nonexistent market intelligence, inadequate funding stream, and obstacles in nurturing relationships with technology providers (Karakosta, 2016). In the absence of efficient technology transfer strategies, it would be difficult for developing nations to avail of this climate-imperative innovation (Karakosta, 2016).
6. Economic inequality: In many developing countries, high levels of economic inequality can act as a barrier to the achievement of climate neutrality. This situation leads to inequity, as wealthier people invariably enjoy more relative access to resources and agency regarding climate policy development that will likely build social stressors by favouring the interests of a few at the expense of precious opportunities for everyone (Semieniuk & Yakovenko, 2020). Reducing these inequalities is necessary to develop a more socially inclusive climate action that produces benefits for all (Semieniuk & Yakovenko, 2020).
7. Climate finance: Developing countries have long been impeded in terms of having access to climate finance. International climate finance mechanisms do exist, but many countries experience difficulties in accessing and managing the diverse range of funding possibilities (Ha et al., 2015). For example, the divide between developed and developing countries in climate finance discourse adds a challenge to securing funding for adaptation initiatives (Ha et al., 2015). To sum up, developing countries are battling all their problems to maintain zero carbon neutral nations and they need unique ways ideally fitting them. These countries need a combined approach of resource growth, improved governance and climate finance in addition to technology transfer capabilities for them to walk the bridge with developed nations towards sustainability.
8. Investment in clean technologies: Becoming carbon neutral will require developing nations even larger investments in clean technology. Another challenge to the use of renewable energy sources and adoption of household-level energy efficient practices is a lack of technological infrastructure, expertise indeed (Wang, Jiang, et al., 2022). Further, a weak industrial base and the importing of many of the technologies to fulfil this kind of strategic requirement due to limited

domestic production and additional demand on fiscal resources (Li et al., 2018). Also, this problem is further worsened by the implementation of the International Climate Agreement as it regularly fails to address capacity building and knowledge transfer, particularly coming from developing countries while harmonising with other extraneous factors (Zhu & Geng, 2013). Other challenges include the decision to choose between economic growth and sustainability (Harrison et al., 2021; Kuzubov, 2023) since developing nations will have to determine whether they want to act within the global economic context (Tang et al., 2015; Xue et al., 2023). However, inadequate research and development capacity (Chen, Msigwa, et al., 2022; Mishra et al., 2022) also limit the capacity of developing nations to achieve climate neutrality same as insufficient infrastructure (Wang, Liu, & Gu, 2022), shortage of skilled workforce (Shu et al., 2022), limited access to advanced technologies (Colenbrander et al., 2015; Hatmanu & Căutişanu, 2022; Unger, 2024), regulatory and policy frameworks (Bruneckienė, 2023; Hatmanu & Căutişanu, 2022; Karakosta, 2016).

4.4 Strategies for Achieving Climate Neutrality in Developing Nations

Climate neutrality represents a comprehensive challenge that needs distinct approaches in each country, notably for developing countries. The key strategies to help transform developing countries towards climate neutrality are:

1. Invest in renewable energy technologies: Developing countries have to prioritise investments in renewable energy which includes solar, wind, hydro, and biomass. Importation for renewable energy not only decreases fossil fuel dependence, but the use of these technologies can bring jobs and create a cycle within local economies (Chen, Yang, et al., 2022). For example, the adoption of renewables into national grids can both bolster energy security as well as aid in reaching climate aims (Chen, Msigwa, et al., 2022). Financial mechanisms could also support these investments, and green bonds and international climate finance are necessary for providing financial resources to developing countries so that they can shift their energy systems from high- to low-carbon ones (Schmidt et al., 2022). They can also adopt renewable energy technologies installation incentivizing policies like feed-in tariffs, and tax credits to promote investment in clean energy (Chen, Yang, et al., 2022). International cooperation in technology transfer would also be beneficial for developing countries to catch up with the knowledge and technological gaps associated with the most advanced renewable energies (Wang et al., 2021).
2. Enhancing energy efficiency: In addition, to reduce GHG emissions and mitigate potential climate change-induced energy stresses in the future, raising energy efficiency across different sectors is also essential. This may be obtained by means of the use of energy efficient technology and practices in industrial installations, buildings, and transportation systems (Lai, 2022). Households and businesses

alike may need to deploy energy management systems as well as retrofit existing building stock utilised in order that a related portion of greenhouse gas emissions from associated power generation might be avoided (Lai, 2022). Subsidies, tax benefits and mandatory regulations to enforce sustainable practices are tools that governments can use for energy efficiency improvements (Flagg, 2015). This may be realised by adopting energy efficiency levels in buildings, appliances, and industrial processes (Chen, Msigwa, et al., 2022). By offering financial stimulus packages for refurbishment of the existing infrastructure and employing energy efficiency technology, governments can encourage sustainable practices. The Green Deal, the European Union's project to make Europe the first climate-neutral continent by 2050 (Leuzzo & Mangano, 2023) pays particular attention to Energy Efficiency.

3. Sustainable land management practices: Implementing sustainable land management practices (SLMP) is essential for achieving climate neutrality, particularly in agricultural sectors. Strategies such as agroforestry, conservation agriculture, and improved livestock management can enhance carbon sequestration while maintaining agricultural productivity (Duffy, Prudhomme, Duffy, O'Donoghue, et al., 2022; Yangka et al., 2018). Additionally, the development of bio-based industries and bioenergy infrastructure linked with carbon capture capabilities can further support emissions reduction efforts (Duffy, Prudhomme, Duffy, Iannetta, et al., 2022). SLMP strategies can bolster climate neutrality by combating poverty and unemployment by increasing food security (Mechiche-Alami et al., 2022) which is associated with other benefits such as enhancing soil fertility and biodiversity (Hassen et al., 2021). Besides, the adoption of reduced use of soil conservation techniques can increase the levels of organic carbon in the soil (Vittorio, 2024). These practices can reduce emissions from soil degradation and they also offer useful co-benefits that will boost crop productivity through the use of geographic information systems (Mema & Jojic, 2024), which is essential for food security in many developing countries (Compaoré, 2024).
4. Capacity building and technology transfer: The education and training of local human capacities offer an essential link for the effective implementation of climate-neutral technologies (Schmidt et al., 2022). Instead of industries, developing nations should build people to run and manage high-tech. The transfer of technology from developed countries also brings access to innovative solutions which are critical for achieving climate neutrality (Bruneckienė, 2023; Karakosta, 2016). Technology transfer mechanisms can be enhanced through collaborative partnerships between governments, private sectors, and international organisations (Karakosta, 2016).
5. Stakeholder engagement and public awareness: Stakeholder engagement, involving local communities as well as the business and civil society sectors is vital to support climate neutrality efforts. Public awareness drives could inform the public about climate action and promote green habits among people at a personal or community level (Lai, 2022). There is also scope for including stakeholders in processes of decision-making, and the ability to confirm that policies are comprehensive, representing diverse groups; so they cater to necessities (Jäger

et al., 2023). Many have made the same argument; no one can expect to approach constituents without engaging their local community or raising general awareness toward climate and sustainability to survive any public-based attack that would undermine even the best of intentions from a global perspective. The effect of the initiatives can be strengthened by involving more local communities in decision-making and providing them with some education, detailed as good practices for sustainability (Mechiche-Alami et al., 2022). Educational channels can also encourage people to embrace sustainable living practices and get involved in climate action (Udas et al., 2018). Involving communities in decision-making and supporting grassroots actions within developing countries helps to increase a sense of ownership over avoiding climate neutrality (Udas et al., 2018).

6. Policy frameworks and regulatory support: Developing nations need to develop stable policy frameworks that help them achieve net zero carbon support (Shu et al., 2022). Comprehensive climate action plans that set clear targets and timelines for emissions reductions should be developed by governments. These include charging for carbon, creating emissions trading systems and establishing green industrial policies designed to create economic pressure on owners of capital assets or users of low-efficiency processes (Gao, 2023). Moreover, having national policies by international climate agreements (e.g. the Paris Agreement) may improve accountability and commitment to achieving a robust set of global policy targets on climate change as well (Shu et al., 2022). Bruneckienė (2023) Proper governance structures, regulatory frameworks and policy coherence to enable climate action in developing countries. Translating the principles of sustainable development into national plans and mainstreaming climate policies in well-defined sectors, as well as engaging relevant authorities, such as governmental departments from different ministries, and private companies in decision-making processes to help achieve broader integrated measures on effective homeland security (Bruneckienė, 2023).
7. Research and development (R&D) initiatives: Establishing a zero carbon infrastructure in the developing world demands high-impact action from across the spectrum: decisive investments in renewables, energy efficiency improvements, sustainable land management measures and clear R&D pathways accompanied by targeted capacity building and stakeholder engagement, enabling policy frameworks (price signals), among others. R&D is necessary to enable new solutions for addressing the challenges of climate change. Governments should work with the private sector to facilitate more research into new sustainable technologies climate adaptation measures, and emissions reduction solutions (Chen, 2023). Developing nations can discover ways to coalesce around effective pathways to climate neutrality, by creating innovation-enablement environments (Chen, 2023). Some of these measures will be carried out by developing countries, while others require action on the part of advanced nations to support their establishment and/or free up resources for use in combating climate change within a development context. During this transition, there is a clear need to establish global partnerships between international trading market countries and developing ones through cooperation among institutional researchers as well as

private sectors with tailored needs due to the differing context of climate changes for those nations (Wang et al., 2021). Making sure that these countries prioritise R&D means they can build technical capacity to develop and implement effective climate strategies suitable for their context (Wang et al., 2021).

8. Sustainable agriculture practices: Agriculture is one of the highest sources of several greenhouse gases, especially in developing countries. Sustainable agricultural practices including agroforestry, organic farming and precision agriculture can decrease emissions while increasing food security (Kingwell, 2021). Reforestation: to plant trees that absorb carbon dioxide from the atmosphere, supporting climate neutrality goals (defining as "no net greenhouse gas emissions"); and avoiding deforestation, carbon sequestration through the planting of new forests may provide potential mechanisms for reducing atmospheric CO_2 concentrations (Kingwell, 2021). Developing nations can also use low-carbon pricing mechanisms (Bruneckienė, 2023), request finance from international donors (Ford et al., 2014), and engage in proper urban planning and sustainable cities (Gouveia et al., 2021; Lai, 2022). Above all, data collection and analysis for the evaluation of these strategies need to be invested by developing countries to make informed decisions (Chen, Msigwa, et al., 2022).

4.5 Implications

The ramifications of the charge towards climate neutrality in developing countries are profound for all parties involved, namely governments and companies as well as communities or even nature itself. Such implications must therefore be addressed so decision-makers have the information they need to inform decisions, guide business strategies, and incentivize collaboration for sustainability goals. This would rely on hard but enforceable climate-neutral policies. Governments need to be more responsive not only in terms of updating their nations on the international agreements they sign, but also merge each nation's part with every other country as a sovereign plan. It would be a declaration of high standards on emission reduction, incentives to transition into clean energy and provide food for the future sustainably. They also need to cooperate with stakeholders to ensure that regulations are not only executed practically yet likewise accommodate the peculiarities of private states.

A climate-neutral economic transition can stimulate economic growth and job creation in developing countries. In addition, the circular economy can accelerate innovation and entrepreneurship—opening opportunities for new markets/industries. Developing countries need to invest in green techs and sustainable practices, creating jobs while solving the environmental crisis. Incorporating multiple voices of stakeholders—including businesses, NGOs, and local communities—is critical to driving innovation across a wider range of possible solutions. Developing nations can address climate challenges and support sustainable development by working together to co-create solutions. Indeed, stakeholder participation is the backbone of trust and sustainability for all climate change measures. Education, provided through capacity building is essential for ensuring that individuals and organisations have the necessary knowledge and tools to support the shift towards a climate-neutral economy, hence, citizens should have adequate access to it. The benefits of climate neutrality and sustainable practices should be the message that resonates with consumers, raising awareness on a local level is crucial to pushing behaviour change in both consumer behaviour as well as business practice. The government should invest in education to help our world understand climate change and encourage the rest of them to work hand-in-hand with initiatives that are trying their best for it.

Becoming carbon neutral is going to be a capital-intensive proposition. It is equally important for developing countries to promote research and development, so that new potentially sustainable technologies may be formed. We need collaboration between governments and private sectors to invest in new ideas or proven projects/ initiatives that improve resource efficiency and reduce emissions. It is an investment that is important for the creation of a robust and sustainable economy. A just and inclusive climate-neutral transition must be represented in a way that results in marginalised communities benefiting from it. Policymakers need to establish specific programmes that help at-risk groups, they must prevent the transition from a climate-damaging economy into a carbon-free way of production—from further dividing already disadvantaged people. Foster a more equal, just society.

Moreover, climate neutrality is particularly timely given the global challenges faced by developing countries. Countries can better secure themselves against environmental change by taking the lessons of circular economy, and general sustainable principles as a whole—mitigate consumption to ease pressure on finite resources. The circular economy is built on a restorative model where the drivers of growth are decoupled from scarce natural capital.

4.6 Conclusion

Reaching climate neutrality in developing countries is a huge challenge on many levels and will have to be tackled from the technological, economic, social as well as institutional angles. Amidst this delicate dance of developing against sustainability, certain strategies appear fundamental to the journey these nations undertake. Fossil fuels need to be switched out in favour of renewable energy sources, as this is crucial if they are serious about both lowering reliance on fossil fuels and reducing the impact these greenhouse gases have. Through the development of solar, wind and biomass energy these countries can fully utilise their natural resources (where they are abundant) while creating jobs as well as encouraging economic growth (Duffy, Prudhomme, Duffy, Iannetta, et al., 2022). More broadly, improving energy efficiency in multiple sectors can help decarbonize while at the same time reducing costs for consumers of that or other types of systems: this represents a double-win policy approach (Brazzola et al., 2022; Oladele et al., 2023; Oyenuga et al., 2023).

Second, tree-planting practices that focus on sustainable land management will play an important role in increasing carbon sequestration and adaptation to climate change. Developing countries could increase their carbon sink by allowing degraded lands to be restored, promoting agroforestry and implementing climate-smart agriculture practices with self-sufficient food production (Peeters et al., 2016). The inclusion of local knowledge and community engagement together with these practices will create a stake, which is necessary for sustaining practice in climate initiatives (Ajani et al., 2013).

Third, it needs strong institutional frameworks and governance structures for climate. Policies to support sustainable practices or promote technology transfer, as well as involving stakeholders in decision-making will all be part of this (Zhao et al., 2022). In addition, global cooperation and climate finance are essential to help developing countries move toward achieving climate neutrality by granting them the tools required for their transformation (Zamarioli et al., 2020). In addition to these, mechanisms for monitoring and reporting the progress of climate neutrality objectives need to be established. Also, in the future decisions that their strategies are variables for developing nations have to keep monitoring emissions reductions and sequestration data so they can evaluate what percentage of performance on a strategy towards carbon neutrality (Dagiliūtė, 2024). But now, the road to climate neutrality is much longer and rockier for developing countries when we come back full circle—leaving behind anecdotical translating here as a success story but also development opportunities galore—where innovation boosts prosperities without exclusivity. They can take strong steps toward carbon neutrality and support worldwide climate change mitigation, by practising sustainability henceforth imbibing the spirit of solidarity in maximising international backing. Hence, achieving climate-neutrality is possible for developing nations.

References

Ajani, E., Mgbenka, R., & Okeke, M. (2013). Use of indigenous knowledge as a strategy for climate change adaptation among farmers in sub-Saharan Africa: Implications for policy. *Asian Journal of Agricultural Extension Economics & Sociology, 2*(1), 23–40. https://doi.org/10.9734/ajaees/2013/1856

Borghi, J. (2024). Climate finance opportunities for health and health systems. *Bulletin of the World Health Organization, 102*(5), 330–335. https://doi.org/10.2471/blt.23.290785

Brazzola, N., Patt, A., & Wohland, J. (2022). Definitions and implications of climate-neutral aviation. *Nature Climate Change, 12*(8), 761–767. https://doi.org/10.1038/s41558-022-01404-7

Bruneckienė, J. (2023). Rethinking national competitiveness for Europe 2050: The case of EU countries. *Sustainability, 15*(13), 10697. https://doi.org/10.3390/su151310697

Cevik, S., & Jalles, J. (2020). An apocalypse foretold. *Imf Working Paper, 20*, 231. https://doi.org/10.5089/9781513560403.001

Chen, C. (2023). Global carbon neutral pathways and China's strategies under Paris agreement climate goals. *Highlights in Business Economics and Management, 3*, 86–94. https://doi.org/10.54097/hbem.v3i.4668

Chen, F., Yang, Q., Zheng, N., Wang, Y., Huang, J., Xing, L., & Kleissl, J. (2022). Assessment of concentrated solar power generation potential in China based on geographic information system (GIS). *Applied Energy, 315*, 119045. https://doi.org/10.1016/j.apenergy.2022.119045

Chen, L., Msigwa, G., Yang, M., Osman, A., Fawzy, S., Rooney, D., & Yap, P. (2022). Strategies to achieve a carbon neutral society: A review. *Environmental Chemistry Letters, 20*(4), 2277–2310. https://doi.org/10.1007/s10311-022-01435-8

Colenbrander, S., Gouldson, A., Sudmant, A., Papargyropoulou, E., Chau, L., & Ho, C. (2015). Exploring the economic case for early investment in climate change mitigation in middle-income countries: A case study of Johor Bahru, Malaysia. *Climate and Development, 8*(4), 351–364. https://doi.org/10.1080/17565529.2015.1040367

Compaoré, C. (2024). Agriculture and climate change: A socio-economic dynamics of climate-smart practices in Oubritenga province, Burkina Faso. *IJAE, 9*(2), 46–58. https://doi.org/10.11648/j.ijae.20240902.12

Dagiliūtė, R. (2024). Impact of land-use changes on climate change mitigation goals: The case of Lithuania. *Land, 13*(2), 131. https://doi.org/10.3390/land13020131

Diffenbaugh, N., & Burke, M. (2019). Global warming has increased global economic inequality. *Proceedings of the National Academy of Sciences, 116*(20), 9808–9813. https://doi.org/10.1073/pnas.1816020116

Digitemie, W. (2024). Assessing the role of climate finance in supporting developing nations: A comprehensive review. *Finance & Accounting Research Journal, 6*(3), 408–420. https://doi.org/10.51594/farj.v6i3.926

Duffy, C., Prudhomme, R., Duffy, B., Iannetta, P., O'Donoghue, C., Ryan, M., & Styles, D. (2022). Randomized national land management strategies for net-zero emissions. *Nature Sustainability, 5*(11), 973–980. https://doi.org/10.1038/s41893-022-00946-0

Duffy, C., Prudhomme, R., Duffy, B., O'Donoghue, C., Ryan, M., & Styles, D. (2022). Goblin version 1.0: A land balance model to identify national agriculture and land use pathways to climate neutrality via backcasting. *Geoscientific Model Development, 15*(5), 2239–2264. https://doi.org/10.5194/gmd-15-2239-2022

Financial Times. (2024). *How each country's emissions and climate pledges compare* (ft.com).

Flagg, J. (2015). Aiming for zero: What makes nations adopt carbon-neutral pledges? *Environmental Sociology, 1*(3), 202–212. https://doi.org/10.1080/23251042.2015.1041213

Ford, J., Berrang-Ford, L., Bunce, A., McKay, C., Irwin, M., & Pearce, T. (2014). The status of climate change adaptation in Africa and Asia. *Regional Environmental Change, 15*(5), 801–814. https://doi.org/10.1007/s10113-014-0648-2

Fu, X., Pietrobelli, C., & Soete, L. (2011). The role of foreign technology and indigenous innovation in the emerging economies: Technological change and catching-up. *World Development, 39*(7), 1204–1212. https://doi.org/10.1016/j.worlddev.2010.05.009

Gao, J. (2023). A comparative study of China's carbon neutrality policy and international research keywords under the background of decarbonization plans in China. *Sustainability, 15*(17), 13069. https://doi.org/10.3390/su151713069

Giri, S., Prabhakar, A., Malla, R., Oli, S., Poudel, S., & Khanal, A. (2023). Climate change mitigation and adaptation in Nepal and South Asia: Challenges, progress, and recommendations. *Journal of Sustainability and Environmental Management, 2*(2), 133–140. https://doi.org/10.3126/josem.v2i2.55206

Gouveia, J., Seixas, J., Palma, P., Duarte, H., Luz, H., & Cavadini, G. (2021). Positive energy district: A model for historic districts to address energy poverty. *Frontiers in Sustainable Cities, 3*, 648473. https://doi.org/10.3389/frsc.2021.648473

Guntuka, L. (2024). From carbon-neutral to climate-neutral supply chains: A multidisciplinary review and research agenda. *The International Journal of Logistics Management, 35*(3), 916–942. https://doi.org/10.1108/ijlm-03-2023-0116

Ha, S., Hale, T., & Ogden, P. (2015). Climate finance in and between developing countries: An emerging opportunity to build on. *Global Policy, 7*(1), 102–108. https://doi.org/10.1111/1758-5899.12293

Harrison, M., Cullen, B., Mayberry, D., Cowie, A., Bilotto, F., Badgery, W., & Eckard, R. (2021). Carbon myopia: The urgent need for integrated social, economic and environmental action in the livestock sector. *Global Change Biology, 27*(22), 5726–5761. https://doi.org/10.1111/gcb.15816

Hassen, G., Bantider, A., Legesse, A., Malesu, M., & Likissa, D. (2021). Land use and land cover change for resilient environment and sustainable development in the Ethiopian Rift Valley region. *Environmental Protection and Natural Resources, 32*(2), 24–41. https://doi.org/10.2478/oszn-2021-0007

Hatmanu, M., & Căutişanu, C. (2022). Investigating the relationships between economic growth and environmental degradation: Evidence from EU15 countries. *Technological and Economic Development of Economy, 29*(1), 192–216. https://doi.org/10.3846/tede.2022.17874

Hertel, T., & Lobell, D. (2014). Agricultural adaptation to climate change in rich and poor countries: Current modelling practice and potential for empirical contributions. *Energy Economics, 46*, 562–575. https://doi.org/10.1016/j.eneco.2014.04.014

Jabakhanji, S., Arnold, S., Aunan, K., Chersich, M., Jakobsson, K., McGushin, A., & Stanistreet, D. (2022). Public health measures to address the impact of climate change on population health—Proceedings from a stakeholder workshop. *International Journal of Environmental Research and Public Health, 19*(20), 13665. https://doi.org/10.3390/ijerph192013665

Jäger, J., Brutschin, E., Pianta, S., Omann, I., Kammerlander, M., Vishwanathan, S., & Ruijven, B. (2023). Stakeholder engagement and decarbonization pathways: Meeting the challenges of the COVID-19 pandemic. *Frontiers in Sustainability, 3*, 1063719. https://doi.org/10.3389/frsus.2022.1063719

Jeudy-Hugo, S., Lo Re, L., & Falduto, C. (2021). *Understanding countries' net-zero emissions targets*. OECD/IEA. https://doi.org/10.1787/8d25a20c-en

Kaewunruen, S. (2024). Comparisons of stakeholders' influences, inter-relationships, and obstacles for circular economy implementation on existing building sectors. *Scientific Reports, 14*(1), 11046. https://doi.org/10.1038/s41598-024-61863-0

Karakosta, C. (2016). A holistic approach for addressing the issue of effective technology transfer in the frame of climate change. *Energies, 9*(7), 503. https://doi.org/10.3390/en9070503

Kingwell, R. (2021). Making agriculture carbon neutral amid a changing climate: The case of South-Western Australia. *Land, 10*(11), 1259. https://doi.org/10.3390/land10111259

Kuzubov, A. (2023). Strengthening the environmental, energy and food security of the state as part of the development of critical technologies in the agro-industrial complex. *E3s Web of Conferences, 431*, 01011. https://doi.org/10.1051/e3sconf/202343101011

Lai, Y. (2022). Urban intelligence for carbon neutral cities: Creating synergy among data, analytics, and climate actions. *Sustainability, 14*(12), 7286. https://doi.org/10.3390/su14127286

Leuzzo, A., & Mangano, G. (2023). Advanced sustainable design and experimental assessment to address climate neutrality in Mediterranean areas. *Renewable Energy and Environmental Sustainability, 8*, 10. https://doi.org/10.1051/rees/2023005

Li, S., Juhász-Horváth, L., Pintér, L., Rounsevell, M., & Harrison, P. (2018). Modelling regional cropping patterns under scenarios of climate and socio-economic change in Hungary. *The Science of the Total Environment, 622*, 1611–1620. https://doi.org/10.1016/j.scitotenv.2017.10.038

Ma, J. (2023). China carbon neutrality policies: Economic impact on corporations, challenges, and recommendations. *Advances in Economics Management and Political Sciences, 52*(1), 155–163. https://doi.org/10.54254/2754-1169/52/20230712

Mechiche-Alami, A., O'Byrne, D., Tengberg, A., & Olsson, L. (2022). Evaluating the scaling potential of sustainable land management projects in the Sahelian great Green Wall countries. *Environmental Research Letters, 17*(8), 084016. https://doi.org/10.1088/1748-9326/ac8111

Mema, O., & Jojic, E. (2024). Mapping spatial variability status of criteria influencing crop-land suitability using geographic information system. https://doi.org/10.17758/uruae23.ua0524301

Mishra, R., Singh, R., & Govindan, K. (2022). Net-zero economy research in the field of supply chain management: A systematic literature review and future research agenda. *The International Journal of Logistics Management, 34*(5), 1352–1397. https://doi.org/10.1108/ijlm-01-2022-0016

Ojha, H., Ghimire, S., Pain, A., Nightingale, A., Khatri, D., & Dhungana, H. (2015). Policy without politics: Technocratic control of climate change adaptation policy making in Nepal. *Climate Policy, 16*(4), 415–433. https://doi.org/10.1080/14693062.2014.1003775

Oladele, T. O., Oyenuga, M. O., & Adoga, G. J. (2023). Green management practices and organisational efficiency in higher institution. *Budapest International Research and Critics in Linguistics and Education Journal, 6*(3), 276–290. https://doi.org/10.33258/birle.v6i3.7716276

Oyenuga, M. O., Marcus, G. O., & Ahungwa, A. I. (2023). Do consumers care about green marketing practices? Insight from a developing nation. *Budapest International Research and Critics Institute Journal, 6*(3), 1424–1436. https://doi.org/10.33258/birci.v6i3.7665

Peeters, P., Higham, J., Kutzner, D., Cohen, S., & Gößling, S. (2016). Are technology myths stalling aviation climate policy? *Transportation Research Part D Transport and Environment, 44*, 30–42. https://doi.org/10.1016/j.trd.2016.02.004

Rogelj, J., Geden, O., Cowie, A., & Reisinger, A. (2021). Net-zero emissions targets are vague: Three ways to fix. *Nature, 591*(7850), 365–368. https://doi.org/10.1038/d41586-021-00662-3

Rozanov, V. (2024). Land degradation neutrality in the Penza region. *Bio Web of Conferences, 84*, 01018. https://doi.org/10.1051/bioconf/20248401018

Schmidt, A., Köster, D., & Strube, J. (2022). Climate neutrality concepts for the German chemical-pharmaceutical industry. *PRO, 10*(3), 467. https://doi.org/10.3390/pr10030467

Semieniuk, G., & Yakovenko, V. (2020). Historical evolution of global inequality in carbon emissions and footprints versus redistributive scenarios. *Journal of Cleaner Production, 264*, 121420. https://doi.org/10.1016/j.jclepro.2020.121420

Seyfried, S., Nagel, L., & Weigold, M. (2023). Empirical investigation of climate neutrality strategies of companies in industrial production. In *Lecture notes in mechanical engineering* (pp. 999–1007). Springer. https://doi.org/10.1007/978-3-031-28839-5_111

Sharma, S. (2023). Developing climate neutrality among supply chain members in metal and mining industry: Natural resource-based view perspective. *The International Journal of Logistics Management, 35*(3), 804–832. https://doi.org/10.1108/ijlm-03-2023-0108

Shively, G., Ward, P., & Diffenbaugh, N. (2012). Vulnerability, income growth and climate change. *World Development, 40*(5), 916–927. https://doi.org/10.1016/j.worlddev.2011.11.015

Shu, Y., Chen, J., & Fu, W. (2022). Sustainable approaches to realize carbon neutrality in China: A case study of Zhejiang province. *Journal of Marine Science and Engineering, 10*(10), 1351. https://doi.org/10.3390/jmse10101351

Tang, L., Wu, J., Yu, L., & Bao, Q. (2015). Carbon emissions trading scheme exploration in China: A multi-agent-based model. *Energy Policy, 81*, 152–169. https://doi.org/10.1016/j.enpol.2015.02.032

The Net-Zero Standard. (2023). *Science based targets*. Retrieved December 13, 2023.

Udas, E., Wölk, M., & Wilmking, M. (2018). The "carbon-neutral university"—A study from Germany. *International Journal of Sustainability in Higher Education, 19*(1), 130–145. https://doi.org/10.1108/ijshe-05-2016-0089

UNEP. (2024). *Responding to climate change. Responding to climate change | Africa*. UNEP - UN Environment Programme.

Unger, B. (2024). Determinants of firms' initiative and inertia in pursuing climate neutrality strategies—Theoretical explanations and empirical evidence. *Business Strategy and the Environment, 33*(5), 4086–4107. https://doi.org/10.1002/bse.3698

Vittorio, A. (2024). Soil management practices can contribute to net carbon neutrality in California. *Environmental Research Letters, 19*(6), 064034. https://doi.org/10.1088/1748-9326/ad4b41

Wang, C., Jiang, H., Wu, Y., Hou, J., Su, Z., Lin, Y., & Zhang, R. (2022). Design of project-based learning activities for carbon dioxide capture under the vision of carbon neutralization. *Journal of Chemical Education, 99*(11), 3662–3671. https://doi.org/10.1021/acs.jchemed.2c00217

Wang, F., Harindintwali, J., Yuan, Z., Wang, M., Wang, F., Li, S., & Chen, J. (2021). Technologies and perspectives for achieving carbon neutrality. *The Innovation, 2*(4), 100180. https://doi.org/10.1016/j.xinn.2021.100180

Wang, S., Chen, J., Ter-Mikaelian, M., Levasseur, A., & Yang, H. (2022). From carbon neutral to climate neutral: Dynamic life cycle assessment for wood-based panels produced in China. *Journal of Industrial Ecology, 26*(4), 1437–1449. https://doi.org/10.1111/jiec.13286

Wang, X., Teigland, R., & Hollberg, A. (2022). A pathway to climate neutral buildings: Definitions, policy and stakeholder understanding in Sweden and China. *IOP Conference Series: Earth and Environmental Science, 1078*(1), 012122. https://doi.org/10.1088/1755-1315/1078/1/012122

Wang, Y., Liu, Y., & Gu, B. (2022). Cop26: Progress, challenges, and outlook. *Advances in Atmospheric Sciences, 39*(8), 1209–1216. https://doi.org/10.1007/s00376-022-2097-z

Wolf, S., Teitge, J., Mielke, J., Schütze, F., & Jaeger, C. (2021). The European green deal — More than climate neutrality. *Intereconomics, 56*(2), 99–107. https://doi.org/10.1007/s10272-021-0963-z

Xue, Y., Ma, T., Zhuang, T., & Ma, Y. (2023). The development of green finance under the goal of carbon neutrality: A review from China's perspective. *Frontiers in Earth Science, 11*, 23020. https://doi.org/10.3389/feart.2023.1123020

Yangka, D., Rauland, V., & Newman, P. (2018). Carbon neutral policy in action: The case of Bhutan. *Climate Policy, 19*(6), 672–687. https://doi.org/10.1080/14693062.2018.1551187

Zamarioli, L., Pauw, P., & Grüning, C. (2020). Country ownership as the means for a paradigm shift: The case of the green climate fund. *Sustainability, 12*(14), 5714. https://doi.org/10.3390/su12145714

Zhao, L., Du, M., Du, W., Guo, J., Liao, Z., Kang, X., & Liu, Q. (2022). Evaluation of the carbon sink capacity of the proposed Kunlun Mountain National Park. *International Journal of Environmental Research and Public Health, 19*(16), 9887. https://doi.org/10.3390/ijerph19169887

Zhu, Q., & Geng, Y. (2013). Drivers and barriers of extended supply chain practices for energy saving and emission reduction among Chinese manufacturers. *Journal of Cleaner Production, 40*, 6–12. https://doi.org/10.1016/j.jclepro.2010.09.017

Chapter 5
Climate Vulnerability and Green Growth in Sub-Saharan Africa: The Roles of Resilience and Governance

Abubakar Jamilu Baita, Dian Masyita, and Rizky Wisnoentoro

Abstract Sub-Saharan Africa dwelled the world's vulnerable economies in respect of climate change. This has implication for green growth transition. Thus, this research examined the effect of climate vulnerability on green growth in Sub-Saharan Africa. Besides, it examined the moderating effect of resilience and governance in mitigating the undesirable consequences of climate vulnerability on green growth. Methodologically, the study collected secondary data between 2000 and 2021, and employed method of moment quantile regression. The findings established that lower values of resilience and governance have negative moderating effects on vulnerability-green growth connection. However, the inhibiting effect of vulnerability reduced drastically as resilience and governance increase in the sampled countries. The chapter provided social and policy implications for promoting green transitions in SSA.

Keywords Green growth · Climate vulnerability · Governance · Climate Resilience · SSA

A. J. Baita (✉) · D. Masyita · R. Wisnoentoro
Faculty of Economics and Business, Universitas Islam Internasional Indonesia, Depok, Indonesia
e-mail: abubakarbaita@gmail.com

D. Masyita
e-mail: dian.masyita@uiii.ac.id

R. Wisnoentoro
e-mail: rizky.wisnoentoro@uiii.ac.id

A. J. Baita
Department of Economics, Northwest University, Kano, Nigeria

D. Masyita
Padjadjaran University, Bandung, Indonesia

R. Singh and D. Crowther (eds.), *Climate Neutrality and Sustainable Ecosystems*, Approaches to Global Sustainability, Markets, and Governance,
https://doi.org/10.1007/978-981-96-7181-6_5

5.1 Introduction

Volze and Ahmed (2020) contended that climate vulnerability might accentuate transition risk as countries transition to low-carbon economy. This transition risk has severe consequences on sustainable development by constraining adaption to climate change and climate resilience. The authors identified seven transmission channels through which climate risk poses threat to sovereign risk. Among these channels are three outstanding aspects that directly affect sustainable green growth. These channels, as highlighted by Volze and Ahmed (2020), include degrading the natural ecosystem, financial costs of climate risk on public budget, and its macroeconomic implications. Natural resources depletion is an outcome of economic activities through resource extraction in order to generate revenues. Hence, unsustainable depletion of natural endowments and climate risks are interconnected. Furthermore, persistence of extreme weather events increase fiscal costs via rise in public expenditure to mitigate climate challenges.

Xiao and Fei (2024) argued that promoting technological innovativeness reduces harmfulness of vulnerability by making country or region more resilient. The inhibiting implication of climate vulnerability for green-centric innovation could trigger unsustainable development due to a number of factors. Climate vulnerable regions such as Sub-Saharan Africa (SSA) find it difficult to invest in green development as a result of low financial capacity. In addition, infrastructural deficits constrains country's capacity to halt the risk of vulnerability. Invariably, low quality educational provisioning affect the quality and quantity of manpower for promoting research and innovation in climate science (Xiao & Fei, 2024). Furthermore, green transition could upscale economic and environmental benefits to countries in the lower income category (Daniel, 2005). As well, green technical development, environment and industrial activities are interconnected. Specifically, green technology reduces industrial waste, which enhances productivity, environmental quality and efficient resource use (Daniel, 2005).

Africa is the most vulnerable to the vagaries of anthropogenic effects of pollution despite its minimal contribution to climate change. Climate change coupled with underdevelopment could trigger high level of vulnerability, which pose substantial adverse ramifications (Welborn, 2018). Sound institution is critical and inevitable in bolstering Africa's adaptive capacity to facilitate transition to inclusive green economy. To buttress its criticality, Welborn (2018) observes that good governance is efficacious for forecasting climate riskiness, and formulating and executing reliable adaptive system in extremely climatic and economic vulnerable societies. Africa confronts hurdles in its green growth agenda. It is projected that Africa's climate vulnerability will severely rise within 2030–2050 period. In the present, physical climate hazards, which are precarious to natural ecosystem and development path have become Africa's "new normal" (AfDB & GGGI, 2022).

Green growth is not achievable unless climate resilience is integrated in the green transition (AfDB & GGGI, 2022). In terms of biodiversity, Africa accounts for one-quarter of the global natural ecosystem, and it has ample untapped renewables for green transition. However, it lags behind in most of the components of green growth. In comparison to other global regions, it outperforms Middle East and South Asia but falters behind other four regions. It experiences progression in access to energy and other social inclusion indicators between 2010 and 2021, though the pace is inadequate compared to other global regions. Besides, there was positive correlation between green growth and climate resilience in African region within 2010–2021; however, it associates negatively with vulnerability to climate change (AfDB, 2023). Besides, accelerating green would enable SSA nations to cut energy intensity by optimising energy resources to achieve economic welfare.

Theoretical and conceptual studies have acknowledged the need for resilience in curtailing climate vulnerability. However, empirical literature in this domain is scanty. Recent discussions (e.g. Lee et al., 2017, 2021; Mustun, 2022) have limitations, which require extending the research frontier. Mustun (2022) focused on Saudi Arabia. The study could potentially provide insight for oil exporting African nations. Nevertheless, the author used annual data spanning 1998–2019, which is insufficient for a rigorous time series analysis. Thus, the broad goal of the chapter is the examination of the outcome of climate vulnerability on green growth transition in SSA. Specific research objectives are twofold. First, it aims to examine the conditioning effect of resilience in downgrading the inhibiting role of climate vulnerability on green growth. Second, it evaluates the marginal effect of vulnerability to climate change on green growth due to improvement in governance.

The chapter provides empirical and policy-oriented contributions to green transition in SSA region and other developing economies. Empirically, it contributes to the existing literature on the implication of climate risk in the course of green transition. Climate factors can enhance or stifle green growth transition, depending on the channel of transmission. The policy contribution is important as well. The author expects the outcomes of the research to influence policy cycle in order to contribute towards achieving progress in sustainable development. Specifically, it will expand research frontier in respect of climate action (SDG 13) and biodiversity conservation (SDG 15). SDG 13 aims to promote climate resiliency (SDG 13.1), making climate an integral part of national policies (SDG 13.2), and enhance climate awareness and promote institutions for tackling climate change (SDG 13.3).

5.2 Empirical Review

5.2.1 Climate Vulnerability and Green Growth

Climate policy makers agreed that vulnerabilities to climate change have adverse economic and environmental consequences (Singh et al., 2011; Thornton et al., 2014). It could lead to fall in productivity, loss of productive resources and reallocation of capital to finance climate adaptation (Bowen et al., 2012). Thus, preservation of natural capital could facilitate sustainable growth. Bowen et al. (2012)

buttressed the relevance of natural capital in sustaining growth. They contended that ensuring better condition of natural capital facilitate growth trajectory that is sustainable, and improve the adaptation state of the biodiversity. Thornton et al. (2014) observed that anticipated rise in rainfall variation could have large effect on the capacity of ecosystem to ensure sustainable environmental services. As stated by Singh et al. (2011 p. 114), vulnerability due to high incidence of climate hazards and economic disadvantage triggers suffering to regional grouping including SSA. Similarly, limited accessibility to resources accentuate vulnerability at both country and regional levels. Besides, there is direct connection between climate change and socioeconomic vulnerability (Kelley, 2016), as well as among climate stress, underdevelopment and vulnerability (Welborn, 2018).

Empirical literature documented connection between climate vulnerability and sustainability. In this respect, Lee et al. (2017) analysed the governance-vulnerability-climate interface in 122 economies covering 2000–2012 periods. Due to cross-sectional dependence issue, the researchers employed Driscoll-Kraay panel estimation. The findings established that governance and climate vulnerability have significant cointegration relationship with sustainability. Based on full sample results, governance has favourable implications for environmental quality, while climate vulnerability hampered it; similar results were obtained for the sample of high-income economies. However, only vulnerability to climate change significantly influenced sustainable environment in sub-sample of developing economies. The authors attributed this scenario to the fact that while developed nations have strong environmental policies, the developing world struggled with weak environmental regulations. This research pointed to the confluence of weak environment governance and high vulnerability in retarding sustainable environment in many developing economies. However, the study left substantial time gap of 12 years. Beyond 2012, there were many global vulnerability issues such as Trump's policy of anti-sustainable energy transition policy, Covid pandemic, and geopolitical crises. These events have mixed effects on vulnerability and sustainability; thus, extension of the research to current period is timely in order to confirm or rebut the previous research findings.

Furthermore, Lockitch (2009) stressed the need for comprehensively appraising the efficacy of green economy policies on climate vulnerability by looking beyond the climatic parameters. Sound climate policies should incorporate both political and economic considerations. The author criticised the orchestration of green climate policies in controlling climate vulnerability as it restricted political freedom and industrial development. Besides, the social science perspective on climate vulnerability, though of critical importance, is often neglected in climate change discussion. As well, adaptive capacity play substantial role in influencing the direction of vulnerability; high adaptive capacity lowers vulnerability (Thomas et al., 2019). Thus, enhancing country's adaptive capacity is key to curtailing socioeconomic vulnerability (Thornton et al., 2014).

Likewise, there existed synergistic relationship between climate vulnerability and economic growth in developing economies. Bowen et al. (2012) explained that growth strengthened country's capacity to adapt to climate risks and redirected economy's sensitivity to climate shocks in either positive or negative directions; however, the final effect was indeterminate. More so, some strands of single country literatures investigated the response of real sector to climate risks. Khurshid et al. (2022) applied non-linear ARDL to inspect the influence of climate change shock on Pakistan's economic performance from 1980 to 2021. The authors identified CO_2 emission and temperature as proxies of climate shocks. Ultimately, they uncovered equilibrium relationship among climatic factors and growth over the longer horizon. The results of short- and long-term analyses revealed significant emission-growth connection, and temperature-growth linkage. Essentially, positive emission led to lower growth, positive average temperature propelled growth, while negative mean temperature stifles economic growth.

Jirophat et al. (2022) used temperature and precipitation to uncover the efficacy of climate challenges on the Thailand's economy. The findings confirmed the reductionist effects of climatic shocks feeding into the real economic activities in the short- and medium-terms respectively. In terms of asymmetries, positive climate shock was beneficial to real growth while negative climate risks dampened real output. At the sectoral level, the asymmetric effect was more glaring in the agriculture relative to industry and service sectors. Berihun and Steven (2022) established the substantial depressive implications of climate variability on economic performance in Ethiopia. In particular, temperature has the largest deleterious consequence on real output growth. Both rise and fall in temperature resulted in decrease in real GDP in the long term. Besides, the outcome of positive rainfall variation has more influence on growth than its negative outcome over the long term. The findings showed compelling evidence for accelerating mitigation strategies to contain the adverse outcomes of climatic shocks. Similarly, Castro et al. (2020) explored the consequence of climate challenge on Brazilian agricultural output from 1990 to 2012. The results showed predominant adverse outcome of average temperature on the productiveness of agriculture.

Considering cross-country literature on the effect of climate shocks, Alagidede et al. (2016) surveyed 18 nations in SSA, between 1970 and 2009, to examine the economic repercussions of climate change. The results revealed a significant short-term dampening effect of temperature on growth. In the long-term, temperature exhibited significant non-linearity where low-to-moderate temperature promoted growth, whereas high temperature depressed growth in the region. Similarly, Meyghani et al. (2022) established the role of climate change in explaining growth dynamics in the MENA region. The authors showed climate change shock and growth were in equilibrium over the long period. They, also, found significant decreasing effects of temperature and precipitation on growth, with temperature accounting for larger effect. Dell et al. (2012) inspected the risks of temperature variability in influencing economic outcomes between 1950 and 2003. The findings documented depressive effect of rising temperature on rates of growth in poor economies, while economic growth rates in high-income countries were not susceptible to temperature changes. However, variations in precipitation have significant modest impacts on growth across high- and low-income economies. The authors' findings suggested deleterious consequences of climate shocks for both the level of output and the growth rates.

5.2.1.1 Resilience, Vulnerability, and Green Growth

Using global evidence, Lee et al. (2021) analysed the moderating role of climate resilience for modelling connectivity of vulnerability and sustainable environment by covering 2000–2014 period. They decomposed vulnerability into two indices including social vulnerability and climate vulnerability. Based on the global sample of 128 economies, the findings documented positive effects of social vulnerability on environmental quality, while climate vulnerability demonstrated significant deleterious outcome on sustainability in the sample countries. Note that Lee et al. (2021) study contributed to the scant literature on the interface between vulnerability and sustainability by providing global evidence. As well, it recognised the roles of institutional factors including climate governance resilience in uncovering the limit of vulnerability in evaluating sustainable environment. However, the study has certain limitations in terms of time scope, indicators of governance and the proxy of sustainability. First, the study was limited to the year 2014 implying a 9-year gap. Second, the study adopted environmental pollution (air quality) as a measure of sustainability. Nevertheless, this single indicator failed to capture other salient features of green growth. Hence, this study utilises adjusted net savings, which is multidimensional. Thus, the study formulated the following hypothesis.

H1: Climate vulnerability weakens green growth, but the effect is lower as resilience improves.

5.2.2 Governance and Green Growth

Researchers have acknowledged the criticality of governance towards sustainable transition to green growth. Degbedji et al. (2024) scrutinised how institutions promote green growth transition in eight members of West African economic and monetary union between 2002 and 2017. The authors applied fully modified OLS and the results demonstrated facilitative impact of institutions on green growth in the member countries as union. However, the findings from the individual countries revealed heterogeneous implications of institutions by exerting positive and negative effects depending on the country's level of institutional enforceability. The results of granger non-causality analysis found significant one-way causality running from institutions to green growth. Outcomes of Dada et al. (2024)'s research demonstrated both enhancing and inhibiting effects of institutions on sustainability in Africa based on two alternative long run panel regression models. Based on the outcomes of panel quantile modelling, significant deleterious effects of institution on sustainable environment was confirmed between fourth and ninth quantiles.

Ahmed et al. (2022) analysed the institution-environment connection by delving into how institution drive sustainable growth across five South Asian economies from 2000 to 2018. The outcomes of both dynamic and fully modified OLS established significant positive influence of institutions in driving sustainable economic growth. Egbetokun et al. (2020) evaluated the consequence of growth and quality of regulations on sustainable environment in Nigerian context covering 1970–2017. Based on ARDL model, the findings showed mixed results in influencing environmental pollution. Across the six pollution indicators, it showed significant positive and negative effects respectively. Though it displayed increasing influence on carbon emission, the relationship was not statistically significant.

Acheampong and Dzator (2020) inspected the influence of institutions in driving environmental quality by surveying 45 countries in SSA between 2000 and 2015. The findings could not established significant institution-environment relationship, except in the political stability dimension. However, the sub-sample analysis demonstrated that institutional factors upscaled sustainable environment in both former British and French colonies. Wang et al. (2021) dissected African economies based on oil-resource presence to examine the effect of institutions on sustainability of environment covering 1999–2017 period. Based on the African sample, the findings demonstrated significant of institutions in enhancing sustainability. Based on the results of sub-sample analysis, quality of institution in non-oil resourceful African countries exert greater impact on sustainable environment than their counterparts do. Besides, panel causality relationship revealed a one-way long-term causality flowing from institution to environment in African region. Furthermore, there was long run two-way causality relationship between institution and sustainability in both oil-resourceful and non-oil rich economies in Africa. As well, results of non-oil rich countries displayed short-term one-way causality running from institution to environment. Hosseini and Kaneko (2013) examined 129 countries between 1980 and 2007 to show the linkage between spatial spillovers of environmental pollution as conditioned by quality of institutions. At the global level, decrease in civil liberties propelled carbon intensity, while the other two institutional proxies were insignificant. Based on regional analysis, civil liberty significantly reduced environmental quality while democracy promoted sustainable environment in Africa, Asia and Europe.

In the context of G7 economies, JinRu and Qamruzzaman (2022) employed cross-sectional ARDL for period spanning 1980–2020 to uncover the link between quality of institutions and sustainability. The authors documented significant effect of institution in promoting sustainability. The findings remained consistent in both short long periods. In addition, nexus test showed unilateral causality, which flow from institution to environment sustainability. Uzur (2021) explored the connection between institutions and environmental unsustainability among Emerging Seven (E-7) economies from 1992 to 2015. The author found quality of institution significantly enhanced sustainable environment over long term. As well, there was significant causality between institutions and environment, which uni-directionally run from institutions to environment. At country level, the results remained consistently significant for five countries out of emerging seven economies. Danish and Ulucak (2020) sampled 18 Asia-Pacific member countries covering 1992–2015 period. The analysis provided compelling evidence that quality of institution significantly induced sustainable environment. Likewise, it confirmed a facilitative role of renewables in fostering sustainable environment. Panel causality established one-way causal relationship, which flowed from institution to environmental sustainability. Again, institution unilaterally granger causes renewable energy development. Similarly, Khandy et al. (2023) examined finance-institution-ecology nexus in India between 1980 and 2019. The findings documented how deleterious were financial development and institutions in stifling sustainable environment in the short span. However, institutions significantly improved environmental quality over the long run.

5.2.2.1 Governance, Vulnerability, and Green Growth

Xiao and Fei (2024) is a pioneering global study on the linkage between vulnerability to climate change and green innovativeness, which covered 131 economies between 1995 and 2019. They documented inhibiting outcome of vulnerability on green innovativeness. In addition, the authors introduced political stability and level of corruption as governance moderators. The results indicate that governance factors are crucial for reducing the deleterious effect of vulnerability on green innovativeness. Lee et al. (2021) used effective government as proxy of governance as well as examined its consequences on sustainability. They documented significant enhancing role of effective government on sustainable environment. However, utilising a whole set of governance indicators will provide a more robust picture of the implications of institution in the sustainability analysis.

H2: Governance mitigates undesirable effect of climate vulnerability on green growth.

5.2.3 Research Gap

Vulnerability-sustainability discussions (e.g. Lee et al., 2017, 2021) utilised emission/ecological indicators to assess the drivers of green growth transition. Lee et al. (2017) integrated climate vulnerability and institutions to uncover their roles in promoting green transition in both advanced and emerging worlds. The research was novel; however, it has substantial time gap, as it was limited to 2012 due to data availability. Although there were better proxies of green growth, the authors used air quality as proxy of green growth. This indicator predominantly aligned with environmental pollution due to its health consequences rather than overall green growth. As well, Mustun (2022) utilised World Bank indicator of green growth to examine the effect of vulnerability to climate and institutional factors in Saudi Arabia. Nevertheless, the author used annual data spanning 1998–2019, which is insufficient for a rigorous time series analysis.

5.3 Modelling

5.3.1 Data and Sample

The study collects secondary data from the World Bank's World Development Indicators (WDI) database and Notre Dame Global Adaptation Initiative (ND-GAIN). WDI provide datasets for various macroeconomic, financial, environmental, social, and demographic indicators. This research collects datasets for green growth, GDP per person, and population density. Additionally, ND-GAIN is a data repository for climate vulnerability, resilience and adaptation. The datasets are freely available from 1995 and 2021; however, the study restricts its scope to 2000–2021 period. During this period, many developing countries started implementing international environmental agreements such as Kyoto Protocols. Besides, the study collects governance dataset from World Governance Indicators (WGI). Similar to WDI, World Bank compiles and manages WGI database.

The population of study covers all the 49 countries in SSA, which extends over different geographical parts of Africa. However, the final sample is 39 nations based on data availability. The sample countries include Angola, Benin, Botswana, Burkina Faso, Burundi, Cabo Verde, Cameroon, Comoros, Congo, Congo DR, Cote d'Ivoire, Djibouti, Eswatini, Ethiopia, Gabon, Gambia, Ghana, Guinea, Guinea-Bissau, Kenya, Lesotho, Madagascar, Mali, Mauritania, Mauritius, Mozambique, Namibia, Niger, Nigeria, Rwanda, Senegal, Sierra Leone, South Africa, Sudan, Tanzania, Togo, Uganda, Zambia, and Zimbabwe.

5.3.2 Variables Description

The research variables include green growth, climate vulnerability, climate resilience, governance, and set of control variables. Adjusted net savings is the proxy of green growth, which nets out the costs of biodiversity and environmental damages from the country's economic savings and incorporates the benefits of educational spending. Climate vulnerability is an index of various components including food, habitat, health, infrastructure and water. Food category covers climate impact on cereal yield, dependence on food importation, population change, agricultural capacity, size of rural dwellers, and child malnourishment. Habitat comprises of likelihood of extreme heat induced by climate change, expected flood hazard, urban concentration, proportion of age dependency, quality of trade and transport infrastructure, and proportion of the paved roads. Health component covers expected death due to climate induced diseases, access to better sanitation, number of people living in slum, size of medical workers, dependence on external financing for health services, and expected change in vector-carrying diseases. Infrastructure category include the capacity of hydropower generation, expected impact of sea level rise, energy dependency from importation, people living below 5 m above sea level, access to electricity, and preparedness for disaster. Water is measured by climate change impacts on water runoff and discharge, accessibility and reliability of drinking water, ratio of water dependency, withdrawal of freshwater.

Resilience, also called climate readiness, is a multidimensional construct, which composes of many categories such as economic, social, and governance dimensions. ND-GAIN provides data on climate readiness index, which measures country's resilience to climate change. The economic resilience relates to the doing business indicators, which reflect the extent of conducive environment for businesses to thrive. The social aspect of resilience captures progress in reducing social inequality, improving digital infrastructure, educational attainment, and technological innovation. Though governance forms an integral part of climate resilience, we separate it from the climate resilience index and construct a comprehensive governance index. We construct climate resilience based on economic and social dimensions. It functions as moderating variable, which potentially reduces the unwanted influence of vulnerability on green growth trajectory. In the context of ND-GAIN dataset, governance deals with stable polity and zero violence, minimal corruption, effective regulation, and rule of law. However, this study incorporated governance as moderating variables by utilising all the six governance indicators produced by WGI database. Table 5.1 presents summarised definitions of these research variables.

5.3.3 Econometric Specification

Equation (5.1) expresses the overall econometric model as follows:

$$\mathrm{GG_{it}} = \alpha_\mathrm{i} + \beta_i \mathrm{CV_{it}} + \beta_i \mathrm{CV_{it}} \times M_\mathrm{it} + \beta_i W_\mathrm{it} + \varepsilon_\mathrm{it} \tag{5.1}$$

Note: *GG* stands for green growth; *CV* is the climate vulnerability; *M* is the vector of moderator variable, and *W* represents a set of control variables consisting of energy intensity, per capita GDP, and population density (PD); α_i is the intercept; β_i represents the parameter estimates; ε_{it} is the error term.

Specifically, Eqs. (5.2) and (5.3) represent the empirical models.

$$\mathrm{GG_{it}} = \beta_0 + \beta_1 \mathrm{CV_{it}} + \beta_2 \mathrm{GOV_{it}} + \beta_3 \mathrm{RES_{it}} + \beta_4 \mathrm{EI_{it}} + \beta_5 \mathrm{PCI_{it}} + \beta_6 \mathrm{PD_{it}} + \varepsilon_\mathrm{it} \tag{5.2}$$

$$\begin{aligned} \mathrm{GG_{it}} = \alpha_0 + \alpha_1 \mathrm{CV_{it}} + \alpha_2 \mathrm{GOV_{it}} + \alpha_3 \mathrm{RES_{it}} + \alpha_4 \mathrm{CV} \times \mathrm{RES_{it}} \\ + \alpha_5 \mathrm{CV} \times \mathrm{GOV_{it}} + \alpha_6 \mathrm{EI_{it}} + \alpha_7 \mathrm{PCI_{it}} + \alpha_8 \mathrm{PD_{it}} + \varepsilon_\mathrm{it} \end{aligned} \tag{5.3}$$

where α_0 and β_0 represent the individual intercepts for each model; $\beta_1 - \beta_6$ and $\alpha_1 - \alpha_8$ are the parameter estimates for the explanatory variables; *i* stands for individual countries; *t* represents the time dimension. Resilience [RES] and governance [GOV] represent M_it. In line with Xiao and Fei (2024), if both α_1, α_4 and α_5

Table 5.1 Summary of variables and data sourcing

Variable	Code	Description	Data
Green growth	GG	Adjusted net savings as % of GNI	WDI
Climate vulnerability	CV	A composite indicator of country's deficiencies in food, habitat, ecosystem, health, infrastructure and water.	ND-GAIN
Resilience	RES	Arithmetic average of economic and social resilience indicators. Economic resilience measures country's conducive business environment for businesses to address climate change. Social resilience is an index of social inequality, digital infrastructure, education, and technological innovation.	Authors' own and ND-GAIN
Governance	GOV	An index of six indicators, which track country's progress in polity and institutions.	WGI
Energy intensity	EI	Energy supply as % of GDP	
Per capita income	PCI	Log of real per person GDP (2015 in USD)	WDI
Population density	PD	Log of population per square kilometre of land area	WDI

Source: Authors' compilation

have similar sign, the moderator increases the marginal effect. If the signs of α_1, α_4, and α_5 differ, the conditioning effect decreases.

Furthermore, the research employs method of moment quantile regression (MM-QR) to uncover the dynamic relationship. The study employs panel MM-QR to track the effects of vulnerability and governance across different quantiles of the green growth. This is important considering the fact that African region has heterogeneous economies with varying progress towards transition to green economy. Researchers attest the suitability of MM-QR model in the event of heterogeneous sample units, the conditional distribution of data, endogeneous covariates, its compatibility with non-linear models, and robustness to heteroscedasticity (Ameer et al., 2023; Emenekwe et al., 2023; Khaddage-Soboh et al., 2023). Besides, it removes the distributional effect of potential outliers, applicable to non-normally distributed data, and incorporate asymmetries (Ameer et al., 2023).

5.4 Results and Discussion

5.4.1 Preliminary Assessments

The research presents descriptive statistics to evaluate the average values of the research variables as well as measure the dispersion. This is followed by testing for normality and collinearity to determine whether the variables are normally distributed and free from multicollinearity. Table 5.2 displays the summarised report along with normal distribution and collinearity checking.

In Table 5.2, mean green growth (GG) is very low in SSA representing 1.2% of GNI. Its standard deviation (SD = 14%), indicate high variability across the countries under investigation. It is as low as 58.5% of GNI in Congo as at 2001, and as high as 38.2% of GNI in Botswana during the year 2001. On average, climate vulnerability (CV) is as high as 52.5%, which is beyond regional average in developing economies. It has low variability (SD = 5.6%), revealing that many SSA nations are climate vulnerable at various stages. Its maximum score is nearly 70% in the region, while the minimum score is about 40%. In addition, the region's overall mean governance (GOV) is −0.601, which highlights large governance deficit needed to improve GG. Its minimum and maximum values, −1.85 and 0.87 respectively, reflect low quality of institutional framework needed for effective climate governance. Mean resilience index (RES) is 30.2% with low SD of 6.4%. It is as high as 56.8% and as low as 12.7%. The average energy intensity (EI) is 6.7%, confirming low energy use in region. The minimum and maximum values of EI are 1.9% and 26.9% respectively.

Furthermore, the results of normal distribution based on Shapiro-Wilk test and skewness-kurtosis test reveal that the variables do not have normal distribution at 0.05 significance level. Nevertheless, the collinearity test using variance inflation factor (VIF) confirms that the variables are free from multicollinearity because none of the VIF values exceeds 10. Beside the descriptive summary and preliminary assessments, the study estimated the green growth equation by employing MM-QR regression

Table 5.2 Descriptive summary and preliminary assessments

Variables	Mean	SD	Min.	Max.	Shapiro-Wilk		Skewness-Kurtosis		VIF
					z	Prob.	Adj. chi2	Prob.	
GG	0.012	0.142	−0.585	0.382	6.696	0.000	53.74	0.000	–
CV	0.525	0.056	0.385	0.658	4.078	0.000	8.01	0.018	3.40
GOV	−0.601	0.578	−1.853	0.870	6.725	0.000	28.33	0.000	1.69
RES	0.302	0.064	0.127	0.568	7.608	0.000	85.86	0.000	1.54
EI	0.067	0.041	0.019	0.269	10.686	0.000	192.07	0.000	1.56
PCI	3.089	0.375	2.407	4.040	7.259	0.000	41.00	0.000	4.39
PD	3.881	1.286	0.793	6.452	6.375	0.000	13.40	0.001	1.38

Source: Authors' compilation

technique. The estimation technique is built on nine quantiles ranging from first to ninth quantiles. Table 5.3 presents the estimated results without interaction effect, while Table 5.4 introduced resilience and governance indices as moderators in the green growth equation.

Table 5.3 Green growth results using MM-QR estimation

Variable	Quantiles								
	0.10	0.20	0.30	0.40	0.50	0.60	0.70	0.80	0.90
CV	−0.02	0.02	0.04	0.07	0.09	0.11	0.13	0.15	0.17
GOV	0.16[a]	0.14[a]	0.12[a]	0.10[a]	0.08[a]	0.07[a]	0.06[a]	0.04[a]	0.02
RES	−0.07	0.05	0.15	0.23[c]	0.31[a]	0.37[a]	0.44[a]	0.52[a]	0.62[a]
EI	−0.04	−0.11	−0.18	−0.23	−0.27[c]	−0.31[c]	−0.36[b]	−0.41[b]	−0.46[b]
PCI	−0.03	−0.01	0.00	0.02	0.03	0.04	0.05[c]	0.07[b]	0.08[a]
PD	−0.02[a]	−0.02[a]	−0.02[a]	−0.02[a]	−0.02[a]	−0.02[a]	−0.02[a]	−0.02[a]	−0.02[a]
Const.	0.18	0.11	0.05	−0.00	−0.05	−0.08	−0.13	−0.17	−0.23
OBS.	736	736	736	736	736	736	736	736	736

Source: Authors' compilation
[a,b,c] 1%, 5%, and 10% significance levels

Table 5.4 Green growth results using MM-QR estimation

Variable	Quantiles								
	0.10	0.20	0.30	0.40	0.50	0.60	0.70	0.80	0.90
CV	0.25	−0.28	−0.58	−0.91[c]	−1.21[b]	−1.42[a]	−1.69[a]	−1.97[a]	−2.40[a]
GOV	−0.55[a]	−0.39[a]	−0.30[a]	−0.20[c]	−0.11	−0.05	0.03	0.11	0.24[c]
RES	−1.16	−1.78	−2.14[b]	−2.53[a]	−2.88[a]	−3.13[a]	−3.44[a]	−3.77[a]	−4.28[a]
CV*RES	2.71	4.04[c]	4.79[b]	5.62[a]	6.37[a]	6.90[a]	7.55[a]	8.26[a]	9.33[a]
CV*GOV	1.42[a]	1.06[a]	0.85[a]	0.63[a]	0.42[a]	0.27	0.09	−0.10	−0.40
EI	0.71[a]	0.44[b]	0.29	0.13	−0.02	−0.12	−0.26	−0.40[b]	−0.61[a]
PCI	0.06	0.07[b]	0.07[b]	0.07[a]	0.07[a]	0.08[a]	0.08[a]	0.08[b]	0.08[b]
PD	−0.01[b]	−0.01[a]	−0.01[a]	−0.02[a]	−0.02[a]	−0.02[a]	−0.02[a]	−0.02[a]	−0.02[a]
Const.	−0.39	−0.09	0.08	0.27	0.45	0.57[b]	0.72[b]	0.88[a]	1.13[a]
OBS.	736	736	736	736	736	736	736	736	736

Source: Authors' compilation
[a,b,c] 1%, 5%, and 10% significance levels

5.4.2 Panel Quantile Regression Results

In here, we present the results of green growth equations by employing method of moment panel quantile regressions (MM-QR) analysis. Table 5.3 reports the baseline results, while Table 5.4 incorporate the interaction effect of RES. Table 5.5 presents the authors' own computation of interaction effects for selected values of RES and GOV in the fourth and median quantiles.

In Table 5.3, CV has no effect on GG in SSA across all the quantiles. GOV significantly enhances GG across all the quantiles with exception of ninth quantile. However, the magnitude of GOV effects decrease at higher level of green growth. As well, RES promotes GG between median and highest quantiles. Its effect is weak in the fourth quantile, and it turns out increasingly crucial for improving GG in the median quantile ($\beta = 0.3$) and reaches a peak ($\beta = 0.61$) in the ninth quantile. However, it does not influence lower levels of GG (0.1, 0.2 and 0.3 quantiles respectively). Results of Table 5.3 proves that RES and GOV can serve as crucial moderators, which mitigate the deleterious effect of CV. Figure 5.1 depicts the quantile plots of the explanatory variables and their relationships with GG.

In Fig. 5.1, the plots reveal that CV is compatible with GG; however, the relationship was not significant. The effect of GOV on GG lies within 0.16–0.02 range, implying that GOV should be more critical in countries with lower GG. The RES trajectory is upward sloping, reflecting the necessity of upscaling it to improve green transition in SSA. To examine the conditioning effects of RES and GOV, the study presents Table 5.4.

Table 5.4 incorporates the interaction effects of RES and GOV. The main effect shows that CV has significant adverse consequence on GG; the effect is larger (in absolute terms) at higher quantiles of GG. Between median and highest quantiles, GG declines by around 1.21–2.4% respectively due to adverse effect of vulnerability to climate risk. The implication of CV is weak in the fourth quantile. However, GG is insensitive to vulnerability in the lower quantiles. This findings support previous

Table 5.5 Computation of moderating effects

Quantile	Range	CV	Moderators	Moderating effect
$q = 0.40$	(0.127, −1.85)	−0.91	(5.62, 0.63)	[−0.91 + 5.62 (0.127) + 0.63(−1.85)] = −1.36
$q = 0.40$	(0.301, −0.60)	−0.91	(5.62, 0.63)	[−0.91 + 5.62 (0.301) + 0.63(−0.60)] = 0.40
$q = 0.40$	(0.568, 0.87)	−0.91	(5.62, 0.63)	[−0.91 + 5.62 (0.568) + 0.63(0.87)] = 2.87
$q = 0.50$	(0.127, −1.85)	−1.21	(6.37, 0.42)	[−1.21 + 6.37 (0.127) + 0.42(−1.85)] = −1.18
$q = 0.50$	(0.301, −0.60)	−1.21	(6.37, 0.42)	[−1.21 + 6.37 (0.301) + 0.42(−0.60)] = 0.46
$q = 0.50$	(0.568, 0.87)	−1.21	(6.37, 0.42)	[−1.21 + 6.37 (0.568) + 0.42(0.87)] = 2.77

Source: Authors' compilation

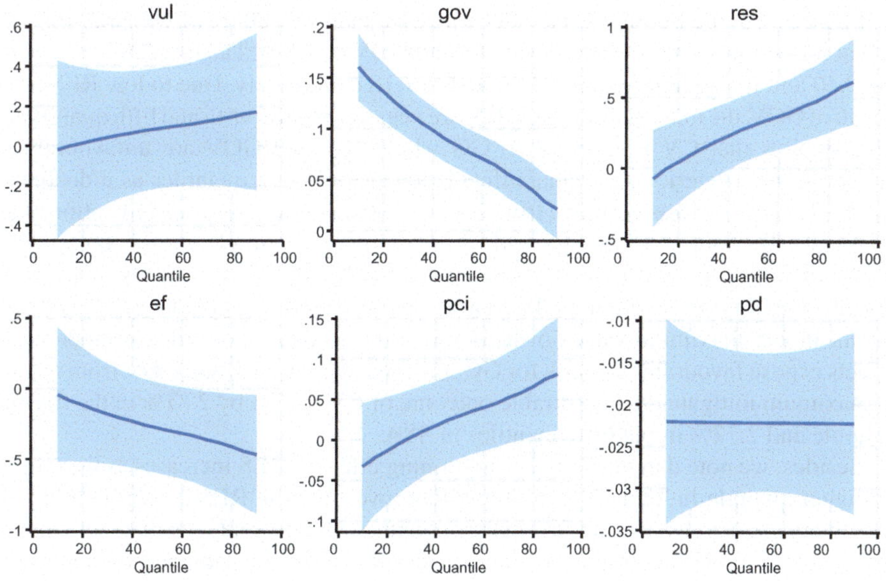

Fig. 5.1 Quantile plots

literatures (Bowen et al., 2012; Lee et al., 2017; Singh et al., 2011; Thornton et al., 2014) which found detrimental effect of vulnerability on sustainable green transition. The moderating effects of RES and GOV favour green transition. RES overwhelmingly mitigates the undesirable consequence of CV on GG. Though GOV reduces the unwanted outcome of CV on green transition in the lower and median outcome, it appears less important in the upper quantiles.

More so, the results of the control variables document an overwhelming positive effect of PCI on GG, which revolves around 0.07–0.08% points. The facilitative connection is significant at 0.05 level. This highlights the efficacy of real per capita income for green transition in SSA. It reflects the ability of high GG nations to consolidate economic progress without compromising environmental conservation. In contrast, EI and PD dwarf the GG trajectories. Energy intensity (EI) significantly decreases GG at the higher quantiles (eighth and nineth) by 0.40% and 0.61% respectively. Theoretically, high EI leads to more energy usage, which hampers green transition. It is worth noting that SSA has low production and consumption of renewables. Upscaling renewable content in the energy mixture can reduce EI, thus, favouring GG. Likewise, population density (PD) distracts GG by 0.02% point. Although PD decrease GG, its size has negligible consequence.

Furthermore, Kingsley et al. (2017) suggest computing interaction effect over certain values of the moderator. In computing the conditioning effect of RES and GOV, we utilise the minimum, mean and maximum values of RES and GOV fourth and fifth quantiles, because they simultaneously have significance of CV, RES, and GOV parameter estimates. Table 5.5 presents the results.

In Table 5.5, the findings establish depressive marginal effect of vulnerability on GG at minimum values of RES (=12.7%) and GOV (−1.85). The marginal effects in the 0.40 and 0.50 quantiles are −1.36 and − 1.18 respectively. Due to low RES and negative GOV, the marginal effects of CV are negative in the fourth and fifth quantiles. This implies that CV is inimical to GG when RES and GOV are unfavourable. However, the deleterious marginal effect is lower for higher quantiles as it declines from −1.36 to −1.18 as we move from 0.40 to 0.50 quantile. This is an indication that marginal effect improves as RES and GOV improve. The implication is that green transition falters if countries do not invest enough resources to upscale resilience and prioritise quality of governance. In contrast, the conditioning effects are higher for the mean and maximum values of RES and GOV. In fact, the results of moderator effects expose favourable outlook for GG. Improvement in RES and GOV from mean to maximum mitigate the undesirable outcome of CV on GG by 2.87% in the fourth quantile and 2.77% in the fifth quantiles in SSA.

Besides, we note that the partial conditioning effect of RES increases as we move to higher quantile in Table 5.4, implying that investment in RES is worthwhile for advancing green transition in SSA. In fact, countries with larger GG will benefit more from promoting RES because they have invested in climate adaptation to downsize vulnerability to climate change. In addition, the partial moderating effect of GOV is indicative of its significance as facilitative mechanism for accelerating GG. Based on the findings, improving GOV landscape will enhance GG trajectory in SSA region. The average GOV score is negative in SSA. Consequently, the region should upscale GG transition by leveraging GOV quality and instituting effective environmental governance mechanisms at both national and regional levels. The positive effect of GOV in driving GG corroborates previous researches (Acheampong & Dzator, 2020; Ahmed et al., 2022; Dada et al., 2024; Degbedji et al., 2024; Lee et al., 2017; Wang et al., 2021). However, Dada et al. (2024) documented deficient effect of GOV between median and upper quantiles, while Egbetokun et al. (2020) found no significant implication of GOV on sustainable environment in Nigeria.

In the light of the findings, we test the hypotheses to confirm their feasibility with the reality. The first hypothesis claims that improving resilience lowers the marginal effect of climate vulnerability on green growth in SSA. The results reveal that low RES is detrimental to CV-GG linkage. Nevertheless, increasing RES beyond median (57%) can accelerate green transition by reducing the adverse outcome of CV. Based on this analysis, the hypothesis is partially supported because the positive moderating effect of RES holds over certain range. Regarding the second conjecture about the conditioning effect of GOV on green transition, the findings confirm the enhancing role of sound governance in achieving higher GG in SSA. From the findings, we observe that the size effect of GOV diminishes as the rates of GG increase, which goes contrary to expectation. We expect countries with higher GG to have better governance structure, including climate governance. Intuitively, the overall GOV could benefit economic and environmental performance without strict vigilance of climate governance. In other words, countries suffering low GG should be more concerned about local climate governance in order to conserve environmental resources.

5.5 Conclusion

The chapter assessed the conditioning role of resilience and governance in forging connection between vulnerability and green growth in Sub-Saharan Africa by covering 2000–2021 period. The analysis documented negative effect of vulnerability on green growth at low levels of resilience and governance, while increases in resilience and governance dwarfed the outcome of vulnerability on green transition. Thus, raising resilience and instilling good governance could lead to positive green growth.

The findings have policy and social implications. In policymaking circle, the research highlights the need for tackling vulnerability to climate change in SSA. Previous analysis confirms the inimical outcome of climate vulnerability on green transition. Note that the main areas of concern for vulnerability include food, habitat, ecosystem, health, infrastructure, and water. Therefore, it is pertinent to dissect vulnerability into sub-components to uncover their implications in the realm of green growth. This will ensure the design of sector-specific policies in addressing its unwanted consequences. First, excessive dependence on food importation risks food insecurity, leading to decline in agricultural capacity. This scenario may induce people to practice unsustainable agricultural methods. Policymakers should harness investing in domestic food production and encourage sustainable agriculture practices. This will facilitate food security, increase farm income, and reduce environmental damage.

Second, habitat and ecosystem are critical vulnerability redlines. They need protection to prevent people from destroying natural capital such as forest resources and marine biodiversity in their quest for survival. Here comes the role of governance. Effective enforcement of habitat safety and protection will ensure minimal vulnerability to climate crisis. Besides, effective climate policy plays beneficial roles of protecting the biomes, reducing flood hazards and ecological deficits, as well as foster international collaboration to mitigate the vagaries of climate change. Thus, resilience is required to ensure climate friendly ecosystem and natural habitats.

Third, water and health are interconnected and both subject societies to climate vulnerability. Policymakers should align health and water policies. For example, provision of sanitation facilities and protection against water-borne diseases should synchronise with clean and reliable water accessibility. Fourth, provision of quality and reliable infrastructure reduces country's vulnerability, which facilitates green transition. The impact of sea level rise attenuates vulnerability to climate crises. Absence of infrastructure for climate adaptation will deepen vulnerability and hinder green growth. However, investing in climate-resilient infrastructure could improve climate resilience and reduce vulnerability leading to improved green growth. Adoption of artificial intelligence (AI) gadgets is another means of improving infrastructure. Leveraging AI for early warning signalling will facilitate preparation for unexpected disasters, which are harmful to the environment. Such action ensure reduction in vulnerability through technological innovation, which upscale green growth in SSA.

Besides fixing vulnerability in the policy issues, resilience is another crucial perspective, which is inevitable for successful green transition in SSA and other developing economies. Stakeholders (such as governments, private sector, donors, and international partners) should invest in climate resilience projects as they are beneficial for green growth. In promoting economic resilience, economic activities should be in harmony with environmental protection goals. These include drastic cut in carbon emission, conservation of biodiversity, protection of forest through radical reduction in deforestation, and sustainable utilisation of natural resources. Again, downscaling climate-damaging economic activities invariably affect jobs in SSA. To avoid the trade off and ensure a win-win strategy between economic activities and green transition, governments should incentivise business community to adopt sustainable business practices such as promoting green lifestyles among employees and customers, adoption of sustainable technologies, utilising renewable energy, and increase energy efficiency.

Furthermore, social resilience facilitates sustainable environment as it empowers the socially vulnerable segments of the society by reducing social inequality, promoting access to education, digital inclusion and access to modern technology. Social resilience helps reduce vulnerability such as displacements and loss of jobs. The vulnerable group living near rivers could be displaced by floods; however, they could neither build new resilient shelters nor afford to relocate elsewhere. Social investment programmes targeting the vulnerable groups will foster social resilience by reducing social inequality among the residents. This avails the vulnerable residents an opportunity to afford flood-resistant dwellings. Additionally, encircling rivers with high walls serves as social protection for the dwellers, which make protected from and resilient to the flood hazards.

Finally, instituting sound governance framework is required in SSA for reducing climate hazards. In particular, enforcing environmental laws is sine qua non for halting deforestation, and reducing the cost of environmental damage. In addition to environmental regulations, policymakers should engage the business community. In the process, regulators should educate business people about the harmful consequences of unsustainable business practices, which damage the ecosystem and makes the habitat uncomfortable. Similarly, governance should target the vulnerable groups by providing them with protection from harsh outcomes of climate crisis.

References

Acheampong, A. O., & Dzator, J. (2020). Managing environmental quality in sub-Saharan African Africa: Does institutional quality matter? In C. Hussain (Ed.), *Handbook of environmental materials management*. Springer. https://doi.org/10.1007/978-3-319-58538-3_215-1

AfDB. (2023). *African economic outlook 2023*. African Development Bank Group.

AfDB and GGGI. (2022). *Africa green growth readiness assessment*. African Development Bank Group.

Ahmed, F., Kousar, S., Pevaiz, A., & Shabbir, A. (2022). Do institutional quality and financial development affect sustainable economic growth? Evidence from south Asian countries. *Borsa Istanbul Review, 22*(1), 189–196. https://doi.org/10.1016/j.bir.2021.03.005

Alagidede, P., Adu, G., & Frimpong, P. B. (2016). The effect of climate change on economic growth: Evidence from sub-Saharan Africa. *Environmental Economic Policy Studies, 18*, 417–436.

Ameer, W., Ali, M. S., Farooq, F., Ayub, B., & Waqas, M. (2023). Renewable energy electricity, environmental taxes, and sustainable development: Empirical evidence from E7 economies. *Environmental Science and Pollution*. https://doi.org/10.1007/s11356-023-26930-5

Berihun, D., & Steven, P. V. (2022). Climate variability and macroeconomic output in Ethiopia: The analysis of nexus and impact via asymmetric autoregressive distributive lag cointegration method. *Environment, Development and Sustainability, 24*, 4064–4087. https://doi.org/10.1007/s10668-021-01604-9

Bowen, A., Cochrane, S., & Fankhauser, S. (2012). Climate change, adaptation and economic growth. *Climate Change, 113*, 95–106. https://doi.org/10.1007/s10584-011-0346-8

Castro, N. R., Spolador, H. F. S., & Marin, F. R. (2020). Assessing the economy-climate relationship for Brazilian agriculture. *Empirical Economics, 59*, 1161–1188. https://doi.org/10.1007/s00181-090-01711-7

Dada, J. T., Ajide, F. M., & Al-Faryan, M. A. S. (2024). The moderating role of financial development in energy poverty-sustainable environment linkages: Evidence from Africa. *Management of Environmental Quality: An International Journal*. https://doi.org/10.1108/MEQ-08-2023-0268

Daniel, P. (2005). Technology revolutions and social development: Prospects for a green technoeconomic paradigm in lower income countries. *International Journal of Social Economics, 32*(5), 454–482.

Danish, A., & Ulucak, R. (2020). The pathway toward pollution mitigation: Does institutional quality make a difference? *Business Strategy and the Environment*. https://doi.org/10.1002/bse.2597

Degbedji, D. F., Akpa, A. F., & Chabossou, A. F. (2024). Institutional quality and green economic growth in West African economic and monetary union. *Innovation and Green Development, 3*, 100108. https://doi.org/10.1016/j.igd.2023.100108

Dell, M., Jones, B. F., & Olken, B. A. (2012). Temperature shocks and economic growth: Evidence from the last half century. *American Economic Journal: Macroeconomics, 4*(3), 66–95. https://doi.org/10.1257/mac.4.3.66

Egbetokun, S., Osabuohien, E., Akinbobola, T., Onanuga, O. T., Gershon, O., & Okafor, V. (2020). Environmental pollution, economic growth and institutional quality: Exploring the nexus in Nigeria. *Management of Environmental Quality*. https://doi.org/10.1108/MEQ-02-2019-0050

Emenekwe, C. C., Onyeneke, R. U., Nwajiuba, C. U., Anugwa, I. Q., & Emenekwe, O. U. (2023). Determinants of consumption-based and production-based carbon emissions. *Environment, Development and Sustainability*. https://doi.org/10.1007/s10668-023-04311-9

Hosseini, H. M., & Kaneko, S. (2013). Can environmental quality spread through institutions? *Energy Policy, 56*, 312–321. https://doi.org/10.1016/j.enpol.2012.12.067

JinRu, L., & Qamruzzaman, M. (2022). Nexus between environmental innovation, energy efficiency, and environmental sustainability in G7: What is the role of institutional quality? *Frontiers in Environmental Science, 10*, 860244. https://doi.org/10.3389/fenvs.2022.860244

Jirophat, C., Manopimoke, P., & Suwanik, S. (2022). *The macroeconomic effects of climate shocks in Thailand*. Puey Ungphakorn Institute for Economic Research.

Kelley, C. (2016). On sustainability, vulnerability, climate, and conflict. *The Journal of Population and Sustainability, 1*(1), 35–44. https://doi.org/10.3197/jps.2016.1.1.35

Khaddage-Soboh, N., Safi, A., Rasheed, M. F., & Hasnaoui, A. (2023). Examining the role of natural resource rent, environmental regulations, and environmental taxes in sustainable development: Evidence from G-7 economies. *Resources Policy, 86*, 104071. https://doi.org/10.1016/j.resourpol.2023.104071

Khandy, I. N., Wani, I. U., & Tarique, M. (2023). Do financial development and institutional quality matter for ecological sustainability in the long run? Evidence from India. *Management of Environmental Quality, 34*(6), 1668–1689. https://doi.org/10.1108/MEQ-01-2023-0002

Khurshid, N., Fiaz, A., Khurshid, J., & Ali, K. (2022). Impact of climate change shocks on economic growth: A new insight from non-linear analysis. *Frontiers in Environmental Science*. https://doi.org/10.3389/fenvs.2022.1039128

Kingsley, A. F., Noordewier, T. G., & Bergh, R. G. V. (2017). Overstating and understating interaction results in international business research. *Journal of World Business, 52*, 286–295. https://doi.org/10.1016/j.jwb.2016.12.010

Lee, T., Chang, Y., & Park, D. (2017). *Governance, vulnerability to climate change, and green growth: International evidence*. Kiel Institute for the World Economy.

Lee, T., Chang, Y., & Park, D. (2021). Governance, environmental vulnerability, and PM2.5 concentrations: International evidence. *The Energy Journal, 42*(2), 24. https://doi.org/10.5547/01956574.42.6.thle

Lockitch, K. H. (2009). Climate vulnerability and the indispensable value of industrial capitalism. *Energy and Environment, 20*(5), 733–745.

Meyghani, S., Mashhadi, M. K., & Salehnia, N. (2022). Long-term effects of temperature and precipitation on economic growth of selected MENA region countries. *Environment, Development and Sustainability, 25*, 7325–7343. https://doi.org/10.1007/s10668-022-02330-6

Mustun, Z. K. (2022). Sustainable development, climate change vulnerability, governance and oil rent: The case of Saudi Arabia. *Journal of Emerging Economies and Islamic Research, 10*(3), 35–62. Retrieved from https://myjms.mohe.gov.my/index.php/JEEIR

Singh, N. P., Bantilan, M. C. S., Kumar, A. A., Janila, P., & Hassan, A. W. R. (2011). *Crop adaptation to climate change* (1st ed.). John Wiley & Sons.

Thomas, K., Hardy, R. D., Lazrus, H., Mendez, M., Orlove, B., Rivera-Collazo, I., et al. (2019). Explaining differential vulnerability to climate change: A social science review. *WIREs Climate Change, 10*, 565. https://doi.org/10.1002/wcc.565

Thornton, P. K., Ericksen, P. J., Herrero, M., & Challinor, A. J. (2014). Climate variability and vulnerability to climate change: A review. *Global Change Biology, 20*, 3313–3328. https://doi.org/10.1111/gcb.12581

Uzur, U. (2021). The relationship between institutional quality and ecological footprint: Is there a connection? *Natural Resources Forum*. https://doi.org/10.1111/1477-8947.12235

Volze, U., & Ahmed, S. J. (2020). *Macrofinancial risks in climate vulnerable developing countries and the role of the IMF: Towards a joint V20-IMF action agenda*. InsuResilience Global Partnership and German Cooperation.

Wang, M. L., Ntim, V. S., Yang, N. J., Zheng, Q., & Geng, L. (2021). Effect of institutional quality and foreign direct investment on economic growth and environmental quality: Evidence from African countries. *Economic Research, 35*(1), 4065–4091.

Welborn, L. (2018). *Africa and climate change: Projecting vulnerability and adaptive capacity*. Institute for Security Studies.

Xiao, Q., & Fei, L. (2024). How does climate vulnerability impact green innovation? A hindrance to sustainable development. *Innovation and Green Development, 3*, 100169. https://doi.org/10.1016/j.igd.2024.100169

Chapter 6
Pathways to Climate Neutrality: Integrating Sustainability in Transport, Energy, and Communication for the Indo-Pacific Region

Narayanage Jayantha Dewasiri, Damith Gangodawilage, and Rubee Singh

Abstract This chapter discusses the pathways to make the Indo-Pacific region climate-neutral, including sustainability within the transport and energy sectors and communications. Thus, the overall purpose of the chapter is to critically evaluate current practices within the region, define specific impediments to attaining sustainable systems, and, more importantly, suggest possible tactical approaches to overcoming such constraints. Based on the Systematic Literature Review (SLR) approach, this chapter adheres to the Preferred Reporting Items for Systematic and Meta-Analytics (PRISMA) protocol to identify the relevant literature and organise the research using the Population, Intervention, Comparison, Outcome (PICO) criteria. The findings showcase a portfolio of challenges unique to the Indo-Pacific region, including the overreliance on fossil energy, technology and infrastructure deficits, and policy bifurcation, but highlight opportunities in renewable energy, electric vehicles (EVs), and green information and communication technology. The evaluated 'case studies' highlighted lessons learned on governmental policy, technological

N. J. Dewasiri
Department of Accountancy and Finance, Faculty of Management Studies, Sabaragamuwa University of Sri Lanka, Belihuloya, Sri Lanka

N. J. Dewasiri (✉)
University Center for Research and Development (UCRD), Chandigarh University, Chandigarh, Punjab, India
e-mail: jayantha@mgt.sab.ac.lk

D. Gangodawilage
Qasim Ibrahim School of Business, Villa College, Malé, Maldives
e-mail: damith.gangodawilage@villacollege.edu.mv

R. Singh
Institute of Business Management, GLA University, Mathura, India

Department of Management Studies, Kumaun University, Nainital, Uttarakhand, India

R. Singh
e-mail: dr.rubeerajput@gmail.com

R. Singh and D. Crowther (eds.), *Climate Neutrality and Sustainable Ecosystems*, Approaches to Global Sustainability, Markets, and Governance,
https://doi.org/10.1007/978-981-96-7181-6_6

advancement, and international cooperation. The implications highlight that attaining carbon neutrality in the context of the Indo-Pacific will require synergy in policies, new funding schemes, and digital solutions for the sectoral decarbonisation of these economies.

Keywords Climate neutrality · Sustainability · Indo-Pacific region · Renewable energy · Decarbonisation

6.1 Introduction

Net-Zero matches greenhouse gas emissions with an equal drawdown level and paves the way towards climate neutrality (Abudu et al., 2024). Climate neutrality is thus crucial to fighting climate change, which has devastating impacts on the biosphere, the economy, and people's quality of life. It entails moving from reliance on non-renewable energy sources to renewable energy systems, increasing energy efficiency and incorporating sustainable. On an international level, the transition to climate neutrality is crucial for implementing the set climate goals, such as those of the Paris Accord, where global temperature increase is limited to below 2% (Meinshausen et al., 2022). This transformation will help prevent extraordinarily adverse weather conditions, minimise environmental pollution, and prevent future generations' suffering and lives from the effects of climate change.

The Indo-Pacific, fused from the eastern shores of Africa and the Middle East to the Pacific Island nations, is a strategic and economic powerhouse with over half the global population. It comprises emerging giants, mainly India and China, plus South East Asia, as well as developed countries like Australia or Japan. The environmental problems affecting this diverse region include climate change effects such as rising sea levels, more frequent disasters, and fossil fuel energy use (Joseph & Sadhasivam, 2023). Indo-Pacific coastal and island states are most sensitive to climate change effects, where many households already bear the brunt of flooding, loss of biodiversity, and displacement, among others (Kumar et al., 2020). Also, it has been established that the region contributes highly to global greenhouse emissions, thus making it an essential cog in addressing the international engagement on climate change. Being a region in exponential industrialisation, the Indo-Pacific has to see how the ever-growing urban populations tangent the economic progressions with ecological concerns.

In the case of the Indo-Pacific region, the transition to climate neutrality is based on the use of sustainable transport, energy, and communication infrastructure. For instance, the transport sector has significant greenhouse emission levels attributed to its extreme dependence on fossil commodities. The reduction of artery pollution can also be achieved by transitioning to EVs, upgrading public transport, and encouraging low-emission logistics. Likewise, the power generation in the region has found its backbone in coal and oil. Therefore, changing to renewable power sources, including solar, wind, and water, is the need of the hour. Others include sustainable communication systems because digital infrastructures and data centres are now vast energy consumers. The region needs to utilise better capacity for communication systems, optimisation, and cleaner technologies. In particular, these sectors operate interdependently, relying on each other to achieve particular goals. By implementing sustainable transport, energy, and communication practices, the Indo-Pacific could form a solid joined-up approach to decarbonisation and for better chances in combating climate change (Dewasiri, 2024).

6.2 Methodology

As this chapter demonstrates, the methodological framework is highly rigorous. It utilises an SLR approach as used by Dewasiri et al. (2024), Rathnasiri and De Silva (2023), Rathnasiri and Gunathilaka (2024), Karunarathna et al. (2024), and Mishra et al. (2024). Further, the study incorporates the PRISMA and PICO guidelines to ensure a robust methodological approach in conducting this study. These strategies were chosen to reflect a comprehensive, systematic, and empirical approach to outlining the sustainability strategies to attain climate neutrality in the Indo-Pacific states. Below is an outline of the methodological steps applicable to this study.

The SLR was carried out according to the aims to identify the state of researched literature concerning the concept of climate neutrality, a subject on which particular emphasis was laid on sustainability in the transport, energy, and communication sectors. Hence, the topics under focus of the review were to discover important themes, challenges, and novelties inherent to the Indo-Pacific region. This was done by identifying the research questions that focused on the current sustainability initiatives in the above sectors, emphasising climate neutrality, the challenges to achieving this agenda, and the tactical approaches and policies that can support the transition to sustainable systems. A rigorous search of Web of Science, Scopus, and Google Scholar databases identified the present literature. Accordingly, the following keywords were used to search for articles, reports, conference proceedings, and government documents in peer-reviewed international journals, databases, and repositories: climate neutrality, sustainable transport, renewable energy, communication sector sustainability, and Indo-Pacific.

While searching for literature, the inclusion and exclusion criteria regarding the type and quality of studies included in this review were used. Seminal articles in English that concentrated on sustainability initiatives in transport, energy, or communication and were published during the last 15 years were considered for the study. On the other hand, papers that did not consider these sectors or where no empirical research was presented were excluded. The fact that this approach was structured made it possible to find several academic articles to be analysed.

The PRISMA framework was used to make the review process more transparent and rigorous, as highlighted by Kelly et al. (2016). This was initiated by searching for specific articles on the subject matter, which was reduced to 317 unique records after eliminating the replicates. These studies underwent initial scrutiny of titles and abstracts to exclude any not strictly speaking research relevant to the formulated questions. One hundred sixty-three studies were eliminated after this step, leaving 154 studies. Full-text review was performed for these articles; only those presenting methodological researchers and empirical relevance were chosen. This process eliminated studies to 101, and 42 were selected and reviewed for analysis.

The PICO framework was applied to guide the review and synthesis of the literature in the present study, as used by Eriksen and Frandsen (2018). The Population (P) studied for this research was identified within the Indo-Pacific region. We have the essential countries: Australia, Japan, India, and Indonesia, as well as the Indo and Pacific countries, and it cover transport, energy, and communication sectors. The Intervention (I) included sustainability measures and technological advancements, including electricity and energy sources, electric and rechargeable vehicles, and energy-efficient communication technology. Comparison (C) considered how these sectors had traditionally operated on a business-as-usual carbon-dense baseline and compared this to a low-carbon transition. Last is the Outcome (O), which focuses on goals such as minimising carbon emissions, improving energy efficiency, and achieving climate neutrality in the region.

Information enrolment comprised a methodical process of recognising and coding information seen in the chosen studies. This involved comparing current practices, policies, technology, by-region issues, and adequate sustainability management case studies. Some topics discussed include carbon footprint in transport, gaining efficiency in energy usage, and the environmental cost of communication technology structures.

Therefore, the combination of SLR, PRISMA, and PICO findings led to better organisation of the findings and an understanding of climate neutrality in the Indo-Pacific region. This methodological framework made it possible to ground the findings in the extensive academic and practical literature, thus providing critical, evidence-based directions for the transition to sustainability in the transport, energy, and communication industries. In this way, this chapter provides practical suggestions to policymakers, businesses, and international organisations that want to take the Indo-Pacific towards climate neutrality.

6.3 Findings and Discussion

6.3.1 Sustainability in the Transport Sector

The Indo-Pacific region boasts huge and diverse transport infrastructure to match the diverse nations in the area. Transportation is significant for economic development, established megacities like Tokyo and Jakarta, or new developing urban centres in India and Southeast Asia. However, this sector has one of the highest emissions of greenhouse gases in the region of the world (Güzel & Alp, 2020). Due to most companies heavily employing road transport using vehicles powered by internal combustion engines associated with gasoline and diesel, the sector ranks among the most significant carbon emitters (Benajes et al., 2024). Most nations across the Indo-Pacific region, especially those cities that are new to urbanisation, are already grappling with traffic jams, pollution, and limited infrastructure in public transport (ESCAP, 2024). Furthermore, the growth of freight transport resulting from the developing trade and industrial demand has increased emissions from heavy-duty vehicles and ships. These factors explain why the transition towards decarbonising the transport sector cannot wait and must be achieved systemically on the path to climate neutrality.

6.3.1.1 Sustainable Mobility Solutions

One potential way to reduce carbon emissions and decarbonise transportation within the Indo-Pacific region is by implementing transitions towards sustainable mobility (Kumar, 2023). Electric Vehicles (EVs) are one of the best starting points for saving the environment from further pollution (Aijaz & Ahmad, 2022). For instance, countries like India, China, and Japan are now actively promoting subsidies, incentives, and infrastructure to increase the use of electric vehicles for cars. The move towards electric buses, cars, and even two-wheelers is gradually taking shape at some city levels, although it remains highly unbalanced worldwide (Noussan et al., 2020). Besides improving the use of battery-operated vehicles, increasing the efficiency of public transport, which is also a critical method of reducing emissions, can be done more quickly. Implementing new metro networks, bus rapid transit (BRT) systems, and tropicalising ongoing and proposed shared mobility service solutions can offer sustainable counterparts to personal transits. Smart city mobility is an emerging idea of using intelligent technologies in combination with urban environment changes to enhance the transport infrastructure (Nikitas et al., 2020). Substantial cuts in emissions may be achieved by improving traffic management, using real-time data to assess and control traffic flow, and implementing the mobility-as-a-service (MaaS) concept (Wong et al., 2020). Moreover, strategies to reduce emissions from freight logistics, like electric delivery trucks and cleaner fuel for shipping, are increasingly important requirements for decarbonising freight transport (Miklautsch & Woschank, 2022).

6.3.1.2 Policy and Technological Innovations

The governments of the Indo-Pacific realise that sustainable transportation requires policy and technological relevance (Abu-Eisheh et al., 2020). Many nations have adopted an environmental policy on using electric cars and other clean fuels. For example, India's National Electric Mobility Mission Plan (Sharma et al., 2023) or Chinese electric vehicle quotas (Li et al., 2018) drive the shift to zero-emission transport. These efforts are also supported by regional projects, including the ASEAN Smart Cities Network that prompts member countries to embrace sustainable urban mobility solutions (Crumpton et al., 2021). Science and technology are also supporting the process of transporting sector decarbonisation in a big way. Technologies to be seen in self-driving cars, hydrogen fuel cells, and battery technologies for energy storage are some of the possible solutions for emissions. Furthermore, the Internet of Things (IoT), Artificial Intelligence (AI), and Big Data in the transport domain facilitate attractive solutions in fleet management and intelligent traffic which are sustainable.

Therefore, though the transportation sector in the Indo-Pacific region continues releasing relatively high emissions levels, the shift to a sustainable mobility future is all the more promising with the right policies and technology available. Indeed, by increasing the rates of EVs, improving the transport system, and encouraging innovation, the region can make significant progress towards transportation-related climate neutrality.

6.3.2 Sustainability in the Energy Sector

6.3.2.1 Energy Demand and Emissions in the Indo-Pacific

The Indo-Pacific region is one of the biggest energy consumers in the world today on account of growth in population, industrialisation, and urbanisation (Bowen, 2022). Global energy consumers such as China, India, and Indonesia exist in this region and it is expected that the energy demand in this region will further increase in the future decades. Sadly, this demand has now been met mainly by fossil fuels and, most significantly, by coal, oil, and natural gas, which are still in huge demand in many Indo-Pacific countries. China and India, for example, use coal frequently as a source of electricity even though it proceeded with high carbon emissions (Sharma et al., 2022). The condition worsened by the high utilisation of the fossil fuel type led to several negative impacts, which include the intensification of climate change, high impact on air quality, health hazards, and environmental depletion. With energy needs increasing over the years, the Indo-Pacific is now in the precarious position of having to meet energy demands while at the same time, cutting greenhouse emissions by half as envisioned in the Paris Agreement.

6.3.2.2 Renewable Energy Integration

Indo-Pacific needs to move away from fossil fuel-based energy supply to attain climate neutrality (Aminjonov & Sim, 2023). Exploring more clean, green, renewable energy such as solar, wind, hydropower, or geothermal is required to minimise the region's carbon footprint. Specifically, the renewable energy source, solar energy, holds great potential because the area receives much sunlight; India and Australia, for instance, have invested big in utility-scale solar development. For instance, India launched the International Solar Alliance (ISA), whose objective is to support the increased deployment of solar energy in the member regions of the alliance

(Shidore & Busby, 2019). Wind energy is another emerging industry, especially in the coasting and islands; Vietnam and the Philippines have considered developing an offshore wind industry (Do et al., 2021). Hydropower is also essential, particularly in Southeast Asia, where countries like Laos and Myanmar have many rivers to generate energy. Also, there is a vast chance of applying geothermal energy, especially in countries such as Indonesia, located in the Pacific Ring of Fire and containing a potential geothermal resource. It minimises emissions and increases energy security by diversification of energy sources with the support of renewable ones integrated into the energy grid.

6.3.2.3 Challenges and Opportunities

Although embracing renewables to meet power requirements in the Indo-Pacific region has a lucrative impact on the natural setting of the earth and economy, it has many difficulties, such as the availability of resources. Further, the existing distribution infrastructure in the region was primarily to support fossil-fueled electrical networks. Integrating more significant amounts of variable renewable generation, mainly through wind and solar, comes with significant capital costs for infrastructure improvements. The problem of financing renewable energy projects persists, though perhaps less so today than in the past. It is particularly acute for developing countries within the region where access to development capital may be more difficult. Therefore, governments must develop policies and regulations that encourage investments in clean energy to expand services (Dewasiri, 2024; Rathnasiri et al., 2024). Also, regional cooperation and knowledge transfer are essential in addressing technical and financial constraints. On the other hand, there are also some opportunities for developing renewable energy in the Indo-Pacific. This region has much potential for clean energy sources, including sunlight, wind, and rivers. Also, investment in batteries as a method of energy storage, improvement in grid systems and the availability of renewable energy technologies are getting cheaper; hence, the trend towards clean power is becoming more achievable.

By and large, the prospects for energy transition in the Indo-Pacific countries remain grim due to the high growth rate of the energy demand and the continent's dependence on fossil fuels. However, there is an opportunity to make this region a leader in the shift towards cleaner energy sources and the consistent policies to support this transition are finally being implemented. Championing renewable energy promotes climate neutrality, boosts energy security, develops green employment, and contributes to economic sustainability in the aftermath of climate change.

6.3.3 Sustainability in the Communication Sector

6.3.3.1 Digital Communication and Its Environmental Impact

The current generation of electronic communication technologies has revolutionised the Indo-Pacific region's economic and social affairs and has posed considerable environmental issues. Globally, with the increased use of the internet, cloud computing and data centres, the energy consumption of information and communication technology is set to rise further (Katal et al., 2023). To excel, perfect, and efficient servers, data centres that are the primary building blocks of the digital economy use a significant amount of electricity, leading to high emission rates of greenhouse gases. More use of 5G networks as it provides quicker and more efficient methods of communication demands much energy to meet the demands of numerous base stations and ceaseless data processing. With digital communication and the internet becoming more and more involved in daily human life when it comes to streaming services or eCommerce platforms, the environmental impact of such technologies also escalates, which places the issue of digital sustainability as an essential part of the climate neutrality agenda.

6.3.3.2 Greening Communication Technologies

Development that seeks to bring about sustainable measures is being advanced and adopted to offset the environmental effects of communication through digital interfaces. Among them is the enhancement of energy efficiency measures, for example, in the 5G and broadband access networks. The management of IT data traffic is improved, and energy consumption is cut through the adoption of software-defined networking (SDN) and network function virtualisation (Kaur et al., 2020). Also, energy-efficient data centres are emerging as critical ones as individuals and organisations are willing to invest in green data centres with the power of renewable sources, efficient cooling systems, and sustainability of the building structure. For example, some data centres are built in cooler regions where the use of air conditioning rates is less likely. Digital decarbonisation also implies lowering the carbon intensity of communication technologies, by utilising low-power semiconductors, recycling the products, and focusing on developing new devices which consume less energy measured in watts. These innovations make the sector less emission-intensive and improve the robustness of digital ecosystems in a climatically changing world.

6.3.3.3 The Role of Communication in Driving Climate Awareness

In addition to minimising its environmental effect, the communication sector plays a significant role in climate change sensitisation and its extension into other sectors.

Web 2.0 technologies, social networks, and communication technologies are effective for creating awareness, sharing information, and mobilising the public to participate in environmental issues, especially climate change. They allow governments, firms, and NGOs to exchange information, attract funding, and organise collaborative actions regarding the climate. For instance, web applications, social media and other technologies can help the public learn the necessary practices and measures for energy saving, efficient use of resources, and environmental conservation. Further, by collecting real-time data, they can assist in sustainable decisions within areas such as agriculture, transport, and power usage, monitoring and ameliorating resource consumption. Therefore, the digital communication sector plays a vital role in the transition towards sustainability of the Indo-Pacific region, and it is an enabler of climate neutrality for the whole region (Carteny & Di Stefano, 2023).

Therefore, there is apparent energy consumption in the region and emissions elevation due to the normalisation of digital communication technologies. However, these technologies present a lot of innovation and climate action prospects. Conclusively, for the Indo-Pacific to achieve its vision of an integrated, climate-neutral future, sustainable practices within the specialised sector and the critical communication platforms for climate must be promoted.

6.3.4 Barriers to Climate Neutrality in the Indo-Pacific

6.3.4.1 Economic and Policy Challenges

This study identified that the economic and policy environment is the most critical challenge facing the Indo-Pacific region in its quest for climate neutrality (Dewasiri, 2024). Most states in the region are developing nations that cannot afford to make significant investments to support sustainable development. Switching to renewable energy systems, upgrading transportation systems, and usually issues. Moreover, there are more barriers in the form of unclear and contradictory measures between countries. Some countries have achieved this through climate action plans or setting carbon reduction targets. On the other hand, many countries largely depend on fossil fuels due to economic implications or the absence of regulations (Rathnasiri & Gunathilaka, 2024). The location of actors within a political space also contributes to the difficulty of working collectively towards climate neutrality regarding resources and governing capabilities. Most Indo-Pacific region states will find it challenging to achieve their climate ambition without comprehensive policy synchronisation and sustainable finance linkages—green financing and international climate finance instruments.

6.3.4.2 Technological and Infrastructural Gaps

The lack of technology and infrastructure is a significant challenge in promoting sustainable interventions across the Indo-Pacific region (Tangney et al., 2021). Taking the cases of renewable energy technologies, electric vehicle transports, and sustainable communication networks, the risks they pose are not distributed uniformly within the region. New technologies, such as green technologies, are yet to surface in developing countries because of their outdated infrastructures and the lack of technological capabilities in developed countries such as Japan and Australia (Rathnasiri & Gunathilaka, 2024). Most cities and countries are not ready to roll out renewable energy or electric cars because they sometimes lack the most basic infrastructure, such as power or charging networks or even innovative city systems (Mastoi et al., 2022). Moreover, there is minimal spending on the innovation of local solutions that address the needs of various regional countries. For instance, small island nations undergoing large-scale utility-scale renewable power plan integration or landlocked or relatively resource-poor countries who face issues in achieving energy mix. Such gaps indicate the need to enhance regional integration and knowledge transfer so that efficient and sustainable deployment of green technologies can be accomplished.

6.3.4.3 Social and Cultural Resistance

Climate neutrality is not only a technological or policy problem but a social and cultural problem (Dewasiri, 2024). This study shows that there are significant differences in knowledge and concern for climate change among the populations of the Indo-Pacific countries. Another emerging challenge tends to emanate from behaviours or changes in practices, such as deciding to minimise energy use or opt for eco-friendly means of transport, which, while essential for sustainability efforts, typically receives pushback because of harmful social norms or reliance on high-emission industries. For example, countries with large-scale industries such as coal mining and the extraction of fossil fuels rely on them for employment and income, which makes change socially and politically challenging (Muttitt & Kartha, 2020). Furthermore, there are cultural influences: for example, Asian culture favours personal car ownership and does not readily accept government interference. According to an Environmental Chief Specialist we interviewed, it is stated that there is also a socio-economic dimension where low-income communities may be unable to afford clean energy or a sustainable means of transport. To remove these barriers, we need awareness and daunting policies to promote socially sustainable transitions for a climate-neutral society without marginalising vulnerable groups.

Hence, while the Indo-Pacific formally committed to peaking emissions before 2030 and achieving net zero by mid-century, the structures and systems that can get the Indo-Pacific region there are not without issues. Each of these calls for intervention that addresses the financial, technological, policy, and cultural nature of the problem. The region can only aspire to attain such climate neutrality objectives and targets through a coordinated partnership approach and persistent and comprehensive interventions at multiple sectoral levels.

6.3.5 Strategic Pathways to Achieve Climate Neutrality

6.3.5.1 Integrated Policy Frameworks

Climate policy strategies for achieving climate neutrality in the Indo-Pacific region call for comprehensive and coherent policies that can work harmoniously, implementing cross-border, and encouraging regionalism (Huang et al., 2024). The member countries of Asia and the Pacific region are so diverse that small island states are nested alongside industrialised giants; therefore, policies that are coordinated to tackle global environmental issues are crucial. Regional cooperation with the climate pacts is possible, establishing carbon trading mechanisms and joint projects for producing renewable energy sources. Bilateral interconnectivity projects for power infrastructure like cross-border renewable electricity networks or regional fast charging networks for electric vehicles bring shared benefits by increasing energy security and decreasing carbon footprints in multiple countries. In addition, there is a need for global cooperation to develop harmonised policy settings that encourage sustainable investment and discourage emissions. This can be in setting high emission cut policies, ratifying and adhering to global climate change pacts such as the Paris Climate Agreement and developing regional knowledge hubs and technology exchange. Thus, integrated policies can define a plan to reach a climate-neutral Indo-Pacific and depict a fair distribution of the notified responsibilities among developed and developing countries (Dewasiri, 2024).

6.3.5.2 Innovative Funding Models

The funds to get to the climate neutrality level are staggering, especially for developing countries in the Indo-Pacific region that do not have access to the necessary funding. Hence, new funding mechanisms are essential for moving the necessary financial resources for green activities (Dewasiri, 2024). One potential solution involves outright support for PPPs so that the government could jointly finance and implement sustainable projects with other private players. These can contribute to sharing the costs and benefits of renewable energies, low-carbon transportation, and other sustainable infrastructure projects. Also, to encourage low carbon emissions by firms, there is a policy tool known as carbon pricing instruments like carbon tax and cap and trade, which can be adapted to support firms' emission reduction and generate revenues for climate investment. Another crucial potential funding source is international climate finance; platforms, for instance, the Green Climate Fund and the Global Environment Facility, among others, provide health finance to developing countries for Climate Change adaptation and mitigation interventions (Schwerdtle et al., 2024). These innovative funding models can thus help the Indo-Pacific to attract the financial means needed to support sustainable change in vital sectors.

6.3.5.3 Technological Innovations and Digitalisation

This study reveals that especially digitalisation contains significant enabling potential for upscaling sustainability in the Indo-Pacific. Some leading innovations critical towards fearless climate-neutral practices include artificial intelligence (AI) and blockchain technology (Dewasiri, 2024). With AI, energy consumption in smart grids can be maximised, accurate weather forecasts for improving the utilisation of renewable energy resources, and more intelligent transportation. For example, deep learning algorithms must be used to improve the network charging infrastructures of electric vehicles or, second, optimise the chains of logistics and supplies to achieve fewer carbon emissions. At its core, blockchain is a highly efficient and secure public ledger, enabling complex carbon trading markets to be more accountable and the whole journey more traceable. However, blockchain can be applied in decentralised recycling and waste management solutions that create circular economy systems that reduce waste and resource use. Furthermore, the surge of the new digital economy being sustainable can lead to lesser adverse effects on the environment by engaging in technologies such as telecommuting, a digital marketplace with green products, as well environmental education through new media platforms., with the help of digitalisation and technological solutions, most of the infrastructural and operation challenges regarding climate neutrality in the Indo-Pacific region can be solved.

6.3.5.4 Other Strategic Pathways

The call for all sectors to adopt sustainable practices that lead to climate neutrality is multifacaceous. As such, decarbonising energy systems and moving away from fossil fuels and towards renewable such as solar, wind and hydropower is a big focus. Replacing fossil fuels with renewable energy can be complemented by improving energy efficiency in buildings, industries, and transport systems, through which energy-saving technologies and intelligent energy management systems can be employed. It is yet another crucial step to electrify transportation and industrial processes with clean, renewable energy.

Transforming the potential transportation sector is one of the critical parts of a climate neutrality strategy. Investments must be made in charging infrastructure and encouraging electric vehicle adoption with policy, including promoting them. Developing and expanding public transportation can help decrease the emissions of reliance on private cars, the other path away from emissions. Urban mobility also has to reduce its carbon footprint, and promoting non-motorised transport options such as cycling and walking is also important. Sustainable fuels like biofuels and hydrogen are also needed to decarbonise complex sectors, such as aviation and maritime shipping.

Waste reduction and resource efficiency, a circular economy, is a very significant strategy towards climate neutrality. When industries and consumers promote recycling, reuse, and waste reduction, it helps lower emissions across the product lifecycle. A circular systems shift also includes introducing decentralised waste management strategies that maximise resource recovery in combination with sustainable production models able to reduce the environmental impact of manufacturing. Moving towards a circular economy allows waste to be minimised and raw material demand to be reduced, lowering the energy intensity of the production processes.

Moreover, Mitigation of emissions and enhancement of carbon sequestration in the agricultural and land use systems require sustainable practices. Improving farming practices that reduce emissions while increasing productivity is a climate-smart agriculture solution. Some examples are the precision farming techniques that optimise resource use and agroforestry that integrates trees into agricultural systems to capture additional carbon. There are also other ways to protect natural carbon sinks and create new opportunities for carbon sequestration: forest conservation and reforestation. It also facilitates improving soil carbon sequestration through sustainable land management practices to allow enhanced carbon storage in the soil and support a climate-neutral world goal.

Behavioural and societal changes are fundamental to achieving climate neutrality along these actions. Public awareness campaigns are needed to encourage people to adopt sustainable lifestyles, from learning to reduce energy consumption to making low-carbon choices in day-to-day life. This process is based on community involvement, optimising the local solutions for specific needs and contexts, and involving the community in climate action by making them the owners of this action. It can significantly encourage consumers to shift their preferences on a large scale. Instead of emitting carbon, they adopt plant-based diets, reduce waste, and purchase eco-friendly products.

A climate-neutral strategy should include monitoring, reporting and accountability. To ensure that countries and corporations are held to account for their climate commitments, the framework must be transparent about where it is going. This identifies gaps in action and maintains the rigour of action in action and climate strategy. Environmental, Social and Governance (ESG) reporting is an essential means to measure the environmental impact of companies and push businesses towards a more sustainable existence in the corporate world.

However, when pursued in concert, these pathways foster a robust strategy to achieve climate neutrality. Solving global warming will require systemic changes in energy, transportation, agriculture, and general behaviour, and by focusing on those changes, the world will move closer to a low-carbon future.

Therefore, to pinpoint the strategic directions by which the Indo-Pacific region can attain climate neutrality, specific policies should be designed together with developing new policies to create imperative new funds and policies that will facilitate and encourage the implementation of new technologies. These elements must all work together for a regional change that will lead to prosperity within a natural, sustainable environment to support a lasting society.

6.4 Case Studies from the Indo-Pacific Region

6.4.1 *Successful Regional Initiatives*

Regarding sustainability, transport, energy, and communication, several countries have continued to implement sustainability projects in the Indo-Pacific region, which have provided some direction towards reaching climate neutrality. For example, Singapore's Smart Mobility 2030 undertaking aims at an enhanced integrated transport system comprising EVs, self-driven public transport, and an intelligent traffic system (Zahraei et al., 2016). Overall, this cuts down the city-state's emissions, or carbon impact, and fashions its urban environment and combat traffic congestion. In the same way, more ambitious renewable energy projects, especially solar and wind power in Australia, have significantly cut the dependency on fossil energy sources. South Australia's Hornsdale Power Reserve is one of the world's most extensive lithium-ion battery facilities, which charges for any extra renewable energy and balances the grid when it is most needed, establishing how others can look to replicate similar storage solutions (Boretti, 2019). In communication, Japan has set the benchmark with green data centres, most of which have renewable energy-sourced options and better cooling technology to curtail carbon output. Such successful initiatives show that the region has the potential to generate innovative, sustainable systems.

6.4.2 *Lessons Learned*

The outcomes from these programs bestowed essential insights for other nations in the Indo-Pacific that aim for climate neutrality. Such projects currently have extensive governmental support, as evidenced by its policies, subsidies, and PPPs. The exemplary Singapore example of innovative city development is the government's vision supported by efficient urban planning that included sustainable transport. Most notably, Australia has received commendation for its advances in renewable energy based on its approach to encouraging more investments in infrastructure and technology for renewable industries. The other crucial lesson for consideration is the relationship between scalable technologies. This is evidenced by Japan's push to have green data centres, demonstrating how beneficial technological applications can be applied across the scale for sustainable purposes. Finally, regional partnerships have been essential because sharing experience and technologies and addressing challenges like cost and infrastructure have been issues. It offers steps for other Indo-Pacific countries to mimic the policy, investment, and innovation in developing climate-neutral societies.

These case studies demonstrate the potential of specific sustainability interventions to change the course in the Indo-Pacific. The basis of such measures can be seen in proper planning, intersectoral cooperation, and an approach ready to promote future and innovative technologies and structures. From such examples, other nations in the region can replicate such efforts up the ladder for a speculative move to climate neutrality.

6.5 Conclusion, Implications, and Future Directions

This chapter examined climate neutrality in the Indo-Pacific region through a thematic sustainability lens, focusing on transport, energy, and communication. Findings showed that existing models operating in these sectors are highly carbon-intensive and contribute to greenhouse gas emissions. Nevertheless, there are opportunities presented through new technologies, new policies, and novel projects being formed at the regional level. Battery-operated automobiles, utilisation of renewable resources, and eco-friendlier information technology are possible solutions towards the climate change problem. However, the chapter also explained some significant limitations, which require further consideration, for example, the lack of financial resources, the problem of technology, and culture. Challenges must be addressed sequentially through strategic pathways such as developing integrated policy to support the system, generating new and alternative funding methods, and utilising technology in artificial intelligence and blockchain. These case studies illustrate that projects on sustainability need not be limited to a specific company but can successfully be replicated at the regional level and can provide important lessons for future planning. Finally, it is concluded that the primary determinant contributing to policy success includes sound policy leadership, regional coordination, and resourcefulness.

6.5.1 Policy Recommendations

The need for consumers to reduce greenhouse gas emissions is accompanied by governments, businesses, and other international actors in the Indo-Pacific region to adopt complex and cooperative strategies for achieving climate neutrality. First, the government needs to ensure and establish transparent and sustainable short-, medium- and long-term regulatory frameworks that would support investments in sustainable technologies and environments. This includes target carbon reduction as much as possible, decommissioning subsidies provided to fossil fuels and supporting renewable resources. Government incentives such as tax subsidies for using electricity in EVs, increased funding for public transport systems, and the development of intelligent mobility platforms are critical in the transport market. There is still more to be done in energy policies to drive higher degrees of renewable power, including wind, solar, and hydro energy, alongside further incentives for grid improvement and energy storage mechanisms. Regarding communications systems, governments must support elaborating low-energy consuming digital equipment such as energy-saver data centres and ecologically friendly 5G networks. Governments should encourage companies to act sustainably through instruments such as carbon trading, green bonds, and the international standard on corporate sustainability reporting. Also, it means that international organisations have to allocate sufficient funds and expertise to help developing countries in the region to achieve climate neutrality.

6.5.2 Roadmap for Implementation

A clear map or strategy must be developed in the Indo-Pacific region to achieve climate neutrality. In the short term, governments must create conditions supporting green technology transition and sustainable development policies. Outcomes on how more can be done include increased investment in renewable power sources, electrification of public transport and the start of green infrastructure demonstration projects. Governments and businesses should also focus on further advancing energy grids for renewable energy and efficient energy industries. In the medium term, it should shift towards popularising these pilot projects and increasing collaboration on regional networks of cross-border infrastructure such as electric car charging and renewable energy frameworks. Strategic objectives should seek to transition the four pillars of society—transportation, energy, and communication—to be carbon-free. It can be sustained through long-term innovation, the widespread implementation of sustainable technologies and products, and the frequent amendments to the policies choking the emission standards. To read more about such separation; Implementing clear deadlines and realistic goals would ensure that a check is done more often to ensure they are on track towards climate neutrality.

6.5.3 Global and Regional Collaboration

The Indo-Pacific cannot become climate-neutral on its own. International and regional cooperation is well needed to improve sustainability in the region. As such, countries within the Indo-Pacific region must build synergies to learn from each other, share the latest technologies, and mobilise adequate funds. The ASEAN, APEC, and the Indo-Pacific Environmental Security Forum are good platforms for countries to work together in climate areas with less duplication. International bodies such as the United Nations, World Bank and the International Renewable Energy Agency (IRENA) can provide leadership in terms of technical support, climate funding and policy advice to support countries in realising sustainable solutions. Additional efforts to advance the specified types of collaboration are large-scale projects, including regional renewable energy grids and low-carbon transportation corridors. However, institutional cooperation in knowledge-sharing and joint research will be necessary in the fight against regionally distinctive problems, including climate vulnerability in the small islands and diversification of energy sources in landlocked countries. When dealing with numerous challenges, including climate change, the

Indo-Pacific region must join forces to force change and guarantee the sustainable development of its nations and partners.

In conclusion, it was found that for the Indo-Pacific region to transition to climate neutrality, strong, effective, and sustainable policies, efficient, effective, and creative sources of revenue, technological developments, and true collaborations with international counterparts must be pursued. Suppose businesses, governments and others join the Great Green Wall Partnership. In that case, the region shall be able to combat global climate challenges and guarantee the prosperity of the people and ecosystems for the future.

6.5.4 Call for Action

The Indo-Pacific region has no choice but to pursue climate neutrality because it can no longer afford to continue as a victim of climate change impacts; it is economically and environmentally influential on the global community. The problem of the climate crisis requires action at least at the government level and at the civil and corporate levels at once. A robust and comprehensive policy should embrace technology development and focus on international cooperation in knowledge and materials sharing. Using a cross-sectional analysis of the case studies, I draw on the concept of the 'sustainability ensemble' and argue that both the public and private spheres require the enabling conditions to be in place to promote sustainability to meet climate goals in a manner that fully integrates social equity and economic rationality. Further, enhancement and piloting would require regional collaboration and global support due to diverse challenges within the region. The Indo-Pacific region is positioned to be the first to propel the world towards climate neutrality. However, this will not come without consistent support from all, risk-taking, and a shared vision for climate change. It is the moment to act together and build a safe and sustainable zone in the Indo-Pacific and the world.

References

Abudu, H., Wesseh, P. K., Jr., & Lin, B. (2024). Climate bonds toward achieving net zero emissions and carbon neutrality: Evidence from machine learning technique. *Journal of Management Science and Engineering, 9*(1), 1–15. https://doi.org/10.1016/j.jmse.2023.10.001

Abu-Eisheh, S., Kuckshinrichs, W., & Dwaikat, A. (2020). Strategic planning for sustainable transportation in developing countries: The role of vehicles. *Transportation Research Procedia, 48*, 3019–3036. https://doi.org/10.1016/j.trpro.2020.08.184

Aijaz, I., & Ahmad, A. (2022). Electric vehicles for environmental sustainability. In P. Agarwal, M. Mittal, J. Ahmed, & S. M. Idrees (Eds.), *Smart technologies for energy and environmental sustainability. Green Energy and technology*. Springer. https://doi.org/10.1007/978-3-030-80702-3_8

Aminjonov, F., & Sim, L. C. (2023). The Gulf states and the energy transition in the indo-Pacific. *Journal of the Indian Ocean Region, 19*(1), 21–39. https://doi.org/10.1080/19480881.2023.2262269

Benajes, J., García, A., Monsalve-Serrano, J., & Guzmán-Mendoza, M. (2024). A review on low carbon fuels for road vehicles: The good, the bad and the energy potential for the transport sector. *Fuel, 361*, 130647. https://doi.org/10.1016/j.fuel.2023.130647

Boretti, A. (2019). Dependent performance of south Australian wind energy facilities with respect to resource and grid availability. *Energy Storage, 1*(6), e97.

Bowen, J. (2022). *Reenergising indo-Pacific relations: Australia's clean energy opportunity.* Climate Council of Australia.

Carteny, A., & Di Stefano, E. T. (2023). The EU and the indo-Pacific. In B. Kratiuk, A. Jaskólska, & Y. Sato (Eds.), *Handbook of Indo-Pacific studies* (p. 405). Routledge.

Crumpton, C. D., Wongthanavasu, S., Kamnuansilpa, P., Draper, J., & Bialobrzeski, E. (2021). Assessing the ASEAN smart cities network (ASCN) via the quintuple helix innovation framework, with special regard to smart city discourse, civil participation, and environmental performance. *International Journal of Urban Sustainable Development, 13*(1), 97–116. https://doi.org/10.1080/19463138.2020.1827411

Dewasiri, N. J. (2024). Sustainable strategies for net zero: An Asia pacific perspective. In R. Singh & D. Crowther (Eds.), *Transition towards a sustainable future. Approaches to global sustainability, markets, and governance.* Springer. https://doi.org/10.1007/978-981-97-5756-5_1

Dewasiri, N. J., Gnanapala, W. K. A. C., Rathnasiri, M. S. H., Yadav, M., Tewari, L. M., & Mishra, U. (2024). The way forward for net zero: Aligning nationally determined contributions (NDCs) and holistic sustainable development in South Asia. In R. Singh & D. Crowther (Eds.), *Transition towards a sustainable future. Approaches to global sustainability, markets, and governance.* Springer. https://doi.org/10.1007/978-981-97-5756-5_5

Do, T. N., Burke, P. J., Nguyen, H. N., Overland, I., Suryadi, B., Swandaru, A., & Yurnaidi, Z. (2021). Vietnam's solar and wind power success: Policy implications for the other ASEAN countries. *Energy for Sustainable Development, 65*, 1–11. https://doi.org/10.1016/j.esd.2021.09.002

Eriksen, M. B., & Frandsen, T. F. (2018). The impact of patient, intervention, comparison, outcome (PICO) as a search strategy tool on literature search quality: A systematic review. *Journal of the Medical Library Association, 106*(4), 420–431. https://doi.org/10.5195/jmla.2018.345

ESCAP. (2024). *Pathways to sustainable development: The nexus of transport, energy, and urban planning in Asia and the Pacific.* Retrieved from https://repository.unescap.org/handle/20.500.12870/6779

Güzel, T. D., & Alp, K. (2020). Modelling of greenhouse gas emissions from the transportation sector in Istanbul by 2050. *Atmospheric Pollution Research, 11*(12), 2190–2201.

Huang, Y., Hossain, M. R., & Haseeb, M. (2024). Energy transition at the crossroads of energy depletion and environmental policy stringency: Energy policy framework for energy giants in the indo-pacific belt. *Energy Policy, 194*, 114311. https://doi.org/10.1016/j.enpol.2024.114311

Joseph, S. V., & Sadhasivam, S. (2023). India as a climate leader in the indo-pacific: Challenges and opportunities. *Journal of the Indian Ocean Region, 19*(3), 255–277. https://doi.org/10.1080/19480881.2024.2314176

Karunarathna, K. S. S. N., Dewasiri, N. J., Singh, R., & Rathnasiri, M. S. H. (2024). What does artificial intelligence–powered ChatGPT bring to academia? A review. In R. Singh, S. Khan, A. Kumar, & V. Kumar (Eds.), *Artificial intelligence enabled management: An emerging economy perspective* (pp. 3–13). De Gruyter. https://doi.org/10.1515/9783111172408

Katal, A., Dahiya, S., & Choudhury, T. (2023). Energy efficiency in cloud computing data centres: A survey on software technologies. *Cluster Computing, 26*, 1845–1875. https://doi.org/10.1007/s10586-022-03713-0

Kaur, K., Mangat, V., & Kumar, K. (2020). A comprehensive survey of service function chain provisioning approaches in SDN and NFV architecture. *Computer Science Review, 38*, 100298. https://doi.org/10.1016/j.cosrev.2020.100298

Kelly, S. E., Moher, D., & Clifford, T. J. (2016). Quality of conduct and reporting in rapid reviews: An exploration of compliance with PRISMA and AMSTAR guidelines. *Systematic Reviews, 5*, 79. https://doi.org/10.1186/s13643-016-0258-9

Kumar, L., Jayasinghe, S., Gopalakrishnan, T., & Nunn, P. D. (2020). Climate change and the pacific islands. In L. Kumar (Ed.), *Climate change and impacts in the pacific*. Springer. https://doi.org/10.1007/978-3-030-32878-8_1

Kumar, T. M. V. (2023). The smart megacity system of indo-Pacific: Emerging architecture and megacities studies. In T. M. Vinod Kumar (Ed.), *Indo-Pacific smart megacity system. Advances in 21st century human settlements*. Springer. https://doi.org/10.1007/978-981-99-6218-1_1

Li, W., Yang, M., & Sandu, S. (2018). Electric vehicles in China: A review of current policies. *Energy & Environment, 29*(8), 1512–1524. https://doi.org/10.1177/0958305X18781898

Mastoi, M. S., Zhuang, S., Munir, H. M., Haris, M., Hassan, M., Usman, M., Bukhari, S. S. H., & Ro, J. S. (2022). An in-depth analysis of electric vehicle charging station infrastructure, policy implications, and future trends. *Energy Reports, 8*, 11504–11529. https://doi.org/10.1016/j.egyr.2022.09.011

Meinshausen, M., Lewis, J., McGlade, C., Gütschow, J., Nicholls, Z., Burdon, R., Cozzi, L., & Hackmann, B. (2022). Realisation of Paris agreement pledges may limit warming just below 2 °C. *Nature, 604*, 304–309. https://doi.org/10.1038/s41586-022-04553-z

Miklautsch, P., & Woschank, M. (2022). A framework of measures to mitigate greenhouse gas emissions in freight transport: A systematic literature review from a manufacturer's perspective. *Journal of Cleaner Production, 366*, 132883. https://doi.org/10.1016/j.jclepro.2022.132883

Mishra, U., Rathnasiri, M. S. H., & Dewasiri, N. J. (2024). Net-zero economy and energy sustainability: A Nepalese perspective. In R. Singh, S. Khan, A. Kumar, S. Luthra, & H. Chokshi (Eds.), *Net zero economy, corporate social responsibility and sustainable value creation. CSR, sustainability, ethics & governance* (pp. 145–161). Springer. https://doi.org/10.1007/978-3-031-55779-8_9

Muttitt, G., & Kartha, S. (2020). Equity, climate justice and fossil fuel extraction: Principles for a managed phase out. *Climate Policy, 20*(8), 1024–1042. https://doi.org/10.1080/14693062.2020.1763900

Nikitas, A., Michalakopoulou, K., Njoya, E. T., & Karampatzakis, D. (2020). Artificial intelligence, transport and the smart city: Definitions and dimensions of a new mobility era. *Sustainability, 12*(7), 2789. https://doi.org/10.3390/su12072789

Noussan, M., Hafner, M., & Tagliapietra, S. (2020). The evolution of transport across world regions. In *The future of transport between digitalization and decarbonization*. Springer. https://doi.org/10.1007/978-3-030-37966-7_1

Rathnasiri, M. S. H., & De Silva, V. (2023). The role of shopping values, cognitive and affective evaluations on the relationship between store environment and store patronage intention: A future research agenda. *Sri Lanka Journal of Management Studies, 5*(1), 100–120.

Rathnasiri, M. S. H., Dewasiri, N. J., & Kumar, A. A. (2024). Policy framework and implementation strategies for Sri Lanka's transition to a net-zero economy. In R. Singh, S. Khan, A. Kumar, S. Luthra, & H. Chokshi (Eds.), *Net zero economy, corporate social responsibility and sustainable value creation. CSR, sustainability, ethics & governance* (pp. 43–60). Springer. https://doi.org/10.1007/978-3-031-55779-8_3

Rathnasiri, M. S. H., & Gunathilaka, D. H. P. M. (2024). Net zero policies for sustainable future: Issues and challenges in the Asia Pacific region. In R. Singh & D. Crowther (Eds.), *Transition towards a sustainable future. Approaches to global sustainability, markets, and governance*. Springer. https://doi.org/10.1007/978-981-97-5756-5_12

Schwerdtle, P. N., Ngo, T. A., Hasch, F., Phan, T. V., Quitmann, C., & Montenegro-Quiñonez, C. A. (2024). Climate change resilient health facilities: A scoping review of case studies in low and middle-income countries. *Environmental Research Letters, 19*(7), 074041.

Sharma, A., Dharwal, M., & Kumari, T. (2022). Renewable energy for sustainable development: A comparative study of India and China. *Materials Today: Proceedings, 60*, 788–790. https://doi.org/10.1016/j.matpr.2021.09.242

Sharma, K., Tiwari, S., Kumar, A., & Sharma, C. P. (2023). Critical review of the electric vehicle ecosystem in India. *Water and Energy International, 66*(8), 42–48.

Shidore, S., & Busby, J. W. (2019). One more try: The international solar Alliance and India's search for geopolitical influence. *Energy Strategy Reviews, 26*, 100385. https://doi.org/10.1016/j.esr.2019.100385

Tangney, P., Nettle, C., Clarke, B., Newman, J., & Star, C. (2021). Climate security in the indo-Pacific: A systematic review of governance challenges for enhancing regional climate resilience. *Climatic Change, 167*, 40. https://doi.org/10.1007/s10584-021-03197-8

Wong, Y. Z., Hensher, D. A., & Mulley, C. (2020). Mobility as a service (MaaS): Charting a future context. *Transportation Research Part A: Policy and Practice, 131*, 5–19. https://doi.org/10.1016/j.tra.2019.09.030

Zahraei, S. M., Choo, C., Cheema, W., & Cheah, L. (2016). Foresight study on Singapore urban mobility: Methodologies and preliminary insights. In M. A. Cardin, S. Fong, D. Krob, P. Lui, & Y. Tan (Eds.), *Complex systems design & management Asia. Advances in intelligent systems and computing* (Vol. 426). Springer. https://doi.org/10.1007/978-3-319-29643-2_10

Part II
Implementing Sustainability for Global Warming

Chapter 7
Climate Neutrality and Global Warming: Problems, Solutions, and the South Asian Path Forward

Mananage Shanika Hansini Rathnasiri, Ali Najeeb, and Lalit Mohan Tewari

Abstract This research focuses on the complexities of attaining climate neutrality in South Asia. Organised using the Preferred Reporting Items for Systematic and Metta Analytics (PRISMA) model of integrated review approaches, the chapter synthesises the literature, case studies, and empirical data to ascertain areas in regions experiencing the effects of climate, challenges to advances, and possibilities. Research has indicated that disaster risks and losses remain high in the South Asia region due to climate change. In contrast, the region sharply continues to depend on fossil energy sources, hence emitting more carbon. However, it underlines potential development measures, including ramping up renewable energy, green urban development, and adopting critical technologies. The findings of this research point to the need for sound regional partnerships, sound policy environments, and the crucial involvement of many players to build the environment for a transition to climate neutrality. Finally, this chapter offers policy and practical implications to those stakeholders who want to ensure sustainable development for South Asia to counter increased vulnerability due to climate change.

Keywords Climate neutrality · Global warming · South Asia · Renewable energy · Sustainable development

M. S. H. Rathnasiri (✉)
Department of Marketing Management, Faculty of Management Studies, Sabaragamuwa University of Sri Lanka, Belihuloya, Sri Lanka
e-mail: shanika@mgt.sab.ac.lk

A. Najeeb
Qasim Ibrahim School of Business, Villa College, Malé, Maldives
e-mail: najeeb@villacollege.edu.mv

L. M. Tewari
Department of Botany, DSB Campus, Kumaun University, Nainital, Uttarakhand, India
e-mail: l_tewari@rediffmail.com

R. Singh and D. Crowther (eds.), *Climate Neutrality and Sustainable Ecosystems*, Approaches to Global Sustainability, Markets, and Governance,
https://doi.org/10.1007/978-981-96-7181-6_7

7.1 Introduction

Climate change and global warming remain some of the biggest concerns of the new millennium, with consequences that affect ecosystems, national and international economies and societies. Climate neutrality, or balancing the carbon input and output load, has become another essential agenda of governments, industries and international organisations. The Paris Agreement signed in 2015 paved for a climate action plan in which countries are to reduce their emissions to keep the global temperature rise below 2 °C and prefer to pursue 1.5 °C (Oztig, 2017). However, the way to climate neutrality, as it seems, is still filled with challenges, even with the commitments made under the Paris Agreement for some regions of the world, such as South Asia. Larger polity, increasing energy requirements, and developmental goals make sustainable development, mitigation, and adaptation to climate change challenging in South Asia. This chapter describes the issues, opportunities, and possible trajectory of climate neutrality and increasing challenges to South Asia.

The rationale behind this chapter is rooted in South Asia's increased climate sensitivity and its importance to climate change characteristics. As a region with a higher population, South Asia remains most vulnerable to climate-related hazards, including floods, droughts, cyclones, and heat waves that negatively impact millions (Patel & Rajak, 2023). In the context of global warming, such the sea levels steadily rising and problematic monsoons define the threat, which includes extreme effects on low-lying countries that are under severe threat of being flooded out of existence, with potential countries such as Bangladesh and the Maldives, monsoons play a pivotal role in agriculture, a crucial facet of most regional economies. Second, CO_2 emissions in South Asia are among the highest globally due to the high demand for electricity generation and industrial uses for fossil fuels, especially coal; hence, conversion to a green economy is challenging but much needed. Climate neutrality is, therefore, both a technical and economic question, and a social and a political question in the region. This chapter will identify numerous challenges that South Asian countries, for example, Bangladesh and Pakistan, experience when addressing climate change, the opportunities to overcome those challenges, and the way to a more sustainable future.

Understanding the climate environment and South Asia's energy status shall be made to comprehend the journey towards climate neutrality. It comprises various climatic topographies, which include high altitude Himalayan region and low land tropical sub-continent with delta-type beaches. Hence, it is vulnerable to global and South Asian regional climatic changes. In the last few decades, the energy demand due to increased urbanisation and industrialisation and burning coal and oil has dominated the region. India, the largest economy in the region, lies in the Carbon emissions list of the world majorly because of its power sector dominated by coal (Ramachandra et al., 2015).

This chapter focuses on these vital issues to comprehensively describe the issues related to attaining climate neutrality while reducing warming in South Asia and present concepts and strategies for the region's progress. Through discussion of best practices for designing clean technologies, learning from successful implementations of clean technology, and understanding of policy environments favourable to clean technology development, this chapter aims to show how South Asia can achieve its climate goals while maintaining its development goals.

7.2 Methodology

This chapter adheres to the PRISMA model, which assists in identifying, selecting, and critically evaluating research studies relevant to the underlying topic, as stated by Karunarathna et al. (2024), Dewasiri et al. (2024), Jayawardena and Dewasiri (2023), Dewasiri et al. (2023), Mishra et al. (2024), Rathnasiri et al. (2024), and Dewasiri (2024). This systematic review of literature, case information, and empirical data helps make the ideas and solutions proposed in this chapter well-researched and supported. The PRISMA model's first step is to search articles through databases such as Google Scholar, Scopus, Web of Science, exceptional environment science, climate policy, and sustainability development journals. Climate neutrality, global warming, South Asia, renewable energy, green technology, and carbon emissions were the keywords selected to find the required materials, studies, reports, and policies.

The second step after a basic understanding of related studies was the screening step, where the title, abstract, and even keywords were scrutinised and determined as fitting to the research questions. Then, inclusion and exclusion criteria were used to filter out available literature, including only the studies that dealt with South Asia and climatic neutrality, global warming, technological solutions, policy frameworks or case studies—the third step involved data extraction from the observed papers. Data germane to climate change, including climate impacts, emission rates, policy measures and technological development, were systematically indexed and classified. The collected data was analysed and combined into general trends such as technology-related issues, financial challenges, and policies. To give a comparative analysis of how other countries in the region are approaching the question of climate neutrality, the cases of Bhutan, India, Sri Lanka, and Bangladesh were examined in detail.

The included studies were critically appraised to assess their quality, validity, and relevance to the chapter's goals. Such studies were evaluated according to the quality of empirical data, credibility of the sources used, and soundness of the conclusion made. When there were concerns about the source of data reported in any study or inconsistencies in the findings, such papers were critically evaluated for selection. This appraisal helped to avoid using low-quality and non-peer-reviewed sources when analysing the chapter. Finally, thirty-six (36) papers were selected as the final sample for this study.

7.3 Findings and Discussion

7.3.1 Current State of Climate Change in South Asia

South Asian community is one of the most sensitive to climatic changes as it has to cope with multiple climate issues that are deteriorating yearly following the increased pace of global warming (Srivastava, 2020). The region is already recording increased and intense frequencies and intensity of extreme weather situations like cyclones, floods, droughts, heat, and icy waves claiming human lives, structures and the economy. Natural hazards such as floods, rivers or streets overflowing with water, usually during the monsoon, that cause houses to be washed away or submerged in water are expected and worsening in Bangladesh, India, Sri Lanka, and many other South Asian countries because of unpredictable movement of the monsoon weather system (Loo et al., 2015). For instance, the floods of 2020 alone inundated millions of people in Bangladesh (Pandey et al., 2022), and recent monsoon-led floods affecting India and Nepal are pretty recent examples of intensifying climate change impacts. Heatwaves have increased with increased frequency and intensity of extreme hot temperatures, with countries like Pakistan and India among those that experienced severe high temperatures. For instance, India is experiencing perilous heat waves impacting humans and animals; heat waves such as these will continue to be felt more often as the temperatures on our planet rise (Mulla et al., 2023).

Among the effects of climate change that are likely to be most devastating for South Asia as a region, one must mention sea-level rise, which threatens the existence of coastal regions. Maldives is a country at the epicentre of this problem (Hinkel et al., 2023). Bangladesh, with a large population along the coastal area and a massive river delta system, is impacted by the threat of flooding and increased salinity in the water, leading to displacement and changes in farming and fishing industries. Since it is a country with 26 atolls, the existence of Maldives is in danger of being plunged underwater if the sea levels rise at the existing pace (Moosa et al., 2020). Nevertheless, the Impact of a one-metre-high sea level is that Bangladesh and the Maldives can be flooded shallowly; this mobility becomes forced, and the loss of income affects millions of people.

Climate change will directly impact agriculture, which forms the base of many South Asian economies. Alteration of monsoon patterns, the upsurge of temperature, and the enhancement of the distressing frequencies of droughts are already diminishing crop productivity and posing formidable challenges to food security in this area (Behnassi et al., 2021). Fickle monsoon rains in India have impaired rice and wheat yields, and the country questions whether it will continue to be able to sustain its increasing population. For instance, in the southern provinces of Pakistan, droughts have recurred and affected the agriculture and drinking water supply. Moreover, glacial melt in the Himalayas, which feeds millions of people through significant rivers such as the Ganges, Brahmaputra, and Indus, is rising due to climate change-induced rising temperatures, which means increased water scarcity in the region (Verma, 2023). These may paint a bleak picture of the region's food and water security situation that demands urgent interventions through effusive actions.

7.3.1.1 Carbon Emissions and Economic Development

South Asia, becoming one of the fastest-developing regions recently, has helped millions of citizens break free from poverty while boosting carbon emissions. Most of the region relies on fossil energy sources for electricity and industrialisation. Gross Domestic Product (GDP) is much more prominent in India than in other South Asian countries, and it is also among the top global providers of carbon emissions, where coal makes up 70% of electricity production. This has resulted from increased energy demand due to industrial growth and urbanisation, with the result now seeking coal-fired power plants. While significant spending has been on solar and wind energy, the shift towards a lower carbon economy has been incremental due to the currently laid down infrastructure and economic development framework.

In Pakistan and Bangladesh, primary energy sources are coal and natural gas, critical greenhouse gas (GHG) emissions sources (Abas et al., 2017). Although these countries have improved their policy and regulation to some extent regarding the use of renewals energy sources for their energy security, their dependence on fossil fuels is still a central sticking point which hinders the work towards attaining climate neutrality. For example, the form of China-Pakistan Economic Corridor (CPEC), which includes the development of new coal-fired power plants to cater to the energy needs of Pakistan while generating electricity, boosts the country's carbon footprint (Munir & Khayyam, 2020). Similarly, while Bangladesh continues to diversify its consumption usage with more electricity generation from natural gas, it becomes challenging to bring down its emissions in the coming years.

The relationship between industrialisation and carbon emission is well illustrated in the South Asia region, and so has the cost of economic development at the expense of environmental conservation. The various industries of the region, such as steel, cement and manufacturing industries, are significant emitters of carbon; major industries within the region continue to use outdated technologies that exploit energy and pollute the environment. Second, transportation has also become a significant source of emissions for developing cities such as Delhi, Mumbai, Dhaka, and Karachi, given the increasing ownership of vehicles. The growth of new personal cars and high dependence on diesel mode worsened air quality and added to the emissions in the region.

This relationship has evolved out of the current picture of economic development, where growth targets are mostly the opposite of emissions goals in South Asia. At the core of the conflict of interest, governments are under pressure to minimise poverty while encouraging economic growth and, at the same time, protecting the natural environment. The desire to improve living standards and industrialise has put the region in a fix to cut or diversify its dependence on carbon-intensive industries. In addition, cheap and easily accessible coal, both locally and internationally, continues to entice countries, especially those that desire to make modest but affordable energy targets a reality.

Still, there are some positive trends visible at the moment. For instance, India has set targets for renewable energy to reach the 450 GW renewable power capacity mark in 2030 (Gulagi et al., 2022). Bangladesh has also established a few policies for spreading solar energy in rural areas, and Sri Lanka has also aimed to shift towards 100 per cent renewable electricity by 2050. However, the process is not very rapid, and the region continues to experience constraints on policy enactment, funding, and technology. Climate neutrality will require a collective effort to decarbonise, upgrade infrastructure and spread technologies that advance development and lower carbon emissions.

However, South Asia needs a transformative change towards a low-carbon economy since its carbon emissions are increasing due to economic development. These features define the need for proactive overall resource management within the region to achieve economic growth and minimise adverse climate impacts. This implies that South Asian countries can only attempt to steer through this multifaceted structure and transition towards climate neutrality with the help of sustained political commitment together with regional and international assistance.

7.3.2 Problems and Barriers to Achieving Climate Neutrality

It is, however, pertinent to note that the processes of attaining climate neutrality in South Asia are accompanied by a host of technological, economic, policy and social impediments, not forgetting political hurdles. Some progress has been made, but challenges still prevent the region from moving further. These barriers indicate the developmental circumstances that typify South Asia as a vulnerable region where even growth, poverty alleviation, and social well-being are overarching factors over environmental preservation. Ignorance of these barriers explains inadequate planning in the pathway towards carbon neutrality.

7.3.2.1 Technological Barriers

Nonetheless, one of the significant challenges to climate neutrality in South Asia is the inability to access some of the exotic green technologies (Amin et al., 2023). The shift from the conventional energy type, for example, the fossil-fueled energy system, to the new natural energy systems like solar energy and wind energy requires new generation technologies that are expensive and difficult to deploy. Although some countries in the region have already made some efforts and started steadily investing in renewable energy, such as India, the general tendencies are slow, mainly because of certain technological constraints. Specifically, coal and natural gas dominate the region's energy systems, and integrating renewable energy technologies into these energy systems entails substantial capital and skill investments.

In addition, the critical component in energy storage technologies, vital for balancing renewable energy generation, is scarce within the South Asian region. Uncertainty in the availability of resources from solar and wind as power sources makes it rather hard for countries to move away from fossil energy sources due to inconsistency. The region also lacks readily available infrastructure to support electric mobility in a way that can lower the emissions from the burgeoning transport sector. As seen in countries of India, electric vehicles (EVs) have started, but the cost of EVs and charging infrastructure remain significant barriers to mass implementation. Also, awareness of green technologies and their implementation and maintenance is challenging because local talent in renewable energy, waste management, and sustainable farming is scarce.

7.3.2.2 Economic Constraints

The immediate problem discussed in this research is that financial constraints in South Asia make it challenging to attain climate neutrality, as Rathnasiri et al. (2024) stated. Green economy transformation mandates investment in discrediting carbon and implementing renewable energy, efficiency, and infrastructure. However, most of the countries in the region have financial problems, which make them unable to support large-scale projects. The affordability of green technologies and innovation remains high, and capital costs for governments and businesses remain high; hence, they cannot afford to invest in sustainable projects. For instance, the initial capital costs for developing solar, wind and power storage batteries are still high despite the long-term socioeconomic returns (Farghali et al., 2023).

Also, most South Asian countries continue to face challenges in eradicating poverty and, indeed, in their overall economic growth, and more often than not, such issues have to take priority over climate issues. The short-term requirements of job creation, standards of living and energy admission, and climate change demands, on the other, are significant challenges for governments. This trade-off has been seen most evidently in the rural setting. Many rely on traditional energy sources such as biomass and coal for cooking and heating in these settings. The switch to clean energy is expensive, and these communities cannot make any changes without sufficient funding.

Multilateral channels, like the Green Climate Fund, offer financial support to developing countries, but the money is rarely enough to cover the needed investment (Seo, 2019). Also, private sector investment in financing climate action is meagre because firms avoid such opportunities due to risks in the green technology market (Kumar et al., 2024). South Asian countries will struggle to generate enough funds to achieve climate neutrality without innovative financing opportunities and favourable public-private partnerships.

7.3.2.3 Policy and Governance Gaps

Several challenges hamper the move towards attaining climate neutrality in South Asia: lack of strong policy measures to support the environment and systematised governance structures (Rathnasiri et al., 2024). While the targets of numerous countries in affluence or cutting down on emissions and boosting renewable energy are set rigid, the actual execution of these policies is insufficient. Standard environmental regulations do not have enforcement capability, and where there is, corruption from the local and national governments exacerbates the problem. For instance, as much as India has envisaged a plan to increase renewable energy sources, it has lately been awarding more coal power projects based on political and economic factors.

In South Asian countries, governance is also divided over various governmental departments; most of the time, they do not work efficiently to implement the policies. Environmental ministries generally do not dispose of the necessary power to compel climate policies, and decentralised governance hampers national, regional, and municipal collaboration. It is crucial to note that some countries' energy, industry and environment ministries have competing agendas, which often blend in the formulation of policies thus making it challenging to rank climate change action. In addition, the concept of regional integration is also absent, and the executive summits refuse to agree on standard climate policies while dealing with problems that can be solved only collectively. For instance, further collaboration on water resources and disaster preparedness is still weak despite several impacts of climate change in the entire region.

7.3.2.4 Social and Cultural Challenges

Lack of social willingness, people's unawareness of environmental problems, and lack of community involvement also prevent South Asia from aspiring to attain climate neutrality (Giri et al., 2023). As has been seen, the traditional ways of using energy are prevalent today, for example, using biomass for cooking or employing coal in industries. Its transition to cleaner technologies is usually opposed because of cost, convenience, and disruption of its existing practice. For example, people of rural India and Bangladesh are reluctant to go for solar home systems or clean cooking solutions because the initial costs are high, and these technologies are either unknown to them or they believe they are unreliable enough.

Also missing is the understanding among the general population that climate change is actual and imminent and that people could benefit from sustainable activities. As environmental movements continue to grow in cities, rural and economically marginalised populations continue to be left out of the fold on climate change. This paper calls for increased focus on achieving sustainable development but emphasises that environmental education as a subcategory remains neglected in school curricula, especially in rurally served schools. Despite all these, there is a lack of appreciation of the dangerous implications of climate change. Thus, building the social and political will, which is critical in facilitating extensive change processes, is not straightforward.

7.3.2.5 Political Dynamics

Political agenda and instability in South Asia make it even more challenging to reach the goal of climate neutrality (Rathnasiri et al., 2024). Political structures in the region are mainly defined by short electoral cycles; therefore, governments are unpredictable in their commitment to dealing with climate change. The strategy that many political leaders choose is near-terms economic and political interests, for instance, the development of industries or supporting fossil fuels. However, as seen in India's case, removing coal from the economies that rely on it for employment and political influence is difficult, as it will face much resistance from industry lobbyists and local communities.

Furthermore, many South Asian regions still have poor political stability, constraining the government's proper and coherent climate policies. Due to unstable governments, especially in Pakistan and Sri Lanka, and recurrent economic instabilities, climate issues have received minimal attention. Sustained regional conflicts and political instabilities also hinder the common cause of climate change, as nations focus on their security and growth to the detriment of climate-friendly policies. This lack of political continuity and commitment diminishes the capacity of the region to plan for substantive, long-term climate change policy for neutrality.

7.3.3 Solutions for Promoting Climate Neutrality

South Asia holds numerous challenges as it works towards climate neutrality. However, there are also opportunities unique to the region, which means the region with the right approach could supercharge the solutions required to shift to a low-carbon economy. For South Asia, climate change urgency juxtaposed with economic growth and social concerns must be approached through policies aligned to variable renewable energy, sustainable urban development, green mobility, policy change, and international support. The following solutions, therefore, are essential to achieve the goal of attaining climate neutrality in the region.

7.3.3.1 Renewable Energy Expansion

Conservation and 'scaling up' of renewable energy technologies like solar, wind, etc., including hydropower in South Asia, indicates one of the region's most potential approaches to carbon neutrality. The region has massive natural resources that could underpin big renewable power projects, especially solar power, due to the region's high sunlight level (Murshed et al., 2021). India and Bangladesh have already started the implementation of the solar power project, as well as have the capacity to increase the included volume due to the technological development and construction of related facilities.

Private-public partnerships (PPPs) have been instrumental in reforming South Asia's electric power and renewable energy initiatives. The government can propose private investments in every country, resulting in more people funding renewable energy infrastructure. At the same time, governments can benefit from the experience and skill of the private players. For instance, the government partnered with private players to achieve India's ambitious target of setting up considerable solar parks/renewable energy, wind farms, and power projects. Bangladesh's solar home system has offered cheaper and decentralised electricity to rural villages using non-governmental organisations (NGOs) and private firms like the latter.

There are specific ways of dealing with the intermittency problem in renewable energy through technological advancements, such as solar panels, energy storage systems, and intelligent grid systems that enhance immediate reliability in a renewable system (Mishra et al., 2024). These technologies help to generate reliable power required to minimise the utilisation of fossil energy sources. Case examples like India's Jawaharlal Nehru National Solar Mission show how policies and financial incentives, followed by technological development, can boost renewable energy and lay a clear path to climate neutrality.

7.3.3.2 Sustainable Urban Planning

The problem is how South Asia can grow as a rapidly urbanising region while attempting sustainable development. A strategy for greening cities can help minimise carbon emissions within the urban built environment, which is a significant factor in climate change. Green infrastructure and integrated resilience strategies, including energy-conserving structures and effective waste disposal techniques, help cities become more resilient to climate change effects and minimise their environmental impact.

Green infrastructure systems refer to systems like parks, green roofs, and urban forests that help lessen the impacts of the urban heat island and energy usage and support biological variety. For example, city designers in Sri Lanka are now inclined to provide greenery in cities to meet the demand for fresh air, plus an enhanced cooling system (Rupasinghe & Halwatura, 2020). Further, while minimising greenhouse gas emissions, energy-efficient buildings are equally important to eliminate emissions from the residential and commercial sectors that consume the most energy in the urban city. Implementing efficient lights, solar power, and intelligent managing energy systems helps the buildings to mitigate their CO_2 emissions intensively.

Another emerging issue in the planning context of the developing cities of South Asia is sustainable waste management. This has been made worse by rapid population growth and urbanisation, which are stretching the available waste management infrastructure. The solutions, including the construction of waste-to-energy plants, recycling bins and composting, would help reduce methane emissions from landfills, and other cities could adopt circular economy principles. It is possible to note that the successful application of the solutions mentioned above would presuppose cooperation between municipal authorities, the corresponding private actors, and civil society organisations to guarantee proper waste management.

7.3.3.3 Green Transportation

Transportation is one of South Asia's most significant sources of carbon emissions, especially in the fast-developing cities (Rathnasiri et al., 2024). Passing over from a traditional public transport system to a green transport system is crucial for attaining climate neutrality in the region. One of the most important solutions is the development of electrification of transport, in other words, overcoming the barriers associated with using electric vehicles (EVs) and necessary charging facilities. India and other countries have also started giving incentives to car manufacturers and buyers and are investing in the charging points of EVs. However, the initial costs associated with purchasing an EV are relatively high while the infrastructure for EVs is still in its early stages.

Enhancing the public transport systems is another critical solution to tackling transport sub-sector emissions. Domestic transportation initiatives such as metro lines, inclusive bus rapid transit or other systems, and electric buses have the potential to offload massive private automobiles, leading to reduced emissions. For instance, both Sri Lanka and Bangladesh are seeking means by which their cities' public transport systems can be enhanced. At the same time, India has used more money and investments in metro lengths in cities, including Delhi and Mumbai (Kanchan et al., 2024). Moreover, increased use of environmentally friendly means of transport like biking and walking can significantly decrease the use of those vehicles caused by fossil fuel.

For these initiatives to work, governments must also put measures that encourage and discourage the use of low and high-emission vehicles, respectively. For instance, the carbon pricing policies and the fuel economy standards bear the risk of changing the consumption patterns regarding transport. When properly implemented, formulating the policy environment encourages public and private entitlement engagement, and South Asia can make considerable progress in cutting transport pollution.

7.3.3.4 Policy Reforms and Regional Cooperation

For the South Asian region, efficient structures and cooperation in policy all contribute significantly to climate neutrality. Policymakers must carry out specification reforms such as encouraging green investments, a market structure that promotes renewable energy and, more importantly, ensuring that environmental policies are complied with (Rathnasiri et al., 2024). For instance, carbon pricing, including carbon taxes or emission trading systems, develops financial virtuous circles for reducing business emissions. Similarly, providing subsidies for renewable energy projects and tax incentives on green technologies will go a long way in creating a low-carbon economy.

Integrated regional cooperation caters for regional commonalities and supports climate intervention at the regional level (Dewasiri, 2024). There is a need to increase cooperation in areas such as water resources management, disaster risk reduction, and energy connectivity concerning the South Asian nations. For instance, cooperation in the construction of hydroelectric power between India, Nepal, and Bhutan is likely to generate clean power for the whole region and, at the same time, cut emissions. Thus, South Asian countries should be able to intensify their efforts in climate change mitigation by sharing knowledge, resources, and technologies.

Other regional associations like SAARC have a central role in proactively participating in regional forums to understand climate problems. Increasing the position of regional organisations to control and facilitate action on climate change and encouraging member-state commitment to the policies would potentially help South Asia develop a united effort to address climate change.

7.3.3.5 International Support and Climate Finance

It is well understood that addressing climate change targets and achieving climate neutrality in South Asia will need international support and climate finance. Some countries with limited fiscal capacity cannot afford big climate projects and need the support of bilateral and multilateral lenders, climate funds, and development partners' aid. Bilateral and multilateral players like the Green Climate Fund (GCF) and Global Environment Facility (GEF) have committed reasonable money to South Asian countries to fund renewable energy projects, climate change mitigation, and enhancement of capacities (Rathnasiri et al., 2024).

However, world bodies should come forward to provide aid in terms of finance and technology to support South Asia in achieving milestones in climate change. Most of the obligations are due from developed countries, which must supply climate finance to support developing nations in shifting to lower-emission economies. Trade and bilateral aid and debt cancellation, along with technology transfer, are necessary for South Asian countries to address global climate change concerns and stimulate their development pathways at the same time.

However, it is crucial to mention many other innovative sources of financing these climate projects, like Green and climate resilience bonds. South Asia can scale up its climate change mitigation efforts and, in the process, move towards achieving climate neutrality by mobilising international finance and engaging critical global financial institutions.

Consequently, South Asia can advance significantly to overcome climate inactivity by discovering solutions like renewable energy growth, climate-smart urbanisation, green mobility, and better policies. They know how to get beyond imposed barriers to do what is essential to assure the region's viability and a transition to sustainable low-carbon societies.

7.3.4 *Case Studies of South Asian Countries*

South Asia is a mixed bag in its journey towards climate neutrality and preparing for climate change. The region embraces countries that are at different levels of development. Therefore, they have different problems regarding their economic status, management of resources, and sensitivity to climate change impacts. South Asian countries' experience in implementing climate policies and achieving general climate goals allows an understanding of potential prospects for becoming climate-neutral in the region. This section focuses on four fundamental case studies: Four countries, which one can learn lessons from, are Bhutan, India, Sri Lanka, and Bangladesh on climate change and sustainability.

7.3.4.1 Bhutan: A Carbon-Negative Country

Bhutan is unique worldwide as much as it can be referred to as a carbon-negative nation; it sequesters even more carbon dioxide than it releases. This remarkable accomplishment can be attributed to sustainability and environmental conservation, which form the principal goals of the country. How Bhutan has adopted Gross National Happiness (GNH) to balance economic and environmental development can help other South Asian countries (Yangka et al., 2018).

It is also important to note that the carbon-negative status of Bhutan is due to the millions and odd tons of carbon dioxide absorbed by forests in the country annually (Soubhari et al., 2024). The country has enshrined in the constitution that at any given time, it must preserve at least 60% of its land as forest land to continue sequestrating carbon emissions. Also, nearly all of the electricity produced in Bhutan comes from hydroelectric power, thus a renewable energy source that has helped the country to phase out the decadent use of fossil energy. Bhutan also sells electricity from hydropower to its neighbouring countries, especially India, making more efforts to reduce carbon footprints.

7.3.4.2 India's Renewable Energy Initiatives

India, one of South Asia's largest and most populous countries, has invested exceptionally well in Renewable Energy, especially Solar energy. India's flagship National Solar Mission, launched in 2010, has set a target of adding only 100 GW to the solar power capacity by 2022, which sums up the country's vision of moving towards low low-carbon economy (Khanna et al., 2023). This constitutes a fair proportion of India's overall energy basket, conceived for curtailing carbon emissions from the power generation sector.

Several policy measures, private sector participation, and foreign direct investments (FDI) have helped India to develop solar energy. Renewable energy centres like the Bhadla Solar Park in Rajasthan, one of the world's most extensive facilities, affirm the country's potential to invest in major solar parks (Jaiswal et al., 2019). India also includes wind energy; the country wants to take the wind energy capacity of the country. All the above-combined efforts are assisting India to diversify away from coal and other more conventional fossil energy sources that have long been preponderant in its energy mix.

However, India remains confronted with several problems in attaining its potential: integration into the grid, financing, and land acquisition. However, the progress made by India in ramping up renewable energy and the increasing emphasis on efficiency in transportation and energy storage, coupled with the introduction of electric vehicles, make India one of the key players in the region's path to net-zero emissions.

7.3.4.3 Sri Lanka's Environmental Policies

The following analysis is based on research on environmental policy and carbon emission shifts in Sri Lanka, a climatically sensitive region with newly implemented policies focused on the environmental issue. On the issue of climate change policies, the country, under the Paris Agreement, has pitched its Nationally Determined Contributions (NDCs) an intention to shift towards 100% renewable electricity sources by 2050 (Dewasiri et al., 2024). In energy sector, development plan of Sri Lanka features the use of solar, wind, and hydroelectric to decrease reliance on imported oil.

Sri Lanka's Soorya Bala Sangramaya (Solar Energy Fight) began in 2016 to find households, businesses, and industries to install solar panels (Hemakumara et al., 2023). The initiative has promoted rooftop solar power generation, thus lowering electricity consumption from conventional power plants that use fossil fuels. Secondly, the government has floated large-scale solar projects like the Mannar Wind Power Project, which also boosts the growth of renewable energy in the country.

Sri Lanka has also come up with measures for energy development together with measures for the conservation of biodiversity, comparing the rate of deforestation to reduced and enhanced waste disposal systems. The country's blue-green economy mainly focuses on the conservation and sustainable use of the ocean, promoting environmental conservation. There are still some challenges Sri Lanka has to overcome in achieving its climate goals. Some of them include lack of adequate funding, political instability, and the required robust enforcement of laws governing the environment. However, its long-term goals for producing renewable energy and environmental policies align with other South Asian countries aiming for sustainable economic growth.

7.3.4.4 Bangladesh's Climate Resilience and Adaptation

Bangladesh is one of the most climate-vulnerable countries globally, which suffers from floods, cyclones and a rising sea level. Nevertheless, Bangladesh has developed adequate acclimatisation and risk management plans on climate change and is well recognised as a world icon of the risks posed by climate change. Bangladesh's

National Adaptation Programme of Action (NAPA) and the Bangladesh Climate Change Strategy and Action Plan (BCCSAP) are strategic plans for measuring climate change impacts and how to mitigate its effects while aiming at lower emissions of carbon (Rahman et al., 2021).

Bangladesh is one example of a country that has made significant efforts to improve climate resilience through disaster risk reduction: Cyclone and flood early warning systems have greatly diminished the number of deaths. The country also has climate-proof measures like building houses on stilts and flood-proof roads for communities vulnerable to sea-level rise and storm surges. Also, Bangladesh has undertaken mangrove afforestation to protect against natural disasters such as storm surges and coastal erosion in the country's coastal regions.

Bangladesh's solar home system (SHS) programme is perhaps one of the most successful in the world, delivering off-grid renewable energy to rural Bangladeshis (Hellqvist & Heubaum, 2023). By using solar energy in areas that lack grid power connection, Bangladesh has managed to reduce its incessant dependence on fossil power sources and, at the same time, increase energy access to its populace. The programme has ensured private/public participation and cooperation with development partners; partnership effectiveness is paramount in climate change resilience.

However, there are new challenges of poverty, population growth in Bangladesh, and poor financial capabilities for considerable-scale climate change adaptation. Nevertheless, the climate resilience strategy, emphatic on community-based adaptation, gives the country much worth emulating by other climate-affected countries within South Asia.

7.4 Conclusions, Implications, and Future Directions

Climate neutrality is one of the most significant endeavours and a chance to redesign South Asia's economy and environmental course. Now, climate change is no longer a subject of debate or uncertainty for millions of people in the region but a daily reality with increased seasonal temperatures affecting food production, severe floods and droughts eradicating livelihoods, and continuing sea-level rise. The climate neutrality process is a collective undertaking that national governments, international organisations, businesses and civil society activities must undertake comprehensively beyond national borders, innovative and integrating sustainable development.

As in the previous chapter, this chapter has aimed to demonstrate the current situation and outlook for climate change in South Asia with rich insights into regional climate change conditions and extensive dependence on carbon-based energy resources. South Asian economies are at this crossroads where energy choices and urban and industrial development decisions will determine the emissions path for the next few decades. The obstacles—technical imperatives, lack of funding for policy chasms, and social opposition—are all natural but surmountable. It will not take conscious effort, precise action, wise investment, and a new policy direction in addressing climate change issues to attain climate neutrality.

The solutions proposed—including reinforcement of renewable power, sustainable cities and green mobility—show how South Asian countries can divorce growth and carbonisation. However, the successful implementation of this solution depends on good governance and regional integration. Climate policies should be considered a priority at the national level, effective and integrated regulatory legislation should be put into practice, and new forms of international cooperation should be developed to form new mechanisms to shift towards the low-carbon economy systematically. International climate finance and technological advancement will be vital to surpassing economic and technological constraints that pose challenges.

The analysis of the South Asian countries can serve as an example of how progress has been made despite these challenges. The other nations have many lessons to learn from Bhutan, whose carbon count is negative, and India, which has set ambitious targets in renewable energy sources, Sri Lankan environmental policies, and Bangladeshi measures on climate change resilience. These examples demonstrate that it is possible to develop successful country-specific strategies backed up by well-directed and sufficiently funded local initiatives. They also posit that the capacity of the successful scaled practices should also be introduced and underscore the value of regional knowledge sharing and successes for all South Asian nations.

Regarding innovative and technological development, South Asia has a latent capacity to move to a higher level of sustainable development. From artificial intelligence in energy consumption to using blockchain in carbon trading and ideas such as digital solutions in waste management, technology can go a long way to closing that yawning gap between climate change targets and implementation. Additionally, the knowledge and beliefs of indigenous people, as well as their traditional practices, can enrich and sometimes replace modern technology and present culturally adequate yet environmentally friendly solutions.

However, the problems are numerous, and paradoxically, the solutions are also attainable in achieving climate neutrality in South Asia. South Asia can be the first region to turn the demographic dividend into a demographic dividend for a sustainable and climate-neutral future safely, sustainably, and prosperously if policy innovation is complemented by technology innovation and regional cooperation. The time for action is today.

References

Abas, N., Kalair, A., Khan, N., & Kalair, A. R. (2017). Review of GHG emissions in Pakistan compared to SAARC countries. *Renewable and Sustainable Energy Reviews, 80*, 990–1016. https://doi.org/10.1016/j.rser.2017.04.022

Amin, N., Shabbir, M. S., Song, H., Farrukh, M. U., Iqbal, S., & Abbass, K. (2023). A step towards environmental mitigation: Do green technological innovation and institutional quality make a difference? *Technological Forecasting and Social Change, 190*, 122413. https://doi.org/10.1016/j.techfore.2023.122413

Behnassi, M., Rahman, M. A., D'Silva, J., Ramachandran, G., Gupta, H., Pollmann, O., & Ramachandran, N. (2021). Enhancing resilience for food and nutrition security within a changing climate. In M. Behnassi, M. Barjees Baig, M. El Haiba, & M. R. Reed (Eds.), *Emerging challenges to food production and security in Asia, Middle East, and Africa*. Springer. https://doi.org/10.1007/978-3-030-72987-5_1

Dewasiri, N. J. (2024). Sustainable strategies for net zero: An Asia pacific perspective. In R. Singh & D. Crowther (Eds.), *Transition towards a sustainable future. Approaches to global sustainability, markets, and governance*. Springer. https://doi.org/10.1007/978-981-97-5756-5_1

Dewasiri, N. J., Dharmarathna, D. G., & Choudhary, M. (2024). Leveraging artificial intelligence for enhanced risk management in banking: A systematic literature review. In R. Singh, S. Khan, A. Kumar, & V. Kumar (Eds.), *Artificial intelligence enabled management: An emerging economy perspective* (pp. 197–213). De Gruyter. https://doi.org/10.1515/9783111172408

Dewasiri, N. J., Karunarathne, K. S. S. N., Menon, S., Jayarathne, P. G. S. A., & Rathnasiri, M. S. H. (2023). Fusion of artificial intelligence and blockchain in the banking industry: Current application, adoption, and future challenges. In A. Saini & V. Garg (Eds.), *Transformation for sustainable business and management practices: Exploring the spectrum of industry 5.0* (pp. 293–307). Emerald Publishing Limited. https://doi.org/10.1108/978-1-80262-277-520231021

Farghali, M., Osman, A. I., Chen, Z., Abdelhaleem, A., Ihara, I., Mohamed, I. M., Yap, P. S., & Rooney, D. W. (2023). Social, environmental, and economic consequences of integrating renewable energies in the electricity sector: A review. *Environmental Chemistry Letters, 21*(3), 1381–1418. https://doi.org/10.1007/s10311-023-01587-1

Giri, S., Prasai, A., Khanal, A., Khamcha, R., Tiwari, S., Khadka, S., Thapa, Y., Rai, N., & Pantha, M. (2023). Assessing climate change challenges and adaptation strategies in south Asian countries: A review. *Journal of Sustainability and Environmental Management, 2*(2), 141–149. https://doi.org/10.3126/josem.v2i2.55207

Gulagi, A., Ram, M., Bogdanov, D., Sarin, S., Mensah, T. N., & Breyer, C. (2022). The role of renewables in the rapid transitioning of the power sector across states in India. *Nature Communications, 13*(1), 1–19. https://doi.org/10.1038/s41467-022-33048-8

Hellqvist, L., & Heubaum, H. (2023). Setting the sun on off-grid solar? Policy lessons from the Bangladesh solar home systems (SHS) programme. *Climate Policy, 23*(1), 88–95.

Hemakumara, G. P. T. S., Athukorala, S. U., & Yapa, L. G. D. S. (2023). "Battle for solar energy" program: An environment-friendly sustainable approach for carbon neutrality in Sri Lanka. In F. Quoquab & J. Mohammad (Eds.), *Sustainability and social marketing issues in Asia* (pp. 23–38). Emerald Publishing Limited. https://doi.org/10.1108/978-1-80071-845-620231002

Hinkel, J., Garcin, M., Gussmann, G., Amores, A., Barbier, C., Bisaro, A., Le Cozannet, G., Duvat, V., Imad, M., Khaleel, Z., & Marcos, M. (2023). Co-creating a coastal climate service to prioritise investments in erosion prevention and sea-level rise adaptation in the Maldives. *Climate Services, 31*, 100401. https://doi.org/10.1016/j.cliser.2023.100401

Jaiswal, A., Joshi, M., & Kwatra, S. (2019). Climate consequences: If India sneezes. *Jindal Global Law Review, 10*, 19–34. https://doi.org/10.1007/s41020-019-00089-y

Jayawardena, N. S., & Dewasiri, N. J. (2023). Food acquisition and consumption issues of south Asian countries: A systematic literature review and future research agenda. *FIIB Business Review*. https://doi.org/10.1177/23197145231194113

Kanchan, A., Nitivattananon, V., Tripathi, N. K., Winijkul, E., & Mandadi, R. R. (2024). A spatio-temporal examination of land use and land cover changes in smart cities of the Delhi–Mumbai industrial corridor. *Land, 13*(7), e957. https://doi.org/10.3390/land13070957

Karunarathna, K. S. S. N., Dewasiri, N. J., Singh, R., & Rathnasiri, M. S. H. (2024). What does artificial intelligence–powered ChatGPT bring to academia? A review. In R. Singh, S. Khan, A. Kumar, & V. Kumar (Eds.), *Artificial intelligence enabled management: An emerging economy perspective* (pp. 3–13). De Gruyter. https://doi.org/10.1515/9783111172408

Khanna, M. K., Malik, S., & Kumar, H. (2023). Indian solar panel initiatives in reducing carbon dioxide emissions. *Energy and Power Engineering, 15*(4), 191–203. https://doi.org/10.4236/epe.2023.154009

Kumar, K., Singh, R., Sharma, A., & Lehri, B. P. S. (2024). Climate change adaptation strategies for achieving net-zero economy. In R. Singh & D. Crowther (Eds.), *Transition towards a sustainable future. Approaches to global sustainability, markets, and governance*. Springer. https://doi.org/10.1007/978-981-97-5756-5_8

Loo, Y. Y., Billa, L., & Singh, A. (2015). Effect of climate change on seasonal monsoon in Asia and its impact on monsoon rainfall variability in Southeast Asia. *Geoscience Frontiers, 6*(6), 817–823. https://doi.org/10.1016/j.gsf.2014.02.009

Mishra, U., Rathnasiri, M. S. H., & Dewasiri, N. J. (2024). Net-zero economy and energy sustainability: A Nepalese perspective. In R. Singh, S. Khan, A. Kumar, S. Luthra, & H. Chokshi (Eds.), *Net zero economy, corporate social responsibility and sustainable value creation. CSR, sustainability, ethics & governance* (pp. 145–161). Springer. https://doi.org/10.1007/978-3-031-55779-8_9

Moosa, A., Do, K., & Jonescu, E. (2020). Design response to rising sea levels in the Maldives: A study into aquatic architecture. *Frontiers of Architectural Research, 9*(3), 623–640. https://doi.org/10.1016/j.foar.2020.04.002

Mulla, S., Ahmed, R., Singh, K. K., Singh, S. K., Deshmukh, N., & Inamdar, F. K. (2023). Climate change effects on-climate parameters like temperature, rainfall and water resources sectors in India. In C. B. Pande, K. N. Moharir, & A. Negm (Eds.), *Climate change impacts in India. Earth and environmental sciences library*. Springer. https://doi.org/10.1007/978-3-031-42056-6_2

Munir, R., & Khayyam, U. (2020). China-Pakistan economic corridor and the impact of coal-based energy projects on tropospheric ozone in Pakistan. *Polish Journal of Environmental Studies, 29*(5), 3729–3747. https://doi.org/10.15244/pjoes/112895

Murshed, M., Abbass, K., & Rashid, S. (2021). Modelling renewable energy adoption across south Asian economies: Empirical evidence from Bangladesh, India, Pakistan and Sri Lanka. *International Journal of Finance & Economics, 26*(4), 5425–5450. https://doi.org/10.1002/ijfe.2073

Oztig, L. I. (2017). Europe's climate change policies: The Paris agreement and beyond. *Energy Sources, Part B: Economics, Planning, and Policy, 12*(10), 917–924. https://doi.org/10.1080/15567249.2017.1324534

Pandey, A. C., Kaushik, K., & Parida, B. R. (2022). Google earth engine for large-scale flood mapping using SAR data and impact assessment on agriculture and population of the ganga-Brahmaputra basin. *Sustainability, 14*(7), 4210. https://doi.org/10.3390/su14074210

Patel, S. K., & Rajak, R. (2023). Effects and resilience to climate crisis among adolescents: A narrative review on south Asian countries. *Climate and Development, 15*(8), 692–703. https://doi.org/10.1080/17565529.2022.2149255

Rahman, M. M., Siddika, F., Ahmed, T., & Hadi, T. (2021). Lessons from climate change adaptation actions in Bangladesh. In R. C. Brears (Ed.), *The Palgrave handbook of climate resilient societies*. Palgrave Macmillan. https://doi.org/10.1007/978-3-030-42462-6_106

Ramachandra, T. V., Aithal, B. H., & Sreejith, K. (2015). GHG footprint of major cities in India. *Renewable and Sustainable Energy Reviews, 44*, 473–495. https://doi.org/10.1016/j.rser.2014.12.036

Rathnasiri, M. S. H., Dewasiri, N. J., & Kumar, A. A. (2024). Policy framework and implementation strategies for Sri Lanka's transition to a net-zero economy. In R. Singh, S. Khan, A. Kumar, S. Luthra, & H. Chokshi (Eds.), *Net zero economy, corporate social responsibility and sustainable value creation. CSR, sustainability, ethics & governance*. Springer. https://doi.org/10.1007/978-3-031-55779-8_3

Rupasinghe, H. T., & Halwatura, R. U. (2020). Benefits of implementing vertical greening in tropical climates. *Urban Forestry & Urban Greening, 53*, 126708. https://doi.org/10.1016/j.ufug.2020.126708

Seo, S. N. (2019). Economics of the green climate fund, Paris agreements, and global funds and currencies: An overview. In *The economics of global allocations of the green climate fund*. Springer. https://doi.org/10.1007/978-3-030-18274-8_1

Soubhari, T., Nanda, S. S., & Shobha, C. V. (2024). Revisiting climate crisis for carbon neutrality through education? A sustainable nudging approach. In P. Singh, S. Daga, & K. Yadav (Eds.), *Nudging green: Behavioral economics and environmental sustainability. World sustainability series*. Springer. https://doi.org/10.1007/978-3-031-65972-0_14

Srivastava, R. K. (2020). Manifestations of climate change in South Asia. In *Managing urbanization, climate change and disasters in South Asia. Disaster studies and management*. Springer. https://doi.org/10.1007/978-981-15-2410-3_4

Verma, S. (2023). Climate change in the Himalayas: Impacts on lives and livelihoods. *International Journal of Science and Research Archive, 9*(1), 419–425.

Yangka, D., Rauland, V., & Newman, P. (2018). Carbon neutral policy in action: The case of Bhutan. *Climate Policy, 19*(6), 672–687. https://doi.org/10.1080/14693062.2018.1551187

Chapter 8
The Role of Human Capital and Emerging Technologies to Achieve the Climate Neutrality Goals in Tourism Industry: A Review

Ali Yüce

Abstract Climate change poses a significant threat to all life on earth. Climate neutrality has emerged as a crucial objective to ensure the long-term survival of the earth's natural resources. Local and national authorities have implemented recent legislation to mitigate carbon emissions by compensating for alternative methods in many industries. Despite being referred to as the "industry without chimney," the tourism sector produces a significant amount of greenhouse emissions that surpasses earlier projections. This study focuses on three primary factors: human capital, emerging technology, and climate neutrality. This study examines the impact of human capital and emerging technology on achieving climate neutrality, with a specific focus on the tourism industry. The existing body of research asserts that there is a connection between human capital, developing technology, and achieving climate neutrality. A study report has stated that human capital and developing technology might have both positive and negative effects on the ecological system. The concept of climatic neutrality requires a thorough understanding, considering its interrelation with the other attributes examined in this study. In addition, despite several studies in the literature on the tourism industry, there has been limited investigation into the relationship between human capital and developing technologies in relation to climate neutrality. This study tries to evaluate and synthesize the relationship between human capital, new technologies, and climate neutrality in a concise manner. This review study aims to demonstrate an association among three essential components of sustainable development activities crucial to the ecosystem through an analysis of the existing literature. Some of the terms mentioned here are sustainability, carbon footprint, tourism industry, and awareness. Furthermore, this study asserts a robust correlation between human capital, developing technologies, and climate neutrality. Finally, the research discusses recommendations as well as theoretical and practical contributions.

A. Yüce (✉)
Cappadocia University, Nevşehir, Turkey

R. Singh and D. Crowther (eds.), *Climate Neutrality and Sustainable Ecosystems*, Approaches to Global Sustainability, Markets, and Governance,
https://doi.org/10.1007/978-981-96-7181-6_8

Keywords Sustainable tourism · Environment sustainability · Emerging technology · Green technology · Climate technology · Climate neutrality · Carbon neutrality

8.1 Introduction

Climate change poses a significant hazard to all life forms on Earth. Hence, climate change is a frequently utilized term that has emerged as a buzzword across different fields in recent decades. Climate change implies periodic temperature fluctuations that lead to erratic atmospheric weather patterns or circumstances (Dietz et al., 2020; Li et al., 2011). In recent years, there has been an increasing emphasis on research that specifically addresses global climate change and environmental degradation (Baloch et al., 2023; Brown et al., 2021; Piguet, 2022). Environmental degradation encompasses a broad spectrum of detrimental consequences, such as the depletion of natural resources (Brown et al., 2021), the decline of biodiversity (Baloch et al., 2023), the exacerbation of air pollution (Zhang et al., 2020), and the accelerating of climate change. The escalation of human activity has led to these consequences, which pose significant challenges to the environmental sustainability (Chen, 2021). Given the imminent threat of resource depletion, organizations are now exploring innovative methods and strategies to reduce their environmental footprint for a more sustainable world (Čuček et al., 2015). Additionally, nations and organizations have implemented further measures to alleviate the adverse impacts of human activities.

The terms "sustainable" and "sustainability" have emerged as significant but pervasive concepts across various scientific fields (Elmo et al., 2020; Zuin et al., 2021). Many authors concisely define sustainability as the conservation of existing natural resources in order to ensure their availability for every living being, both presently and in the future (Hajian & Kashani, 2021; Kumar, 2021; Uralovich et al., 2023; Yuce, 2023). According to the author of this chapter, the term "sustainable" etymologically signifies the ability to conserve natural resources amidst continuously growing populations, in conjunction with increasing output and consumption levels.

Climate neutrality has been a very contentious issue among researchers in multiple disciplines, experts, and politicians since the Paris Climate Summit in 2015 (Huovila et al., 2022; Kılkış et al., 2020; Wolf et al., 2021). This summit was a significant milestone in addressing the unsustainable productivity and consumption of human practices; yet, it was not the inaugural effort to contemplate the future of living and nonliving entities. Climate neutrality, in a simple form, is neutering the anthropogenic activities or practices related emitted all greenhouse gases (Elavarasan et al., 2022; Kılkış et al., 2020). Net zero refers to the equilibrium achieved when greenhouse gas emissions released into the atmosphere are counterbalanced by the adoption and implementation of environmentally sustainable actions, strategies, and rules and regulations. On the other hand, carbon neutrality is another important term that is used interchangeably with climate neutrality. In existing literature, carbon neutrality is used and investigated more often than climate neutrality (Chen, 2021;

Huovila et al., 2022). There is a slight difference between two of them. While carbon neutrality emphasizes reducing the carbon dioxide (CO_2) emission in the atmosphere, climate neutrality refers to combatting all greenhouse gasses including carbon dioxide emitted to the atmosphere.

Although it falls outside the scope and aim of this conceptualized paper, it would be beneficial to provide a brief and concise overview of greenhouse gases. The atmosphere stores gases like carbon dioxide, nitrous oxide, chlorofluorocarbons, and methane, which we refer to as greenhouse gases. In fact, these gases are critical for the earth's sustainability. Without these gases, the Earth would freeze and become lifeless (Panepinto et al., 2021). The primary issue with greenhouse gases stems from the conversion of beneficial gases into harmful ones, a result of human activities that prioritize unrestrained consumption and production methods. When gases accumulate excessively in the atmosphere, the previously balanced system begins to malfunction, leading to an unsustainable environment. Therefore, it is crucial to understand how we can achieve climate neutrality, or at least balance, our daily practices that negatively impact the environment's health due to the excessive greenhouse gas effect.

Human capital can be classified as individuals who possess a high degree of education, environmental awareness, knowledge, and abilities (Ahmed et al., 2021; Carraro et al., 2014; Nakipova et al., 2021). Ahmed et al. (2021) individuals who possess these characteristics play a vital role in order to preserve the natural capital of a country. Research has shown a substantial correlation between environmental sustainability and negative human capital (Ahmed et al., 2021; Liu et al., 2020). The majority of the studies indicate that individuals who possess advanced skills, knowledge, and competence are more tending to conserve natural resources. Nevertheless, it is imperative to acknowledge that this does not always mean that individuals with a high level of education can be considered to have sufficient awareness and maturity to grasp the importance of natural capital. It is also evident that they have the capacity to maintain their unsustainable production and consumption practices. As a matter of fact a study claimed similar argument by stating that "Surprisingly, human capital adds to environmental degradation" (Ahmed et al., 2021) in addition the economic growth and urbanization. Despite all of these studies, impact of human capital on environmental degradation has not been studied comprehensively, particularly in tourism industry. In addition to the existing knowledge, this chapter aims to analyze the literature and present their arguments about the impact of human capital on climate neutrality. This study will also enhance the growing body of research on the relationship between human capital, emerging technology, and climate neutrality in the context of the tourism industry.

Besides human capital (Carraro et al., 2014; Milovanović, 2017), there is also another important asset called natural capital. Natural capital (Collins, 1999), encompassing both renewable and non-renewable resources, constitutes the most essential asset for living organisms. Meraj et al. (2021) defined natural capital as the assets, for example forests, rivers, air and agricultural land, of a nation that are responsible for its economic performance. Bateman and Mace (2020) also argued that there is a significant relationship between natural capital and decision-making process. Since natural resources are not infinite, organizations, individuals and policy makers need to contemplate on how they will use their assets before they make the decisions (Ahmed et al., 2021).

The global ecology is experiencing serious degradation due to various critical factors, including excessive energy use and the utilization of outdated technologies. Rapid and sophisticated technological breakthroughs undeniably offer countless advantages, profoundly impacting all facets of human existence (Yuce, 2021). Emerging technologies are becoming more prevalent and significantly impact several aspects of life constantly (Buhalis & Law, 2008). Whether it is emerging or not, technologies help companies and organizations in a number of ways, such as improving their efficiency and increasing the productivity along with increased consumer and labor force satisfaction (Ahmed et al., 2021; Buhalis & Law, 2008). Yet this does not mean that they are reducing carbon dioxide and help the planet.

By employing innovative solutions, we can mitigate the adverse effects of human activities and promote a more sustainable environment for all. Implementing suitable technologies that exert minimal adverse effects on the environment with various names such as green technologies (Nylund et al., 2022), climate-friendly technologies (Popp, 2011) energy-efficient technologies (Jaffe et al., 1999). In pursuit of sustainable economic growth and responsible usage of natural resources, it is also essential to implement relevant and effective creative solutions alongside suitable technologies. Despite the diversity of terminology, they all share a singular objective: to minimize the adverse effects of human activities and natural resource exploitation while increasing socioeconomic advantages. This method constitutes the most effective option, alongside the responsible utilization of supplies, products, and services, for alleviating detrimental effects on organizations. The current literature includes many studies highlighting the need to adopt suitable solutions to alleviate their negative environmental impacts. Grubb (2004) posited over two decades ago that identifying innovative and appropriate solutions for achieving climate neutrality is essential for environmental sustainability.

Achieving climate neutrality goals requires not only human capital and technological improvements but also the integration of numerous aspects, including financial support for the countries and industries that cannot afford adapting the newer approaches. As Chen et al. (2022) reported that 26th United Nations Climate Change Conference of the Parties (COP26) which was held in Glasgow between 31 October to 12 November 2021 reaffirmed the major goals of Paris agreement. Nearly 200 countries and 120 leaders also agreed on four major headings to reduce the global greenhouse gas emissions (United Nations, 2021). These four key headings were coal, cars, cash, and trees in order to keep the global average temperature below 2 °C, above pre-industrial levels with a target of 1.5 °C as soon as possible.

Adapting to any challenge, including climate change, is crucial for addressing persistent ecological problems. Despite the necessity for proficient workers and sophisticated technology solutions to reverse atmospheric carbonization and environmental degradation, the challenge is exacerbated by individuals' failure to adapt to climate change-related threats (Ahmed et al., 2021). Developing adaptation strategies and procedures to address climate change will mitigate the risks linked to climate change-related challenges, particularly extreme weather phenomena.

The travel industry has been an essential economic asset for many nations globally (Thommandru et al., 2023). Tourism generates foreign exchange and income; however, it also generates a significant amount of waste, such as plastic and culinary waste, which is referred to as the global tourism footprint (Lenzen et al., 2018; Thommandru et al., 2023). Tourism contributes to carbon emissions yet does not produce adequate remedies to alleviate its negative impacts.

In light of the escalating climate change challenges threatening the earth, there is a heightened interest in exploring the correlation between human capital, developing technology, and climate neutrality. Additionally, there is a growing need to investigate novel ideas, approaches, and practices in order to mitigate rapid climate fluctuations. Consequently, it is essential to ascertain the precise principles that can facilitate climate neutrality, as this is crucial for the well-being of individuals, nations, and our ecosystem including the biodiversity. Lastly, despite various studies defining climate neutrality, carbon neutrality, and offsetting, there is a necessity to differentiate these concepts, especially in the tourism industry.

Although tourism has become a major business yielding considerable economic advantages, it is essential for all stakeholders to devise creative measures to mitigate the deterioration of the ecological system (Liu et al., 2020). Despite substantial study on human capital, new technologies, and climate neutrality, there is a scarcity in peer-reviewed papers that expressly examine the convergence of these factors. Fewer studies have highlighted the correlation between human capital, developing technology, and climate neutrality in the tourism industry. This chapter intends to deliver a fundamental analysis through a survey of recent literature. This study aims to examine the following questions in the light of the stated objectives:

(a) What is the influence of human capital on climate neutrality?
(b) What is the effect of new technology on climate neutrality?
(c) In what ways may humans and emerging technologies facilitate climate neutrality within the tourism sector?

8.2 Theoretical Background and Review of Related Literature

8.2.1 Climate Change

The world has witnessed many dramatic disasters including wars, epidemics, and economic turnovers which led all living things hardly adapt the new habitat since the antiquity. The human beings and other living organisms also faced with the extreme weather conditions such as 'temperature, precipitation and storm events' (Dietz et al., 2020) for tens of hundreds of years according to the findings of the geological (Rousseau et al., 2022; Weyl, 1968) and archeological work (Wang & Bresson, 2021) such as Port of Ephesus, even though some argue that sea level is dropped due to the different reasons other than the climate change. Although some critical warnings made by Joseph Fourier and Eunice Newton Foote in between 1820–1850 of the greenhouse gases effect on earth exist, it took a long time to convince scientific community, government and organizational authorities to be greatly interested in the near future (History, 2023). As in the aforementioned definition, climate change refers to uncommon and unpredictable dramatic fluctuations in temperature that causes catastrophic ecological and social outcomes over short or long durations (Dietz et al., 2020; Li et al., 2011). While it is challenging to definitively attribute climate change to specific parameters, research suggest that anthropogenic activities are a significant factor contributing to the climate-related challenges faced by the Earth (Dietz et al., 2020; Zalasiewicz & Williams, 2021). According to Ritchie (2020), who reported on Our World in Data website, individuals emit 50 billion tons of greenhouse gases per year. Ritchie (2020) also contended that numerous sectors are responsible for global greenhouse gas emissions. Moreover, Ritchie (2020) asserted that the energy sector (electricity, heat, and transport) is the largest contributor to greenhouse gas emissions at 73.2%, followed by the chemical and petrochemical industry at 3.6%, and the food and tobacco industry at 1%. Lastly, Ritchie (2020) argued in her article that other industries contribute 10.6% of the greenhouse gas emission.

According to Lenzen et al. (2018), the tourist industry is a trillion-dollar sector that strongly influences global trade and local economies. Tourism contributes 7% of global exports, 10% to global domestic goods and generating approximately 20% of employment (Greif & Houdre, 2024; Lenzen et al., 2018). The tourism industry undoubtedly offers various additional economic advantages throughout the globe. However, this study focuses on the significance of human capital, as most current research prioritizes statistical indicators associated with economic success.

Despite the data being sourced from highly reputable institutions, complete reliance on them is unfeasible due to their primary focus on the results of economic conditions. These institutions often disregard to consider the costs associated with human impact, environmental degradation, and other critical issues. Although the facts may seem accurate, it is essential to examine the concealed costs one must confront. Ultimately, the question should be asked: Was it worth it? If not, how can one ensure that the outcomes are significantly beneficial for all involved? Consequently, local, national, and international organizations must devise new strategies to ensure that outcomes significantly benefit all stakeholders without compromising finite natural resources, in accordance with the definition of sustainability.

An extensive body of literature highlights that, tourism substantially adds to global greenhouse gas emissions, which exacerbate extreme weather occurrences (Greif & Houdre, 2024). The adverse impact of tourism is increasing as an escalating number of individuals visit the destinations around the earth. Although measuring and estimating the actual impact and volume of greenhouse gas emissions generated by tourism activities is challenging, some organizations and research studies asserts different estimates from various sources differ. The World Trade Organization (WTO) reported that tourism accounted for 2.5–3% of carbon dioxide emissions from 2008 to 2010 (Lenzen et al., 2018). It is evident that numerous criteria contributing to the tourism carbon footprint are not incorporated into the aforementioned ratio. Given that tourism operations are intricately linked with numerous other sectors. The diagram above also omits the adverse effects of tourism. According to Greif and Houdre (2024) at weforum.org, tourism accounts for 8% of world carbon dioxide emissions. A substantial number of studies have concluded that transportation is the primary source of greenhouse gas emissions. In the research that was conducted by Lenzen et al. (2018), the authors demonstrated that transportation, which includes hot air balloons and aircraft, is responsible for approximately 49%, followed by goods 12% of the 8% carbon footprint of global tourism.

Lenzen et al. (2018) also argued that each tourism branch approximates an environmental adverse consequence. Previously, the sole method of measuring the total emission of the tourism industry was to concentrate on aviation, as it was recognized as an industry without a chimney. The remaining branches and factors, including the construction of tourism facilities and the provision of accommodation services, were excluded from the equation. Upon the inclusion of the estimated data from the other branches and activities of tourism, it is evident that the greenhouse gas emissions from tourism are more than four times the level that was previously declared (Lenzen et al., 2018).

Research indicates that the tourist sector comprises direct, indirect, and induced contributors (Aynalem et al., 2016). Nonetheless, each classification has a variety of suppliers that release differing amounts of carbon into the atmosphere. According to the study conducted by Lenzen et al. (2018), aircrafts account for 49% of greenhouse gas emissions, making it the predominant source of pollution. The remaining industries comprise products at 12%, food and beverages at 10%, and various categories. These industries can establish objectives to diminish their carbon footprint and mitigate their adverse effects by adopting a climate-neutral policy. Lastly, while tourism is an essential source of income for many destinations, yet, classical tourism activities cannot be sustained. Hence, more eco-friendly approaches such as green businesses and alternative tourism types are needed in order to achieve sustainable tourism goals.

Organizations have primarily addressed issues by identifying the key factors detrimental to their environment. Therefore, corporate and organizational leaders, along with civil society leaders, must collaborate to enhance the environmental sustainability of corporate practices and human activities. A recent study conducted by Berrang-Ford et al. (2011) asserts that human will need to adapt the climate change challenges since it has been unavoidable task. Further, bringing together of the different shareholders into the table can help to analyze the issues with broader understanding and quicker adaptation towards the environmental situation. When resources are used at minimum, less greenhouse gases will be produced and it will be easier to balance the negative impacts. Working as a team will not only generate newer ideas, but it will also help focusing on achieving the common goals of the organizations and society at large. Moreover, collaborating with authorities using the advanced technology will improve the individual and organizational skills to help the society in case of the imminent environmental situations such as storm and float related issues.

Another study by Khan et al. (2024) indicates that carbon dioxide (CO_2) can be reduced using appropriate technological adaptation and effective action approaches such as education of all in a society. Although technology has been seen by some researchers as one of the detrimental factors on environmental related problems earth face today, if it is used wisely, it can play a significant role in reducing greenhouse gases produced due to the anthropogenic activities, in particularly, industrial production, mostly called Fordism.

8.2.2 The Concept of Climate Neutrality, Carbon Neutrality, Offsetting and Net Zero in Literature

Climate change is one of the not only ecologically, but also socially, culturally and politically greatest threats in the age we live in. According to the US National Climate Assessment, "earth's climate is now changing faster than at any point in the history of modern civilization, primarily as a result of human activities" (Jay et al., 2018). Due to increasing human population and consumption, energy consumption also increases significantly. This vicious cycle leads to a significant decrease in limited natural resources of the world and causes constantly mounting greenhouse gases in the atmosphere. Consequently, climate neutrality has emerged as a vital global policy mandating substantial commitment to address this challenge.

As it was mentioned above, climate neutrality has been receiving a popular interest among researchers in recent years as climate change severely influences every aspect of our life. According to Geden and Schenuit (2019) the European Union (EU) and other developed countries, which are the major greenhouse gas contributors, are under pressure to ensure the climate neutrality objectives are achieved by 2050 since they are the frontrunners of the international climate policy. Geden and Schenuit (2019) also asserted that European Union has already its greenhouse gas emission by 23% since 1990. The Paris Climate Summit plays a pivotal role to confront with the ongoing unsustainable anthropocentric practices, approaches, and ideas. The 2015 (December 12) Paris Climate Summit (COP21), also known as the Paris Agreement, was the catalyst for the objective of achieving climate neutrality before by the end (2050) of this century (UNFCCC, n.d.). What makes Paris Agreement important? The Paris agreement is a significant milestone in the fight against climate change, as it is the first mandatory accord that unites all countries to address and mitigate its effects (UNFCCC, n.d.). Paris Agreement required all parties to adhere to the policies that aim to reduce greenhouse gas emissions by a minimum of 55% from 1990 levels by 2030. The COP 21 summit also determined to limit the global temperature rise to fewer than 2 °C and limiting the temperature increase 1.5 °C above pre-industrial levels (UNFCCC, n.d.; United Nations, n.d.).

Furthermore, the Paris Climate Agreement declared that all sectors (Geden & Schenuit, 2019), including organizations, cities, destinations, and governments, can achieve climate neutrality if they endeavor to reduce greenhouse gas emissions to a minimal level. To attain this crucial objective, all stakeholders must overcome obstacles such as financial constraints and the formulation of requisite regulations (Zhou, 2021). On the other hand, to achieve climate neutrality, consumers must adopt sustainable consumption practices, including land use, avoiding unnecessary purchases, reusing repairable items, rejecting non-eco-friendly products and services, and recycling recyclable materials (Geden & Schenuit, 2019; Lenzen et al., 2018; Wolf et al., 2021). Therefore, in addition to adapting to the reality of climate change-related issues, it is crucial to adopt novel insights, approaches, and perspectives to ensure a more sustainable environment.

In reality, what does the term "climate neutrality" mean? Despite the fact that the ultimate objective is to achieve climate neutrality or net zero emissions, the public and mainstream media have confused the two concepts. In research papers, it is also challenging to obtain a concise and lucid explanation of these two terms. It is evident that these concepts are inadequately defined. Climate neutrality can be defined as "greenhouse gas neutrality," as cited by Geden and Schenuit (2019). The common thread between the two is a desire to mitigate the environmental damage caused by human and commercial activities. Climate neutrality is achieved by absorbing the greenhouse gas emission at a specific location in a manner that can be eliminated greatly even if it is not completely through the implementation of avoidance measures by businesses and individuals.

Achieving climate neutrality goals are challenging since the world is demanding a bigger economic growth, more comfortable life style and possessing more or bigger personal assets while the population constantly increases. In order to meet the requirements of demand, suppliers will need to use more raw materials and energy that will cause more global greenhouse gas emissions (Chen et al., 2022). Over-exploitation of raw materials and non-renewable energy sources undoubtedly, will undermine the future generations' living conditions (Chen et al., 2022; Wolf et al., 2021).

Net Zero denotes a concept synonymous with climate neutrality, which seeks to eliminate all greenhouse gases, including CO_2 emissions, from the atmosphere (Greif & Houdre, 2024). The objectives of climate neutrality and Net Zero do not aim to eliminate all greenhouse gas emissions. They ensure that anthropogenic activities associated with hazardous gas emissions are balanced with an equivalent number of emission mitigation strategies. Lastly, in the literature, climatic neutrality and Net Zero are used interchangeably as they both accomplish identical objectives.

Conversely, as Geden and Schenuit (2019) employed the term "greenhouse gas neutrality," this work defines it succinctly as "carbon dioxide neutrality." Carbon neutrality can be defined as carbon emissions are generated by individuals, organizations, and enterprises, but they do so with the aim of the adverse effects of the carbon dioxide (CO_2) emissions through the application of mitigation techniques (Chen, 2021; Sun et al., 2021). These methods involve implementing advanced technologies to reduce energy and water consumption that will help carbon emission levels to the minimum.

The extent of which sectors, destinations, and individuals are impacted by the greenhouse gas or carbon dioxide effect is contingent upon a variety of factors, such as their level of awareness of the ecosystem, the quantity of human and financial capital they possess, and the technology they utilize. Researches also indicate that eco-friendly practices can reduce the impact of individuals' awareness, expertise, and leadership on the quantity of carbon emissions that leads more sustainable work environment (Chen, 2021). Waste management, energy conservation, and water conservation are just a few of the readily implementable new measures that individuals and companies can implement these to mitigate their negative impacts. While it is currently impossible to completely eliminate greenhouse gas emissions, there are numerous measures that businesses and organizations can take to significantly reduce their CO_2 emissions.

8.2.3 The Concept of Sustainable Tourism in Literature

Tourism has been a significant propelling force for global domestic product and has generated a variety of economic benefits worldwide (Khan et al., 2020; Lenzen et al., 2018). In the 1980s, efforts were initiated to promote a tourism industry that is referred to as "Industry without Chimneys" (Kumar, 2021). An industry that lacked a chimney was not required to generate any products; rather, it was to provide visitors with access to the natural resources. However, history has shown that natural resources can suffer harm even in the absence of chimneys. Eventually, tourism evolved into an industry that not only significantly contributes to the global economy (Khan et al., 2020), but also releases greenhouse gases into the atmosphere (Dube & Nhamo, 2021). As previously said, tourism provides substantial prospects for the economic, social, and cultural vitality of destinations, while also exerting a considerable influence on the environment. Estimates indicate that the tourist sector currently accounts for 8% of worldwide carbon dioxide (CO_2) emissions (Baloch et al., 2023; Dube & Nhamo, 2021; Lenzen et al., 2018). The tourism sector has been endeavoring to attain climate neutrality (Lenzen et al., 2018; Sun et al., 2021) by following particular directives set forth by international organizations. Tourism industry highly depends on a skilled workforce that creates the human capital to ensure a quality of service for the visitors (Milovanović, 2017; Nakipova et al., 2021). Moreover, tourism also requires constant follow up with the latest effective technologies for its effective business operation and improve consumer satisfaction in the context of tourism field (Yuce et al., 2020). To mitigate greenhouse gas emissions, the industry can enhance its efficiency and productivity by adopting sustainable practices and advanced digital technologies.

8.2.4 Human Capital, Climate Change, Emerging Technology, and Climate Neutrality

Human capital is a crucial concept for nations, corporations, and organizations. Highly qualified human capital is more likely to result in greater value, enhanced economic growth, and a more secure future for society as a whole. Although there is not a common agreement on the definition of the human capital, it can be defined as the quality of knowledge, heightened awareness, enhanced responsibility, and receptiveness to novel concepts and technology (Carraro et al., 2014). Having a healthy human capital offers numerous benefits from personal development to environmental sustainability (Zhang & Danish, 2019). A study asserts that educational attainment provides numerous benefits such as using the information communication technologies (Ahmed et al., 2021). Human capital also stimulates awareness that positively affects the environmental sustainability (Ahmed et al., 2021; Zhang & Danish, 2019). Ahmed and Wang (2019) posited that human capital has potential to reduce the ecological footprint (Čuček et al., 2015).

Given that human activities profoundly influence climate change, climate neutrality, and sustainability (Ahmed & Wang, 2019; Carraro et al., 2014; Zhou, 2021), it is imperative to address natural capital (Bateman & Mace, 2020; Collins, 1999). The literature has extensively analyzed the impact of human capital on sustainability (Caradonna, 2022; Elmo et al., 2020). Policymakers at local, national, and international levels, along with scholars, prioritize not only financial capital or economic outcomes to attain sustainability objectives, but also assess the tangible effects of human activities, such as tourism. They also ensure the consideration of both human (social) and natural (environmental) capitals. Human capital and policies are essential in influencing the success or failure of sustainability initiatives. As policymakers constitute a segment of human capital, their awareness, knowledge, and competence are crucial. Consequently, it is also essential to examine the scale, intensity, causation, and directionality of the relationship between human capital and climate change.

Extensive studies have investigated the role of the human capital in different concepts. While some of these studies examined the relationship between human capital and environmental sustainability (Ahmed et al., 2021), human capital and tourism development (Hajiyeva & Teymurova, 2019), human capital and tourism education (Kasa et al., 2020) others investigated relationship between human capital and economic growth (Fahimi et al., 2018) and human capital and innovation (Carraro et al., 2014). All these studies indicate that people with high quality of the human capital is crucial for all aspect.

Studies also show that as human capital's value increases, their impact on productivity and efficiency also increases in terms of protecting the natural resources where tourism affairs in high demand (Ahmed et al., 2021; Milovanović, 2017; Nakipova et al., 2021). A study also indicates that human capital plays an important role in adapting the climate change-related issues and acting actively to produce the solutions (Berrang-Ford et al., 2011). Human capital is driving force in terms of more sustainable economic growth (Zhang & Danish, 2019).

Despite the fact that majority of the literature indicate that technology has positive and strong relation to stimulate the individuals' awareness to reduce their carbon footprint (Lenzen et al., 2018), yet some studies indicate technology itself is the major source of the environment degradation (Ahmed et al., 2021).

Technology has significantly transformed the lives of individuals across every aspect (Pencarelli, 2020; Yuce et al., 2020). Moreover, technology profoundly impacts business operational processes. In the tourism sector, technology has revolutionized operations, encompassing reservations, marketing, and the enhancement of business intelligence. Current research posits that the integration of relevant technology in the hospitality sector converts the corporate landscape into an eco-friendly organization (Yuce, 2023). An environmentally sustainable hotel that utilizes advanced technology and proficient personnel reduces its greenhouse gas emissions by conserving energy and water. The organization not only aids in environmental conservation but also diminishes expenses, allowing for expenditures to mitigate their carbon dioxide emissions. Current literature indicates that developing technologies may positively influence the mitigation of the adverse effects associated with rapid population growth, productivity, and consumption patterns. What is crucial to comprehend regarding new technologies? Emerging technologies encompass artificial intelligence, robots, nanotechnology, digital twins, immersive technologies like Immersive Technologies (Yuce, 2022), electronic word of mouth (E-WoM (Abubakar & Ilkan, 2016), Big Data (Samara et al., 2020), and Digital Twin (Yuce, 2023).

While technology significantly influences societal and economic advancement, it concurrently presents substantial environmental hazards due to its enhancement of industrial efficiency, resulting in reduced pricing (Ahmed et al., 2021). As the trend of accelerating the production of electronics at reduced costs intensified, energy consumption increased, resulting in heightened greenhouse gas emissions. Companies are accountable for surpassing carbon emissions at the operational level. Tourism enterprises must function dynamically to satisfy the expectations of continually evolving consumer profiles. Recent research indicates that environmentally sustainable hotels considerably attract more visitors and impact their decision-making. Consumer expectations have significantly evolved over the past century due to heightened understanding of climate change. Consequently, numerous travelers solicit the perspectives of others when investigating destinations and accommodations prior to their vacation. Conversely, websites that endorse hotels and other enterprises within the tourism sector utilize eco labels to distinguish environmentally sustainable hotels from their counterparts.

8.3 Conclusion

Anthropogenic activities are major triggers for the greenhouse gas emissions absorbed in the atmosphere. Increased greenhouse gases cause the malfunctioning of the natural climate cycle that leads climate change-related problems. Unpredictable weather events such as sudden storms, hurricanes, excessive heat or cold, float and more are the major outcomes of the climate change-related problems human challenge in recent years. The fundamental tenet of sustainability is human capital with accountability, awareness, adaptation and using the emerging the technology efficiently to minimize the negative impacts of climate-related issues. Responsible action demands that all endeavors, encompassing policy formulation and technical implementation, are essential for fostering a sustainable environment. Individuals and organizations should adopt and utilize the appropriate technology to operate ethically and optimize the advantages gained from natural resources across the whole production and consumption cycle to extend the longevity of these resources.

It is increasingly possible that tourism will continue to produce an increasing amount of greenhouse gas emissions, as it is expanding at an unprecedented rate. In other words, the complexity of addressing climate-related challenges will also increase. It is essential to take decisive action to reduce the emissions that are the result of global tourism, which contributes to climate change. National governments and international organizations are endeavoring to address this escalating challenge. Conversely, enterprises and individuals within the tourist sector have commenced efforts to mitigate the escalating levels of greenhouse gases accumulating in the environment. Ultimately, business and organizational leaders, in conjunction with civil society leaders, must cooperate to improve the environmental sustainability of corporate operations and human endeavors.

On the other hand, this paper aimed to widen the current understanding and meaning of sustainability by looking into from the etymological perspective. Despite the extensive use of the sustainability not only in public media, but also in literature, it demands clarification about what we are trying to sustain. It is clear that humans cannot sustain the classical attitude, approach, strategy, and method while producing and consuming. Therefore, the question must be asked: what do we need to sustain the natural resources? Hence, this paper discussed the role human capital and newer innovative and appropriate emerging technology on sustaining the natural capital of the earth.

There is parallel relationship between production and consumption that directly influence the destiny of the environment. As mass production phase speeded up, consumption of the products and services were increased. All in all, it is essential to cultivate individuals to ensure they obtain sufficient knowledge, wisdom, and awareness regarding natural resources along with adopting the appropriate technologies that is eco-friendly by using the renewable energies. It is clear that individuals with more income potential exhibit a superior comprehension of the significance of environmental protection.

As a result, this reviewed paper discussed how human capital and technological innovations affected the preservation of natural sustainability. This paper also presents recommendations for researchers in tourism area to investigate the relationship between each dimension of this study in different methodologies to provide more analytical based information. Moreover, this paper also suggests philosophers to discuss about the relationship between human capital's human and anthropogenic activities' anthropos (In Greek anthropos is human). The majority of researchers agree that human activities contribute to climate change, while simultaneously underscoring the importance of human capital in mitigating its causes. In other words, whereas human capital is beneficial, its impact on environmental conservation is harmful. This sounds like an absolute dilemma. Prior to discussing the primary feature of human capital, hence, it is prudent to elucidate the distinction between 'human' and 'anthropos,' despite their synonymous in nature. How can an individual rectify an issue when they are the principal instigator? Therefore, this paradoxical concept requires clarification. Since this chapter does not aim to participate in academic debate in this topic; rather, it offers philosophers the chance to explore whether humanity represents the solution or the dilemma of climate change.

References

Abubakar, A. M., & Ilkan, M. (2016). Impact of online WOM on destination trust and intention to travel: A medical tourism perspective. *Journal of Destination Marketing & Management, 5*(3), 192–201.

Ahmed, Z., Nathaniel, S. P., & Shahbaz, M. (2021). The criticality of information and communication technology and human capital in environmental sustainability: Evidence from Latin American and Caribbean countries. *Journal of Cleaner Production, 286*, 125529.

Ahmed, Z., & Wang, Z. (2019). Investigating the impact of human capital on the ecological footprint in India: An empirical analysis. *Environmental Science and Pollution Research, 26*, 26782–26796.

Baloch, Q. B., Shah, S. N., Iqbal, N., Sheeraz, M., Asadullah, M., Mahar, S., & Khan, A. U. (2023). Impact of tourism development upon environmental sustainability: A suggested framework for sustainable ecotourism. *Environmental Science and Pollution Research, 30*(3), 5917–5930.

Bateman, I. J., & Mace, G. M. (2020). The natural capital framework for sustainably efficient and equitable decision making. *Nature Sustainability, 3*(10), 776–783.

Berrang-Ford, L., Ford, J. D., & Paterson, J. (2011). Are we adapting to climate change? *Global Environmental Change, 21*(1), 25–33.

Brown, D., Boyd, D. S., Brickell, K., Ives, C. D., Natarajan, N., & Parsons, L. (2021). Modern slavery, environmental degradation and climate change: Fisheries, field, forests and factories. *Environment and Planning E: Nature and Space, 4*(2), 191–207.

Buhalis, D., & Law, R. (2008). Progress in information technology and tourism management: 20 years on and 10 years after the internet—The state of eTourism research. *Tourism Management, 29*(4), 609–623.

Caradonna, J. L. (2022). *Sustainability: A history*. Oxford University Press.

Carraro, C., De Cian, E., & Tavoni, M. (2014). Human capital, innovation, and climate policy: An integrated assessment. *Environmental Modeling & Assessment, 19*, 85–98.

Chen, J. M. (2021). Carbon neutrality: Toward a sustainable future. *The Innovation, 2*(3), 100127.

Chen, L., Msigwa, G., Yang, M., Osman, A. I., Fawzy, S., Rooney, D. W., & Yap, P.-S. (2022). Strategies to achieve a carbon neutral society: A review. *Environmental Chemistry Letters, 20*(4), 2277–2310.

Collins, A. (1999). Tourism development and natural capital. *Annals of Tourism Research, 26*(1), 98–109.

Čuček, L., Klemeš, J. J., & Kravanja, Z. (2015). Overview of environmental footprints. In *Assessing and measuring environmental impact and sustainability* (pp. 131–193). Elsevier.

Dietz, T., Shwom, R. L., & Whitley, C. T. (2020). Climate change and society. *Annual Review of Sociology, 46*(1), 135–158.

Dube, K., & Nhamo, G. (2021). Greenhouse gas emissions and sustainability in Victoria Falls: Focus on hotels, tour operators and related attractions. *African Geographical Review, 40*(2), 125–140.

Elavarasan, R. M., Pugazhendhi, R., Irfan, M., Mihet-Popa, L., Khan, I. A., & Campana, P. E. (2022). State-of-the-art sustainable approaches for deeper decarbonization in Europe–an endowment to climate neutral vision. *Renewable and Sustainable Energy Reviews, 159*, 112204.

Elmo, G. C., Arcese, G., Valeri, M., Poponi, S., & Pacchera, F. (2020). Sustainability in tourism as an innovation driver: An analysis of family business reality. *Sustainability, 12*(15), 6149.

Fahimi, A., Saint Akadiri, S., Seraj, M., & Akadiri, A. C. (2018). Testing the role of tourism and human capital development in economic growth. A panel causality study of micro states. *Tourism Management Perspectives, 28*, 62–70.

Geden, O., & Schenuit, F. (2019). *Climate neutrality as long-term strategy: The EU'S net zero target and its consequences for member states*. European Union.

Greif, S., & Houdre, H. (2024). *How travel and tourism can reach net zero*. Retrieved from https://www.weforum.org/agenda/2024/01/travel-tourism-industry-net-zero/#:~:text=There%20are%20a%20variety%20of,8%25%20of%20global%20carbon%20emissions

Hajian, M., & Kashani, S. J. (2021). Evolution of the concept of sustainability. From Brundtland report to sustainable development goals. In *Sustainable resource management* (pp. 1–24). Elsevier.

Hajiyeva, L., & Teymurova, V. (2019). Analysis of the impact of the human capital on tourism development in Azerbaijan. In *Economic and social development: Book of proceedings* (pp. 1579–1588). Varazdin Development and Entrepreneurship Agency.

History. (2023). *Climate change history*. Retrieved from https://www.history.com/topics/natural-disasters-and-environment/history-of-climate-change

Huovila, A., Siikavirta, H., Rozado, C. A., Rökman, J., Tuominen, P., Paiho, S., & Ylén, P. (2022). Carbon-neutral cities: Critical review of theory and practice. *Journal of Cleaner Production, 341*, 130912.

Jaffe, A. B., Newell, R. G., & Stavins, R. N. (1999). *Energy-efficient technologies and climate change policies: Issues and evidence*. Resources for the Future.

Jay, A., Reidmiller, D., Avery, W., & C., Barrie, D., Dave, A., DeAngelo, B., Winner, D. (2018). Overview. In D. R. Reidmiller, C. W. Avery, D. R. Easterling, K. E. Kunkel, K. L. M. Lewis, T. K. Maycock, & B. C. Stewart (Eds.), *Impacts, risks, and adaptation in the United States: Fourth national climate assessment* (p. 34). US Global Change Research Program.

Kasa, M., Kho, J., Yong, D., Hussain, K., & Lau, P. (2020). Competently skilled human capital through education for the hospitality and tourism industry. *Worldwide Hospitality and Tourism Themes, 12*(2), 175–184.

Khan, I., Zhong, R., Khan, H., Dong, Y., & Nuţă, F. M. (2024). Examining the relationship between technological innovation, economic growth and carbon dioxide emission: Dynamic panel data evidence. *Environment, Development and Sustainability, 26*(7), 18161–18180.

Khan, N., Hassan, A. U., Fahad, S., & Naushad, M. (2020). Factors affecting tourism industry and its impacts on global economy of the world. *SSRN Journal*. https://doi.org/10.2139/ssrn.3559353

Kılkış, Ş., Krajačić, G., Duić, N., & Rosen, M. A. (2020). *Advances in integration of energy, water and environment systems towards climate neutrality for sustainable development* (Vol. 225, p. 113410). Elsevier.

Kumar, B. (2021). The role of transportation in tourism development in India. In *Tourism in India-status, challenges and opportunities* (p. 24). Times of India.

Lenzen, M., Sun, Y.-Y., Faturay, F., Ting, Y.-P., Geschke, A., & Malik, A. (2018). The carbon footprint of global tourism. *Nature Climate Change, 8*(6), 522–528.

Li, J., Wang, M.-H., & Ho, Y.-S. (2011). Trends in research on global climate change: A science citation index expanded-based analysis. *Global and Planetary Change, 77*(1–2), 13–20.

Liu, P., Teng, M., & Han, C. (2020). How does environmental knowledge translate into pro-environmental behaviors? The mediating role of environmental attitudes and behavioral intentions. *Science of the Total Environment, 728*, 138126.

Meraj, G., Kanga, S., Kranjčić, N., Đurin, B., & Singh, S. K. (2021). Role of natural capital economics for sustainable management of earth resources. *Earth, 2*(3), 622–634.

Milovanović, V. (2017). The impact of human capital on tourism development. Paper presented at the Tourism International Scientific Conference Vrnjačka Banja-TISC.

Nakipova, G., Arynova, Z., Kunyazov, E., Berezyuk, V., Zhunusova, A., & Syzdykbayeva, B. (2021). Influence of human capital on the development of tourism industry. *Journal of Environmental Management & Tourism, 12*(6), 1694–1706.

Nylund, P. A., Brem, A., & Agarwal, N. (2022). Enabling technologies mitigating climate change: The role of dominant designs in environmental innovation ecosystems. *Technovation, 117*, 102271.

Panepinto, D., Riggio, V. A., & Zanetti, M. (2021). Analysis of the emergent climate change mitigation technologies. *International Journal of Environmental Research and Public Health, 18*(13), 6767.

Pencarelli, T. (2020). The digital revolution in the travel and tourism industry. *Information Technology & Tourism, 22*(3), 455–476.

Piguet, E. (2022). Linking climate change, environmental degradation, and migration: An update after 10 years. *Wiley Interdisciplinary Reviews: Climate Change, 13*(1), e746.

Popp, D. (2011). International technology transfer, climate change, and the clean development mechanism. *Review of Environmental Economics and Policy, 5*(1), 131–152.

Ritchie, H. (2020). *Sector by sector: Where do global greenhouse gas emissions come from?* Retrieved from https://ourworldindata.org/ghg-emissions-by-sector#article-citation

Rousseau, D.-D., Bagniewski, W., & Ghil, M. (2022). Abrupt climate changes and the astronomical theory: Are they related? *Climate of the Past, 18*(2), 249–271.

Samara, D., Magnisalis, I., & Peristeras, V. (2020). Artificial intelligence and big data in tourism: A systematic literature review. *Journal of Hospitality and Tourism Technology, 11*(2), 343–367.

Sun, Y., Duru, O. A., Razzaq, A., & Dinca, M. S. (2021). The asymmetric effect eco-innovation and tourism towards carbon neutrality target in Turkey. *Journal of Environmental Management, 299*, 113653.

Thommandru, A., Espinoza-Maguiña, M., Ramirez-Asis, E., Ray, S., Naved, M., & Guzman-Avalos, M. (2023). Role of tourism and hospitality business in economic development. *Materials Today: Proceedings, 80*, 2901–2904.

UNFCCC. (n.d.). *The Paris agreement*. Retrieved from https://unfccc.int/process-and-meetings/the-paris-agreement

United Nations. (2021). *COP26: Together for our planet. Climate action*. Retrieved from https://www.un.org/en/climatechange/cop26.

United Nations. (n.d.). *The Paris agreement*. Retrieved from https://www.un.org/en/climatechange/paris-agreement

Uralovich, K. S., Toshmamatovich, T. U., Kubayevich, K. F., Sapaev, I., Saylaubaevna, S. S., Beknazarova, Z., & Khurramov, A. (2023). A primary factor in sustainable development and environmental sustainability is environmental education. *Caspian Journal of Environmental Sciences, 21*(4), 965–975.

Wang, T. Z., & Bresson, A. (2021). Guild inscriptions of Ephesus: A comparative study of silversmith's unique social status and economic prestige.

Weyl, P. K. (1968). *The role of the oceans in climatic change: A theory of the ice ages*. Meteorological Monographs.

Wolf, S., Teitge, J., Mielke, J., Schütze, F., & Jaeger, C. (2021). The European green deal—More than climate neutrality. *Intereconomics, 56*, 99–107.

Yuce, A. (2021). The impact of digital transformation and virtual reality in tourism marketing: A conceptualized exploration. In *Handbook of research on IoT, digital transformation, and the future of global marketing* (pp. 1–14). IGI Global.

Yuce, A. (2022). ICT pandemic time adoption and immersive technologies: A comprehensive review. In *Digitalization as a driver for smart economy in the post-COVID-19 era* (pp. 243–255). IGI Global.

Yuce, A. (2023). Digital twins and sustainable developments in the tourism and hospitality industry. In *Handbook of research on applications of AI, digital twin, and internet of things for sustainable development* (pp. 461–472). IGI Global.

Yuce, A., Arasli, H., Ozturen, A., & Daskin, M. (2020). Feeling the service product closer: Triggering visit intention via virtual reality. *Sustainability, 12*(16), 6632.

Zalasiewicz, J., & Williams, M. (2021). Climate change through earth history. In *Climate change* (pp. 49–65). Elsevier.

Zhang, J., & Danish. (2019). The dynamic linkage between information and communication technology, human development index, and economic growth: Evidence from Asian economies. *Environmental Science and Pollution Research, 26*, 26982–26990.

Zhang, N., Ren, R., Zhang, Q., & Zhang, T. (2020). Air pollution and tourism development: An interplay. *Annals of Tourism Research, 85*, 103032.

Zhou, W. (2021). Climate change and low-carbon society: Coping with uncertainty. In *East Asian low-carbon community: Realizing a sustainable decarbonized society from technology and social systems* (pp. 3–36). Springer.

Zuin, V. G., Eilks, I., Elschami, M., & Kümmerer, K. (2021). Education in green chemistry and in sustainable chemistry: Perspectives towards sustainability. *Green Chemistry, 23*(4), 1594–1608.

Chapter 9
Balancing Economic Growth and Climate Neutrality: Solutions for a Warming Planet

Rubee Singh, Akshay Singh, David Crowther, and Raghuveer Katragadda

Abstract This research chapter analyzes how to grow the economy and reduce greenhouse gas emissions to address climate change. It examines decarbonization solutions, from renewable energy transitions to energy efficiency measures to international cooperation, all of which advance sustainable development on a path that doesn't derail economic progress. The study applies a qualitative, case study approach from developed and developing economies to show how growth can be decoupled from emissions through technological progress, policy innovations, and sustainable business models. Yet in such rapidly growing economies, (such as Nigeria with its large oil revenues) financial and infrastructural constraints remain major obstacles. On the other hand, research chapter exhibits the significance of harmonizing the economic growth and climate neutrality for advancing long-term prosperity globally, environmental sustainability, and social equality. It calls for green technology policies, carbon capture and international financial cooperation. To avoid climate neutrality with economic growth, the study concludes that a coordinated effort among governments, businesses, and international organizations is required. Achieving this balance is crucial to dealing with the global warming crisis and to delivering a sustainable future. This research contributes to the discourse on green growth by providing insights to policymakers and businesses from both developed and developing countries how they can be climate neutral without sacrificing economic progress.

Keywords Economic growth · Decarbonization · Sustainable development · Climate neutrality

---

R. Singh (✉)
Institute of Business Management, GLA University, Mathura, India

A. Singh
School of Management Studies, Indra Gandhi National Open University, New Delhi, India

D. Crowther
Social Responsibility Research Network, Derby, UK

R. Katragadda
V.R. Siddhartha Engineering College, Kanuru, India

R. Singh and D. Crowther (eds.), *Climate Neutrality and Sustainable Ecosystems*, Approaches to Global Sustainability, Markets, and Governance,
https://doi.org/10.1007/978-981-96-7181-6_9

9.1 Introduction

Ever growing threat of global warming has pushed governments, businesses and societies globally to rethink on how to find a ground to ensure economic growth and environmental sustainability. The growth in greenhouse gas (GHG) emissions is unprecedented, with economic development having lifted millions out of poverty and improved living standards across the globe but it also sped up climate change. On the other hand, issues relating to the global warming crisis and moving towards climate neutrality have been the core priority today. However, it is not easy to find a balance between it and the economic growth. Sustainable ways to reach climate neutrality for continued economic prosperity, environmental protection, and well-being of our society arecritical.

9.1.1 Overview of the Global Warming Crisis and the Need for Climate Neutrality

Global warming refers to the rise in global temperatures, which is caused by too many heat trapping GHGs in the atmosphere. Repeatedly, the Intergovernmental Panel on Climate Change (IPCC) has stated that global temperatures (relative to preindustrial levels) will rise by more than 1.5 °C if no urgent and major action is taken, with severe and unmitigable consequences including rising sea levels, extreme weather events and biodiversity loss (Change, 2023). Climate change has already started to bring about the adverse effects, and does so disproportionately, mainly to vulnerable people in developing countries where severe weather occurrences and environmental degradation are damaging livelihoods, food security, and infrastructure (Böhringer et al., 2022).

Achieving climate neutrality—or what is also called net zero emissions—by mid-century is becoming a global consensus to reduce these risks (Doda et al., 2019). To be climate neutral, you are balancing the emission of GHGs into the atmosphere with the amount taken out—the latter being taken out by carbon capture or natural sink like forests. The key to achieving this balance is to avoid further increases in temperature, and to avoid the worst impacts of climate change. Countries around the world—including those in the European Union, China, and the United States—have pledged to reach climate neutrality by 2050 or sooner as a means of securing a liveable planet for generations to come. Yet maintaining economic growth, however, entails important challenges for the transition to climate neutrality. Most countries continue to see increased energy consumption and GHG emissions as the natural outcome of economic development, especially for industrialized countries and rapidly growing economies. For most of our history, fossil fuels have underpinned both industrialization and economic expansion, as a source of energy that can power transportation, manufacturing, and infrastructure. As a result, reducing such emissions is difficult and requires radical redesign of energy systems, industrial processes, and consumption patterns.

9.1.2 Challenge of Balancing Economic Growth with Climate Neutrality Goals

One of the most important policy dilemmas of the twenty-first century is balancing economic growth against climate neutrality goals. There is one side of the story that economic growth would increase living standards, create jobs, and address poverty, especially in developing countries. For many of these countries, economic growth is seen as a means to increase equity and meet their people's rising expectations. In contrast, rapid economic growth supported by fossil energy source may cause more carbon emissions, aggravating the climate change problem and hurting climate neutrality action (Zhong & Pei, 2023).

But this balancing act also comes with one of the main challenges, which is that you need a just transition. To many developing countries, economic growth is a tightly joined feature with carbon-intensive industry like manufacturing, mining, and agriculture. If not carefully managed, a swift transition to climate neutral practices could lead to economic disruption, job losses, and growing inequality. Therefore, developing economies face a difficult dilemma as they must grow to meet the needs of their citizens and reduce at the same time their carbon footprints. Similarly, in developed economies with developed industries based on fossil fuels (coal and oil), there is clearly the need for a change in energy production and energy consumption patterns.

The financial and technological investments required to achieve climate neutrality offer another challenge. For achieving climate neutrality, billions of dollars have to be invested in renewable energy infrastructure, green technologies, energy efficiency measures, and carbon capture solutions. Access to financial resources and technological innovation needed to move to low carbon or carbon neutral model of growth is problematic for many countries, mainly those with limited economic capacity. International climate finance attempts to fill this gap, but the scale of the investment needed remains daunting (Fankhauser et al., 2021; Nwokolo et al., 2023).

Adding to that, there is also the fact that the economies of the world are interdependent. Instead, trade, investment, and supply chains crisscross over many countries such that action to reduce emissions in one region can have knock on effects throughout the world. For example, if a particular country introduces climate neutral technologies, that country is likely to see emissions offshoring of emission-intensive activities to countries with lower environmental standards and therefore create "carbon leakage". Meeting climate neutrality goals is only possible if global economic stability is not compromised on while addressing these global spillovers (Cosbey et al., 2018).

9.1.3 Importance of Balancing Economic Growth and Climate Neutrality for Sustainable Development

Balancing economic growth and climate neutrality has to be encouraged to promote sustainable development. According to the United Nations, sustainable development occurs when the development that meets the needs of the present doesn't compromise the ability of future generations to meet their own needs. Prosperity as a phenomenon of economic development, and social inclusion and environmental protection constitute this vision of development.

This vision is supported by balancing economic growth with climate neutrality that also promotes green growth, i.e., a development model focused on generating the economic progress at a cost of minimum environment harm. Moving to climate neutral economies brings new opportunities for innovation, jobs, and modernization of our industrial base. The use of green technologies like renewable energy, electric vehicle and their movement, and sustainable agriculture pave the way for new job creation while decreasing the carbon footprint. Moreover, energy efficiency, resource conservation and circular economy investments can strengthen long-term resilience and minimize climate-related shocks (Breen, 2019; Friedlingstein et al., 2022).

But the pursuit of climate neutrality must be done while promoting social equity at the same time. The poor and marginalized are more vulnerable to climate change as they are least able to adapt to changing environment condition. Consequently, climate neutrality policies need to be fair and inclusive transition policies, supporting those who will bear the burden of moving to low carbon economies (e.g., workers in carbon-intensive industries) (Sovacool et al., 2020).

As such, achieving climate neutrality while maintaining economic growth is in reality a challenge but also an opportunity for countries around the world. The transition to climate neutral economies needs a change in the way we produce energy, how we carry out industrial processes and what we need and consume, but is an opportunity for innovation, creating jobs and sustainable development. To ensure long-term global prosperity, environmental sustainability and social well-being achieving this balance will be critical (Grossman & Krueger, 1995).

9.1.4 Research Objectives

The research objective of this chapter is 'Balancing Economic Growth and Climate Neutrality: The approach to the question posed in 'Solutions for a Warming Planet' is to examine policy innovations, technological breakthroughs, sustainable business models, and global cooperation mechanisms that support continued growth without raising greenhouse gas emissions, and to identify sector-specific solutions that reconcile development and climate neutrality. Therefore, this research chapter objectives for the study are as follows:

1. Analysis of new policy innovations that spur both new economic development through private capital and yet buttress minimum emissions reduction targets.
2. Study part which technological progress can be played to achieve climate neutrality without the inevitable economic growth losses.
3. Examines sustainable business models which support a low carbon economy.
4. Global cooperation mechanisms and outcomes of alignment of economic policies to climate goals should be assessed.
5. Offer sector-specific solutions of how to reconcile their economic growth through industry and climate neutrality.

Consequently, the central focus of this research chapter is on "how can nations attain economic growth without increasing the global warming", and what are the possible solutions to ensure climate neutrality in highly developing in economies, mainly through the decarbonization, use of technology and farming on a sustainable level as well as international cooperation?

9.2 Literature Review: Balancing Economic Growth and Climate Neutrality

Increasingly governments and industries are waking to the need to combat climate change and are faced with the challenge of balancing economic growth and climate neutrality. Economic development was traditionally associated with environmental degradation especially where GHG were emitted. Yet, decoupling growth from emissions has become an increasingly prominent concept in contemporary economic and environmental thought allowing nations to pursue sustainable development. In this literature review, we examine this complex relationship between economic growth and its environmental implications, examine the Environmental Kuznets Curve (EKC) phenomenon, review global efforts towards reaching climate neutrality, including the policy, legislation, and economic tools being used, and look at several case studies that show how growth can be decoupled from growing carbon emissions.

9.2.1 Economic Growth and Environmental Impact

In the past, industrialization is considered to be the source of economic growth and the major culprit for increasing greenhouse gas (GHG) emissions. Over two thirds of global GHG emissions are caused by industrial activity, energy production, and transportation, as stated by the International Energy Agency. Many nations, especially in the Global North, had relied on a growth model that was built on fossil fuels, and hence bestowed greater CO2 concentrations in the atmosphere (Quéré et al., 2019; Sakai & Barrett, 2016). This trajectory of rapid industrialization and environmental degradation has also been followed by early phases of industrialization in developing countries where economic growth has been closely linked to growing environmental degradation (Sovacool, 2021; Sovacool et al., 2020).

As early as the 1970s, McGlade and Ekins (2015) recognized the linear relationship between economic growth and environmental impact in their warning over environmental limits to growth. Unprecedented economic growth based on consumption and industrial output would inevitably end in ecological collapse, they said. While detractors quickly derided Meadows' predictions, they subsequently became the foundation of future investigations of the connection between economic development and sustainability for the environment.

Emissions grew along with the global economy. During 1990–2010, carbon emissions grew globally by 45% with developed nations being the biggest contributors. Yet the tide is beginning to turn, as greater emphasis is placed not only on green growth, but also climate-friendly policies that can bring economic development while saving the environment. This is especially significant given that many countries now have climate neutrality goals, to dial back carbon emissions while not decreasing economic output.

9.2.2 The Environmental Kuznets Curve (EKC) and Its Relevance to Climate Neutrality

The Environmental Kuznets Curve (EKC) hypothesis provides a conceptual framework for understanding the relationship between economic growth and environmental degradation. The EKC is derived from Simon Kuznets' work on the evolution of income inequality and economic development arguing that environmental destruction rises in the early stages of economic expansion, peaks at some stage and then diminishes with the continued development of the economy and the capability to adopt cleaner technology and environmental regulations (Sarkodie & Strezov, 2018).

The curve is in an inverted U shape meaning that environmental degradation is actually a phase of economic development and that when an income level is reached, societies will tend to prioritize environmental quality. Grossman and Krueger (1995) were among the first to empirically examine the EKC in which they find evidence of an EKC for various pollutants. Since then, more recent research has looked as the EKC, in relation to climate change and in relation to GHG emissions—but with mixed results.

Some studies back up the EKC hypothesis, in the sense that richer countries have been able to separate growth from emissions, while others believe that the EKC is too simple and does not take into account the fact that ecological damages are a global problem (Apergis & Payne, 2021). Furthermore, the EKC relies on the assumption that all countries will travel along the same developmental trajectory, which might not be true of countries that depend heavily on carbon-intensive industries.

The fact that the EKC does offer relevance to climate neutrality is predicated upon the proposition that after a certain stage of economic development, economic growth can continue to occur without an adverse impact on the environment. Although critics claim that, economic growth alone will not suffice for climate neutrality (Fankhauser et al., 2021), instead deliberate policy interventions and technological innovation is needed.

9.2.3 *Climate Neutrality and Net-Zero Goals*

The term climate neutrality means that GHG emissions are net zero; we are neither emitting nor absorbing GHGs. In response to increasing climate change, more nations have adopted net zero targets to limit its effect on the environment (Fankhauser et al., 2021). Climate neutrality is also the concept of worldwide initiatives such as the Paris Agreement that limits global warming to 1.5 °C above the preindustrial level.

The most comprehensive international framework to tackle climate change was signed by 196 countries, the Paris Agreement. Signatories are bound to having climate neutral by the second half of the twenty-first century through nationally determined contributions (NDCs). For example, the European Union has committed to reaching the first climate neutral continent by 2050, in support of the European Green Deal and the Fit for 55 legislative package. The world's largest emitter, China, has also promised to reach emissions' peak prior to 2030 and embody a carbon neutral economy by 2060.

Reaching climate neutrality is not only a policy goal—it is a necessity for limiting the worst impacts of climate change. These targets are going to require countries to revamp their energy systems and move away from fossil fuels; instead they must get their energy from renewable sources. The challenges and the opportunities of a climate neutral economy. Others also show that climate neutrality can create green jobs, make companies more competitive profits, save on healthcare expenses powered by low pollution levels (Luderer et al., 2019; Sakai & Barrett, 2016).

9.2.4 *Decoupling Growth from Carbon Emissions*

This debate on economic growth's decoupling from carbon emissions is one of the most important in the debate over climate neutrality. Decoupling means decoupling environmental impact—GHG in particular—from economic growth. Decoupling absolute not only implies economic growth without environmental degradation but also occurs when the magnitude of environmental degradation is falling. Decoupling relative is when the rate of degradation slows more slowly than the rate of the economy. Recent research indicates that decoupling is indeed possible, at least in high-income countries. For example, for example, Quéré et al. (2019) found decoupling in 18 countries (including USA, UK, and several European countries) where absolute decoupling occurred between 2005 and 2015. These countries achieved reduced carbon emissions and economic growth largely by combining policy, greater energy efficiency and transitioning to renewable energy.

There are lessons to be learned from the success of decoupling in these countries as other nations aim to reduce growth and achieve climate neutrality. An important part is the role of government policy in fostering decoupling. The European Union's Emissions Trading System (EU ETS) or other carbon pricing mechanisms, have been central in reducing emissions while stimulating economic growth (Geels, 2021). Renewable energy subsidies and efficiency standards, too, are bringing about newer practices in the private sector which are more green in nature. Despite such successes, critics say that decoupling will not be sufficient to reach global climate targets on its own. But they argue that these decoupling efforts must be accompanied by a profound transformation of economic systems, including a shift in consumer patterns, technological dynamics, and the governance of societies (Hickel & Kallis, 2019). Furthermore, although some high-income countries have experienced decoupling, there are numerous less developed countries that are yet to enter latter stages of the industrialization life cycle and need their industrialization to fall on carbon-intensive industries for growth.

9.2.4.1 Examples of Successful Decoupling: Countries and Industries

Many countries, from coal mining regions to entire nations, and a wide variety of industries have already shown that decoupling economic growth from carbon emissions is not only possible, but also advantageous. One of the best known is Sweden, which has decreased its carbon emissions by around 25% since 1990, and raised its GDP by roughly 60%. Sweden's success was owing largely to its early adoption of carbon pricing, the switch to renewable energy, and expenditure on public transport and energy efficiency.

Danish renewable energy transition proves another compelling case in industrial sector. Beyond, Denmark has made a success of wind energy, with more than 40% of electricity being generated through wind power. This transition has done not only for Denmark, helped reducing the carbon footprint but also has contributed to the economic transition, by creating jobs in the renewable energy sector and export Denmark's wind technology.

Decoupling also offers promising examples from the technology sector. Google and many others (e.g., Google's own slogan: Google Google's mission, 2020) have made commitments to reaching carbon neutrality, a step that Google has already taken, operating on 100% renewable energy. New data analytics, artificial intelligence and cloud computing are being exploited by these companies to increase energy efficiency and lower carbon emissions.

Sustainability means delivering the economic growth required to solve climate change. This is consistent with the historical relationship between economic growth and environmental degradation as observed by the Environmental Kuznets Curve that growth, eventually, leads to reduced emissions. But it's not a necessity that this needs to be decoupled from environmental harm; it's possible if it requires intentional policy and technological innovations. While the Paris Agreement offers a framework to get to climate neutrality, the success of the endeavors will require countries to have the appropriate strategies in place to decouple growth from emissions.

There are plenty of countries—Sweden and Denmark are just two such examples—that prove that you can grow economically alongside a reduction in carbon emissions. But as the world marches towards climate neutrality, it faces huge challenges, particularly for those developing nations that are heavily dependent on carbon-intensive industries. However, ultimately, reaching climate neutrality is going to require global cooperation, technological innovation and policy that incentivizes sustainable practices and that can reap economic growth.

9.3 Conceptual Framework for Studying Balancing Economic Growth and Climate Neutrality

The demand for combating climate change in the world today raises a challenging issue of how to achieve economic development through climate neutrality. This conceptual framework integrates key dimensions such as policy innovations, technological advancements, sustainable business models, and global cooperation to explore how economies can reduce emissions without compromising growth. The framework builds on the sustainability transitions and green growth literatures to show how the economic and environmental agendas are connected and conflicting.

9.3.1 Policy Innovations

The primary driver in the shift from the fossil fuel-based economy to a low carbon economy is the government policies. One of the tenets of this approach is the use of carbon pricing measures such as carbon taxes and emissions trading systems (ETS), which encourage companies to minimize their carbon footprint (Bataille et al., 2020). The policies that should be pursued in this regard should, therefore, be those that should not have adverse effects on the economy but at the same time should encourage investment in clean energy technologies. Also, green finance measures like sustainability bonds, and incentives for renewables are useful in engaging the private sector in the delivery of low carbon infrastructure.

9.3.2 Technological Advancements

It is crucial to achieve climate neutrality while preserving growth by using technological advancement. Solar, wind, and hydroelectric power sources are important in decreasing the usage of fossil fuels and to lessen greenhouse gases that are dangerous to the environment. The framework also underlines the need for investment in research and development (R&D) to enhance the performance of the technologies and enhance their diffusion globally (Bonk et al., 2020). Also, innovations and applications of carbon capture and storage (CCS) technologies provide possible options for industries that are hard to electrify like cement and steel industries.

9.3.3 Sustainable Business Models

Firms are critical in steerage of the change towards sustainability. The conceptual framework is underpinned by the principles of circular economy, which aims to minimize waste and lose materials through thinking differently about the way products and services are designed and used (Geissdoerfer et al., 2020). Green entrepreneurship and startups are pinpointed to be the drivers of change and innovation from where comes clean energy, sustainable agriculture, and waste management. Organizations are urged to implement cleaner production practices, enhance energy efficiency, and work with governments to align business activities to climate change objectives (Johnson & Schaltegger, 2019).

9.3.4 Global Cooperation and Policy Alignment

Climate neutrality cannot be reached without the help of other countries. National commitments, as outlined in the Paris Agreement, require countries to plan and harmonize their actions to achieve agreed-on goals. The developing countries have their own problems such as financial challenges and inability to access green technologies. These countries thus receive much-needed technical and financial support from the international organizations. The framework stresses the importance of international policy coherence so as not to cause carbon leakage and thus not to harm the climate agenda with different national policies (Karlsson et al., 2020).

This conceptual framework captures policy reforms, technological solutions, sustainable business models, and international co-operation as the main strategies for achieving sustainable economic development and climate change mitigation. It recognizes the tensions between economic development and environmental protection, and presents a complex model for addressing these issues. Through partnership between the governments, companies, and the intergovernmental organizations, the economies can attain climate neutrality while concurrently fostering on growth.

9.4 Challenges in Balancing Economic Growth and Climate Neutrality

Many of these challenges result from trade-offs and policy gaps, technological and financial limits, and political and social resistance.

9.4.1 Trade-Offs and Policy Gaps

The trade-off between short-term economic growth and long-term climate goals is one of the main challenges associated with a wide adoption of carbon capture technologies (André & Pirlot, 2024; Keen et al., 2022). Economic pressure on many developing economies to immediately reduce poverty and unemployment typically takes precedence over the long haul sustainability initiatives. This results in a conflict between being climate neutral and becoming prosperous rapidly (Khurshid et al., 2022; Kuehner et al., 2022). For example, while activities that are economically intensive in fossil fuels drive short-term growth, they also adversely affect the reduction of carbon emission and the transition into a low carbon economy (Zhu et al., 2024).

It is also found that the gaps in governance and policy frameworks, in developing economies, may hinder the implementation of climate neutral policies. In the countries where climate-related regulations often falter during enforcement, it is partly due to lack institutional capacity to enforce. Furthermore, existing policies may not sufficiently account for the particular requirements of those industries adopting more environmentally friendly technologies. These gaps act as sources of uncertainty that can slow or prohibit the appropriation of low carbon technologies (Zhu et al., 2024).

9.4.2 Technological and Financial Constraints

A second serious challenge towards getting to climate neutrality is technological and financial capacity to undertake the change towards a low carbon economy. Reducing greenhouse gas emissions requires green technologies including renewable energy systems, and energy-efficient industrial processes. Nevertheless, the roll out and development of such technologies are often limited by high costs, inadequate access to financing, and by a lack of research and development infrastructure in many countries (Krogstrup & Oman, 2019). This issue is clearly illustrated in a study of the sustainability of the green economy in China. This provides not only an illustration of how the interaction between green technology innovation and the digital economy is contributory to high-quality sustainability but also specifies the conditions that need to be made conducive. It emphasizes, however, that regional differences in technological capabilities and financial resources act as a barrier to progress in various fields. For instance, areas of China have developed remarkable progress in green technology, however, areas remain laggard in this respect because of the lack of adequate financial and technological infrastructure (Zhu et al., 2024).

In addition, this problem is compounded by financial constraints mainly in developing nations. Often dependent on external funding sources, such as international financial institutions or carbon markets, these countries can be very volatile, subject to the decisions of others. Stable funding for climate projects is non-existent, allowing green technologies to make a widespread transition, holding these nations to achieve their climate goals (Zhu et al., 2024).

9.4.3 Political and Social Resistance

Another formidable do' is political resistance. Coal mining and oil extraction, industries dependent on the use of fossil fuels, usually are strongly politically influential, lobbying unsuccessfully while trying to resist stringent environmental regulations. Industry groups say such regulations deter jobs and economic stability, especially in places where such industry forms the main economic engine. This creates political inertia, it is hard for governments to pass robust climate policies. For instance, while more and more people recognize the dangers of climate change, many countries around the world are under intense pressure from the fossil fuel industries to put off a move towards greener forms of energy (Tobin et al., 2023).

Also arising from this fear may be social resistance through the fear of job losses in carbon-intensive industries. Where these industries represent a large employment sector, switching to a low carbon economy may be faced with opposition from workers and locals. This kind of resistance can paralyze policy implementation because governments are caught between the need to act on climate change and the political and social fact of keeping people employed and the economy going (Krogstrup & Oman, 2019).

Often it is compounded by a public who is poorly informed about the benefits of developing a green economy. Misperceptions regarding economic impacts of climate policies, coupled with insufficient knowledge of opportunities for creating new green jobs, create a rallying cry among the public against climate efforts (Zhu et al., 2024).

9.5 Overcoming the Challenges

These challenges have to be addressed in a complex way. First, economic growth and climate goals must be seen as non-excluded. Green technologies should be promoted as a driver of economic growth together with job creation in efforts in renewable energy and energy efficiency sectors. With the mass of fossil reliant regions, there can be placed policies filled with retraining programs and financial support for workers changing over to green jobs shrinking social opposition (Tobin et al., 2023). In addition, international coordination, along with financial tools such as climate funds and carbon markets will be necessary to assist developing countries overcome the financial and technological gaps. Those countries who are struggling to make the transition can learn from those who have the best plans and green technologies to learn. Krogstrup and Oman (2019) argue that large climate projects outside of the Western world need international financial institutions' support to supply required funding.

Economic growth and climate neutrality are entangled; a complex challenge, with trade-offs between short- and long-term goals, between technological and financial constraints, and political and social resistance. These challenges, though, require multi-level coordinated efforts, from national governments to international organizations. Policymakers can promote green technologies and encourage sustainable economic development, both by ensuring that workers in carbon-intensive sectors experience equitable transitions and by fulfilling climate goals.

9.5.1 Solutions for Achieving Climate Neutrality without Sacrificing Growth

A goal for the global community is to achieve climate neutrality without sacrificing economic growth. This is based on inventive policy apps, technological developments, sustainable organization models, and international collaboration. I have spent the last few sections looking into possible solutions and strategies in which nations and businesses can shift towards climate neutrality without jeopardizing economic prosperity.

9.5.2 Innovative Policy Approaches

9.5.2.1 Carbon Taxation and Emissions Trading Systems

Carbon pricing mechanisms (e.g. carbon taxes, ETSs) are one of the most effective policy tools in the search to achieve climate neutrality. A carbon tax charges emitters of carbon dioxide a price associated to the pollution their emissions cause, which creates a financial incentive to reduce emissions. The ability of firms to trade their emission allowances creates incentives to optimally minimize the reduction in emissions and allows the market to decide the cost. According to research, carbon pricing can increase greenhouse gas cuts dramatically without slowing economic growth. Haven't countries that have implemented carbon taxes and emissions trading systems; had positive outcomes? The revenues are often reinvested in green technologies and infrastructure. These systems force companies to innovate and employ cleaner technologies with stable and predictable policy environments for them to function in.

9.5.2.2 Green Finance and Investment in Sustainable Infrastructure

Public and private investments in environmentally sustainable projects, called green finance, are concurrently important for economic growth and climate goals. Renewable energy, energy efficiency, and sustainable transport all need this investment, and green bonds and sustainable finance initiatives can channel this investment. We need to go towards sustainable infrastructure, like smart grids and green buildings, both to reduce the carbon footprint, and to reap economic benefits. For example, investments in renewable energy would turn out to be more cost-effective than continuation of funding fossil fuels. In 2020, renewable energy projects created more jobs worldwide than industries relying on fossil fuels, making the economic case for investing in clean energy very strong.

9.5.2.3 Incentives for Private-Sector Engagement in Low Carbon Technologies

The transition to a low carbon economy requires the intervention of the private sector. Tax incentives, grants, and subsidies can help persuade governments to encourage businesses to invest in clean technologies. Additionally, these policies free the burden of the costs of adopting sustainable practices from companies and encourage innovation.

In addition, OECD indicates that inviting constructive contribution from the private sector is fundamental in fighting climate change if this sector is to be allowed to thrive in a favorable business environment. Low Carbon Thinking puts business interests in line with environmental goals, encouraging businesses to adopt low carbon technologies, which in turn helps nations achieve climate neutrality.

9.5.3 Technological Innovations

9.5.3.1 Renewable Energy and Energy Efficiency Improvements

Solar, wind, and hydroelectric power are renewable energy technologies essential to realization of climate neutrality. Moving towards renewable energy is not just necessary for lowering carbon emissions, but it is all necessary for maintaining sustainable economic growth in the long run. Thanks to falling prices for solar and wind energy, increasing performance in the battery storage and grid management spaces, renewable energy is becoming more and more appealing. Furthermore, steps including upgrading industrial processing, better insulating buildings, and utilizing energy saving appliances all help shrink costs while shriveling down emissions. It is shown in studies that countries can save two dollars for every dollar invested in energy efficiency (Rosenow et al., 2022).

9.5.3.2 Carbon Capture and Storage (CCS) Technologies

Carbon capture and storage (CCS) technologies allow capturing industrial process and power plant carbon emissions before they enter the atmosphere. These technologies are of particular importance for the significant decarbonization of difficult to decarbonize industries—cement, steel, and chemicals (Luderer et al., 2019). There is a strong potential for widespread reduction of global carbon dioxide emissions through the use of CCS. If deployed at a large scale, CCS can cut up to 14% of cumulative emissions by 2050, according to a recent study by International Energy Agency. Nations can achieve deeper emissions reductions whilst still maintaining industrial output and economic competitiveness if they invest in CCS.

9.5.4 Sustainable Business Models

9.5.4.1 Circular Economy Principles

This is a circular economy, a model of economy based on minimizing waste and maximizing resources by making them circulate in a loop as long as possible—from products, materials, and services in the circular economy. This model is resource efficient, waste reduction, and recycled. By adopting circular economy principles, businesses can reduce their environmental impact while also cutting costs (Singh et al., 2022). By a 2022 study, it was found that businesses that have adopted circular practices such as product reuse, remanufacturing, and recycling have both greatly improved profitability and also sustainability (Geissdoerfer et al., 2020). A circular economy transition provides a pathway from growth that is decoupled from environmental degradation (Mohan, 2024).

9.5.4.2 Green Entrepreneurship and Startups

Green entrepreneurship is where businesses are created to provide products or services that lower the environmental impact. Sustainability startups, like those creating renewable energy solutions or circular economy practices, are doing the most to make a transition to a low carbon economy. The forces of climate change can be also advanced by these businesses, which can create new jobs and drive innovation and economic growth. For example, renewable energy startups like the solar and wind energy adoption in both developed and developing countries. Green entrepreneurs can be supported by governments through the access to funding, training and market opportunities, to speed up the transition into a low carbon economy (Wagner et al., 2021).

9.6 Global Cooperation and Policy Alignment

9.6.1 Role of International Cooperation

This is an international problem and it requires international effort. Greenhouse gas emissions do not recognize national boundaries, a fact that makes it impossible for one country to go it alone towards climate neutrality. The Paris Agreement and agreements of this type frame global cooperation on emission reductions. Sharing technologies and financial resources as well as expertise will also need international cooperation. Particularly for developing countries, green technologies and financing mechanisms for developing and implementing green technologies can be transferred to the developing countries, in order to pursue economic growth without increasing the carbon footprint. Furthermore, cooperative research and development efforts (e.g. Mission Innovation platform) have promoted partnerships that accelerate clean energy technology innovations (Bataille et al., 2020).

9.6.2 Aligning Economic Policies with Environmental Sustainability

To avoid sacrifice of growth and achieve climate neutrality, it is necessary to ensure that the economic policies conform with the environmental sustainability. Fiscal policies, for instance, subsidies, and tax regimes must ensure that these promote low carbon industries, and discourage carbon-intensive practices. In this record, green stimulus packages have been extensively discussed as a mechanism to promote an economic recovery response to COVID-19, without falling short of climate goals (Andrijevic et al., 2020; Hepburn et al., 2020).

For instance, the European Union's Green Deal is helping prompt climate neutrality by 2050 and economic development at the same time. The agreement in place features policies conducive to investing in renewable energy, green infrastructure and sustainable agriculture; showing how policy convergence can pave the way to an advanced, inclusive economy and support climate resilience. We show that achieving climate neutrality is not only achievable but essential for long-term sustainability while staying on a path to economic growth. This transition is driven by innovative policy approaches, technological advancement, sustainable business models and global cooperation (Antimiani et al., 2016). Solutions that are good for the economy and for the planet include carbon pricing, renewable energy, green finance and circular economy principles. Enacting these strategies enables nations and firms to be part of a low carbon future without sacrificing future economic prosperity.

9.6.3 Policy Recommendations

To meet the dual challenges of economic growth and climate neutrality, clean growth requires coordinated action by governments, businesses, and international organizations. This in turn brings to light the policy recommendations that indicate the ways these entities can contribute to fight climate change and promoting sustainable development.

9.6.3.1 For Governments

Adopt Comprehensive Green Growth Strategies

Green growth strategies in which governments create and carry out strategies that link economic development objectives with emissions reduction targets should be developed. The alignment of the policies and programs of countries with international climate commitments, including under the Paris Agreement, in ways that guarantee economic resilience and job creation is needed. Sectors with the highest emissions should be targeted through a comprehensive design of strategies such as energy, transportation, and agriculture, by introduction of environment-friendly processes and promotion of low carbon technologies.

Current research has focused on the use of national green growth strategies as a means to push economic growth while mitigating environmental damage. Consider Germany or Denmark where gradually integrating renewable energy within their growth models has seen decrease in emissions, and maintaining their economic competitiveness. In addition, clear regulations, tax incentives, and subsidies are needed to back up these strategies for they have to encourage green investments.

Invest in Research and Development (R&D) for Climate Technologies

The development of new climate technologies requires investment in R&D. Research into renewable energy, energy storage, carbon capture, and energy-efficient technologies must be prioritized, and governments must make funding for this research their first priority. When they do that, however, they can help foster innovation and ease industries into the land of sustainability without sacrificing output. Renewable energy technology costs will be reduced in order to become competitive with fossil fuels, but R&D plays a critical part in developing renewable energy technologies, according to IEA. Actively funding climate-related R&D by governments can accelerate the deployment of these technologies and thus make climate neutrality a financially feasible option for all sectors.

Encourage Private Sector Innovation and Investment in Sustainability

Policies that reward private sector investments in sustainable technologies should be developed by the governments. Tax breaks for companies switching to energy-efficient processes, subsidies to renewable energy projects, and carbon pricing mechanisms (including carbon tax or an emissions trading system (ETS)) are part of such policies (Bryant, 2016). In an effort to encourage innovation and reduce the economic risk to phasing out low carbon technologies, governments can help assure the market by creating a favorable investment environment (Buylova et al., 2022). Targeted incentives, such as the EU Green Deal, can also show how private sector involvement in sustainability initiatives can be attained. When government policies support long-term benefits, businesses tend to invest in sustainable practices.

9.6.3.2 For Businesses

Shift to Sustainable Business Practices

Business should begin to move towards sustainable processes and energy use. These can span everything from lessening waste by adopting circular economy models, cutting energy consumption, and moving to clean technologies. Switching to renewable energy sources and reaching sustainability goals not only decreases the emissions but also improves the corporate reputation and its resilience.

As more companies embrace sustainability, companies are discovering that these practices are not only good for the environment but can actually help reduce costs and give them a competitive advantage. For instance, a report from Geissdoerfer et al. (2020) demonstrates that a company adopting a circular economy as a business model has reported lower operating costs and higher customer loyalty—evidence that economic growth and sustainable development can go together.

Collaborate with Governments on Carbon Reduction Initiatives

Governments and businesses should collaborate for the corporate sustainability goals of the businesses to align with national and global climate targets. This could include getting involved in carbon reduction efforts including participation in carbon offset programs or completing public-private partnerships related to advancing low carbon technologies.

Achieving large-scale emission reductions requires public/private sector collaborations. According to studies, businesses that voluntarily embrace government supported climate programs have the potential to curb their carbon footprints and help contribute towards national climate objectives (Wagner et al., 2021).

9.6.3.3 For International Organizations

Support Developing Countries in Their Transition to a Green Economy

As such, international organizations, including the United Nations and World Bank, help developing countries move towards transition to a green economy. These organizations can help developing economies to adopt low carbon technologies while avoiding the trade-off between economic growth and low carbon development through financial assistance, technical support, and capacity building initiatives. On the other hand, low- and middle-income countries require financial assistance, especially the form of climate financing to enable countries to invest in renewable energy and sustainable infrastructure. Asmelash et al. (2020) state that international organizations have played an important role in mobilizing the investment for renewable energy projects in developing countries that have reduced their dependence on fossil fuel.

Strengthen Global Partnerships to Accelerate Climate Neutrality

Climate change is a problem that affects the entire world, therefore partners around the globe are necessary to deal with it. To enable stronger cooperation between countries and knowledge, technology and best practice exchange, the international organization should drive it. Furthermore, these partnerships can also seek to align climate policies in separate jurisdictions or to run growth and protect the environment through growth. Bataille et al. (2020) have shown in a study how global partnerships in climate neutrality efforts are essential to sharing resources and expertise for scaling up the adoption of climate neutral policies. International organizations can facilitate the diffusion of technological innovations and the coordination of policy towards a green economy, resulting in more efficient, more equitable transition to a green economy.

To get to climate neutrality at the same time as growth is a matter of looking at all of this in a coordinated way among governments, businesses, and also international organizations. Governments, instead, should concentrate on wider green growth strategies and encouraging private sector investment in sustainable technologies. Lessons that can be adapted in these endeavors include adopting sustainable practices that businesses need to adopt and working with governments to reduce carbon emissions. Global cooperation is essential for many developing countries, and only International organizations have major roles in supporting such countries. Together, these entities represent a more sustainable and prosperous future.

9.7 Conclusion

The post, future research should prioritize the development of industry-specific net-zero strategies to examine how different sectors such as transportation, manufacturing, and agriculture can transition to net-zero in a sustainable way. Moreover, the contribution that innovation can play in scale-wide climate solutions (such as wind and solar power systems, and carbon capture systems) remains to be explored. Advancements in these areas and the ability to integrate innovative solutions into various economic sectors will govern the pathway to long-term sustainability (Geissdoerfer et al., 2020; Wagner et al., 2021).

Innovative policy approaches, namely, carbon taxation, green finance, and incentives to private sector investments in low carbon technologies, are the key solutions to growth and climate neutrality at the same time. Both goals are supported by technological advancements in renewable energy and carbon capture as well. We also need economic policy at the national and global level to match with environmental objectives, and this can be achieved by taking up sustainable business models like the circular economy and global cooperation. Bataille et al. (2020) show the usefulness for solutions in establishing a reciprocal relationship between governments, businesses, and international organizations that drive long-term sustainability and economic progress.

References

André, P., & Pirlot, A. (2024). When are carbon border adjustment measures just? *Political Studies*. https://doi.org/10.1177/00323217241238125

Andrijevic, M., Schleussner, C., Gidden, M. J., McCollum, D. L., & Rogelj, J. (2020). COVID-19 recovery funds dwarf clean energy investment needs. *Science, 370*(6514), 298–300. https://doi.org/10.1126/science.abc9697

Antimiani, A., Costantini, V., Kuik, O., & Paglialunga, E. (2016). Mitigation of adverse effects on competitiveness and leakage of unilateral EU climate policy: An assessment of policy instruments. *Ecological Economics, 128*, 246–259. https://doi.org/10.1016/j.ecolecon.2016.05.003

Apergis, N., & Payne, J. E. (2021). Natural disasters and monetary policy: Evidence from an augmented Taylor rule. *Journal of Financial Economic Policy, 14*(3), 317–332. https://doi.org/10.1108/jfep-04-2021-0110

Asmelash, E., Prakash, G., Gorini, R., & Gielen, D. (2020). Role of IRENA for global transition to 100% renewable energy. In *Lecture notes in energy* (pp. 51–71). International Renewable Energy Agency. https://doi.org/10.1007/978-3-030-40738-4_2

Bataille, C., Waisman, H., Briand, Y., Svensson, J., Vogt-Schilb, A., Jaramillo, M., Delgado, R., Arguello, R., Clarke, L., Wild, T., Lallana, F., Bravo, G., Nadal, G., Treut, G. L., Godinez, G., Quiros-Tortos, J., Pereira, E., Howells, M., Buira, D., & Imperio, M. (2020). Net-zero deep decarbonization pathways in Latin America: Challenges and opportunities. *Energy Strategy Reviews, 30*, 100510. https://doi.org/10.1016/j.esr.2020.100510

Böhringer, C., Fischer, C., Rosendahl, K. E., & Rutherford, T. F. (2022). Potential impacts and challenges of border carbon adjustments. *Nature Climate Change, 12*(1), 22–29. https://doi.org/10.1038/s41558-021-01250-z

Bonk, A., Braun, M., Sötz, V. A., & Bauer, T. (2020). Solar salt—Pushing an old material for energy storage to a new limit. *Applied Energy, 262*, 114535. https://doi.org/10.1016/j.apenergy.2020.114535

Breen, R. (2019). Education and intergenerational social mobility in the US and four European countries. *Oxford Review of Economic Policy, 35*(3), 445–466. https://doi.org/10.1093/oxrep/grz013

Bryant, G. (2016). Creating a level playing field? The concentration and centralisation of emissions in the European Union emissions trading system. *Energy Policy, 99*, 308–318. https://doi.org/10.1016/j.enpol.2016.06.007

Buylova, A., Fridahl, M., Nasiritousi, N., Overland, I., & Reischl, G. (2022). Climate action in the making: Business and civil society views on the world's first carbon border levy. *Climate Action, 1*(1), 17. https://doi.org/10.1007/s44168-022-00015-4

Change, N. I. P. O. C. (2023). *Climate change 2021—The physical science basis*. IPCC. https://doi.org/10.1017/9781009157896

Cosbey, A., Droege, S., Fischer, C., & Munnings, C. (2018). Developing guidance for implementing border carbon adjustments: Lessons, cautions, and research needs from the literature. *Review of Environmental Economics and Policy, 13*(1), 3–22. https://doi.org/10.1093/reep/rey020

Doda, B., Quemin, S., & Taschini, L. (2019). Linking permit markets multilaterally. *Journal of Environmental Economics and Management, 98*, 102259. https://doi.org/10.1016/j.jeem.2019.102259

Fankhauser, S., Smith, S. M., Allen, M., Axelsson, K., Hale, T., Hepburn, C., Kendall, J. M., Khosla, R., Lezaun, J., Mitchell-Larson, E., Obersteiner, M., Rajamani, L., Rickaby, R., Seddon, N., & Wetzer, T. (2021). The meaning of net zero and how to get it right. *Nature Climate Change, 12*(1), 15–21. https://doi.org/10.1038/s41558-021-01245-w

Friedlingstein, P., O'Sullivan, M., Jones, M. W., Andrew, R. M., Gregor, L., Hauck, J., Quéré, C. L., Luijkx, I. T., Olsen, A., Peters, G. P., Peters, W., Pongratz, J., Schwingshackl, C., Sitch, S., Canadell, J. G., Ciais, P., Jackson, R. B., Alin, S. R., Alkama, R., & Zheng, B. (2022). Global carbon budget 2022. *Earth System Science Data, 14*(11), 4811–4900. https://doi.org/10.5194/essd-14-4811-2022

Geels, F. W. (2021). From leadership to followership: A suggestion for interdisciplinary theorising of mainstream actor reorientation in sustainability transitions. *Environmental Innovation and Societal Transitions, 41*, 45–48. https://doi.org/10.1016/j.eist.2021.10.021

Geissdoerfer, M., Pieroni, M. P., Pigosso, D. C., & Soufani, K. (2020). Circular business models: A review. *Journal of Cleaner Production, 277*, 123741. https://doi.org/10.1016/j.jclepro.2020.123741

Grossman, G. M., & Krueger, A. B. (1995). Economic growth and the environment. *The Quarterly Journal of Economics, 110*(2), 353–377. https://doi.org/10.2307/2118443

Hepburn, C., O'Callaghan, B., Stern, N., Stiglitz, J., & Zenghelis, D. (2020). Will COVID-19 fiscal recovery packages accelerate or retard progress on climate change? *Oxford Review of Economic Policy, 36*(Supplement 1), S359–S381. https://doi.org/10.1093/oxrep/graa015

Hickel, J., & Kallis, G. (2019). Is green growth possible? *New Political Economy, 25*(4), 469–486. https://doi.org/10.1080/13563467.2019.1598964

Johnson, M. P., & Schaltegger, S. (2019). Entrepreneurship for sustainable development: A review and multilevel causal mechanism framework. *Entrepreneurship Theory and Practice, 44*(6), 1141–1173. https://doi.org/10.1177/1042258719885368

Karlsson, M., Alfredsson, E., & Westling, N. (2020). Climate policy co-benefits: A review. *Climate Policy, 20*(3), 292–316. https://doi.org/10.1080/14693062.2020.1724070

Keen, M., Parry, I., & Roaf, J. (2022). Border carbon adjustments: Rationale, design and impact. *Fiscal Studies, 43*(3), 209–234. https://doi.org/10.1111/1475-5890.12307

Khurshid, N., Fiaz, A., Khurshid, J., & Ali, K. (2022). Impact of climate change shocks on economic growth: A new insight from non-linear analysis. *Frontiers in Environmental Science, 10*, 1039128. https://doi.org/10.3389/fenvs.2022.1039128

Krogstrup, S., & Oman, W. (2019). Macroeconomic and financial policies for climate change mitigation: A review of the literature. *SSRN Electronic Journal*. https://doi.org/10.2139/ssrn.3463780

Kuehner, A., Jakob, M., & Flachsland, C. (2022). German stakeholder perceptions of an EU carbon border adjustment mechanism. *Environmental Research Letters, 17*(12), 124007. https://doi.org/10.1088/1748-9326/ac9f23

Luderer, G., Pehl, M., Arvesen, A., Gibon, T., Bodirsky, B. L., De Boer, H. S., Fricko, O., Hejazi, M., Humpenöder, F., Iyer, G., Mima, S., Mouratiadou, I., Pietzcker, R. C., Popp, A., Van Den Berg, M., Van Vuuren, D., & Hertwich, E. G. (2019). Environmental co-benefits and adverse side-effects of alternative power sector decarbonization strategies. *Nature Communications, 10*(1), 5229. https://doi.org/10.1038/s41467-019-13067-8

McGlade, C., & Ekins, P. (2015). The geographical distribution of fossil fuels unused when limiting global warming to 2 °C. *Nature, 517*(7533), 187–190. https://doi.org/10.1038/nature14016

Mohan, P. S. (2024). Stakeholder perspectives on border carbon adjustments: Trinidad and Tobago's balancing act between competitiveness and climate action. *Climate Policy*. https://doi.org/10.1080/14693062.2024.2401855

Nwokolo, S. C., Singh, R., Khan, S., Kumar, A., & Luthra, S. (2023). Africa's path to net-zero. In *CSR, sustainability, ethics & governance*. Springer. https://doi.org/10.1007/978-3-031-44514-9

Quéré, C. L., Korsbakken, J. I., Wilson, C., Tosun, J., Andrew, R., Andres, R. J., Canadell, J. G., Jordan, A., Peters, G. P., & Van Vuuren, D. P. (2019). Drivers of declining CO2 emissions in 18 developed economies. *Nature Climate Change, 9*(3), 213–217. https://doi.org/10.1038/s41558-019-0419-7

Rosenow, J., Gibb, D., Nowak, T., & Lowes, R. (2022). Heating up the global heat pump market. *Nature Energy, 7*(10), 901–904. https://doi.org/10.1038/s41560-022-01104-8

Sakai, M., & Barrett, J. (2016). Border carbon adjustments: Addressing emissions embodied in trade. *Energy Policy, 92*, 102–110. https://doi.org/10.1016/j.enpol.2016.01.038

Sarkodie, S. A., & Strezov, V. (2018). Effect of foreign direct investments, economic development and energy consumption on greenhouse gas emissions in developing countries. *The Science of the Total Environment, 646*, 862–871. https://doi.org/10.1016/j.scitotenv.2018.07.365

Singh, R., Khan, S., & Dsilva, J. (2022). A framework for assessment of critical factor for circular economy practice implementation. *Journal of Modelling in Management, 18*(5), 1476–1497. https://doi.org/10.1108/jm2-06-2021-0145

Sovacool, B. K. (2021). Who are the victims of low-carbon transitions? Towards a political ecology of climate change mitigation. *Energy Research & Social Science, 73*, 101916. https://doi.org/10.1016/j.erss.2021.101916

Sovacool, B. K., Turnheim, B., Hook, A., Brock, A., & Martiskainen, M. (2020). Dispossessed by decarbonisation: Reducing vulnerability, injustice, and inequality in the lived experience of low-carbon pathways. *World Development, 137*, 105116. https://doi.org/10.1016/j.worlddev.2020.105116

Tobin, P., Farstad, F. M., & Tosun, J. (2023). Intermediating climate change: the evolving strategies, interactions and impacts of neglected "climate intermediaries". *Policy Studies, 44*(5), 555–571. https://doi.org/10.1080/01442872.2023.2230899

Wagner, D. L., Grames, E. M., Forister, M. L., Berenbaum, M. R., & Stopak, D. (2021). Insect decline in the Anthropocene: Death by a thousand cuts. *Proceedings of the National Academy of Sciences, 118*(2), e2023989118. https://doi.org/10.1073/pnas.2023989118

Zhong, J., & Pei, J. (2023). Carbon border adjustment mechanism: A systematic literature review of the latest developments. *Climate Policy, 24*(2), 228–242. https://doi.org/10.1080/14693062.2023.2190074

Zhu, T., Zheng, L., Li, F., Liu, J., & Zhuang, W. (2024). Sustainable carbon sequestration via olivine based ocean alkalinity enhancement in the east and South China Sea: Adhering to environmental norms for nickel and chromium. *The Science of the Total Environment, 2024*, 172853. https://doi.org/10.1016/j.scitotenv.2024.172853

Chapter 10
Climate Action and Resilience for a Sustainable Future: The Roles of the UN, WTO, and IMF

Lucy

Abstract As the global community faces escalating climate challenges, the roles of the United Nations (UN), World Trade Organization (WTO), and International Monetary Fund (IMF) are pivotal in advancing sustainable and climate resilience. This paper explores key initiatives from each of these international institutions that contribute to global sustainability efforts. The UN's initiatives include the Paris Agreement, which sets international climate targets; the Sustainable Development Goals (SDGs), providing a comprehensive framework for addressing various global challenges; the Green Climate Fund (GCF), supporting climate action in developing countries; the UN Environment Programme (UNEP), offering environmental leadership and policy guidance; and the Global Compact, promoting corporate sustainability. The WTO's contributions involve the Environmental Goods Agreement (EGA) to promote sustainable trade, integrating environmental policies into trade practices, supporting sustainable development through the Committees on Trade and Environment, and fostering green technology trade through the Technical Barriers to Trade (TBT) Agreement. The IMF's role includes the resilience and Sustainable Trust (RST) to support climate adaptation, integrating climate risks into fiscal policies, providing technical assistance, and supporting debt relief for climate action. Each of these initiatives plays a crucial role in shaping effective responses to climate change and fostering sustainable development. This chapter concludes with recommendations for enhancing the effectiveness of these institutions' roles in sustainability, including climate risks into economic policies, promoting equity in sustainability efforts, and supporting innovation in green technologies. The collaborative efforts of the UN, WTO, and IMF are essential for achieving long-term sustainability and resilience in a rapidly changing world.

Keywords Climate neutrality · Resilience building · Global governance · Sustainable development · Multilateral institutions

---

Lucy (✉)
Amity Institute of Public Policy, Amity University, Noida, India

R. Singh and D. Crowther (eds.), *Climate Neutrality and Sustainable Ecosystems*, Approaches to Global Sustainability, Markets, and Governance,
https://doi.org/10.1007/978-981-96-7181-6_10

10.1 Introduction

Climate change is one of the most pressing global challenges, with widespread environmental, social, and economic consequences. Addressing climate change requires concerted efforts from various international institutions, including the United Nations (UN), World Trade Organization (WTO), and International Monetary Fund (IMF), whose roles in promoting climate neutrality and resilience have become increasingly vital. Climate neutrality, defined as achieving a balance between emitted and absorbed greenhouse gases (IPCC, 2021), has become a global goal, particularly with the Paris Agreement's commitment to limit global, temperature rise to below 2 °C (United Nations, 2015a). In parallel, building resilience to climate impacts, especially in vulnerable countries, is crucial for sustainable development.

The UN, through its numerous agencies, conventions, and frameworks such as the UNFCCC and Sustainable Development Goals (SDG's), provides a comprehensive global framework for addressing climate change. The Paris Agreement ratified under the UNFCC, is a landmark accord that unites nations in their efforts to combat climate change and adapt to its effects. Studies have demonstrated that this agreement significantly shapes climate policy, driving countries towards decarbonization and resilience-building strategies (Gupta & Van Asselt, 2019; Hale, 2016).

The WTO plays a key role in aligning global trade policies with environmental objectives. The organization's work on trade and environmental sustainability includes reducing barriers to climate-friendly technologies and products through negotiations like the Environmental Goods Agreement (Zeng & Chang, 2017). By promoting trade liberalization for green goods, the WTO facilitates the global dissemination of technologies essential for climate mitigation and adaptation (Bacchus, 2016).

Meanwhile, the IMF has increasingly integrated climate risks into its macroeconomic and fiscal policy frameworks. Through initiatives such as the Resilience and Sustainability Trust (RST) and climate vulnerability assessments, the IMF supports countries in adapting to climate risks while promoting fiscal stability (IMF, 2022a). Researchers emphasize that IMF's role in climate finance and policy guidance is crucial for countries, especially developing economies, to navigate the economic shifts driven by the transition to low-carbon economies (Batten, 2018; Lagarde & Gaspar, 2019).

Given the interconnected nature of climate action, trade, and finance, this chapter examines how the collaborative efforts of the UN, WTO, and IMF contribute to climate neutrality and resilience-building. The analysis underscores the importance of multilateral governance in ensuring that climate action is both equitable and effective, especially in addressing the disproportionate impacts on vulnerable nations (Obergassel et al., 2020; Stern, 2015).

10.2 Objectives

1. To examine the role of UN, WTO, and IMF in climate action, sustainable trade, and financial and policy support for climate resilience.
2. Analysis of the interaction among the UN, WTO, and IMF and evaluate the impact on Developing countries and vulnerable countries.
3. Identify the best practices, future direction, and policy recommendations.

10.3 Theoretical Framework: Climate Change, Resilience, and Sustainable Development

Climate change poses a significant challenge to sustainable development, necessitating an integrated approach that addresses environmental, economic, and social dimensions. Sustainable development seeks to meet current needs without compromising the ability of future generations to meet their own needs (World Commission on Environment and Development, 1987). At the intersection of climate change and sustainable development lies the need for policies and practices that simultaneously tackle environmental degradation while promoting economic growth and social equity.

System theory offers valuable insights into this integration by emphasizing the interconnections among ecological, social, and economic systems. It helps us understand how climate change impacts various aspects of sustainability and how integrated approaches can mitigate these effects (Meadows, 2008).

Ecological modernization theory further supports this perspective by suggesting that economic development and environmental protection can coexist through technological innovation and policy reforms, thus enabling societies to achieve sustainable development by adopting eco-efficient technologies and practices (Mol & Spaargaren, 2000).

Resilience theory focuses on the capacity of systems to absorb disturbances and adapt to changing conditions while maintaining essential functions. In the context of climate change, resilience refers to the ability of ecological and social systems to withstand and recover from climate-related shocks and stresses (Holling, 1973). Adaptive capacity, a key concept within resilience theory, is the ability of systems to adjust to potential damage and take advantage of opportunities (Adger, 2006). Panarchy theory complements this by exploring the dynamics of systems through cycles of growth, collapse, and renewal, highlighting the importance of flexibility and learning to enhance resilience (Holling, 2001).

An integrative framework combines these principles to provide a holistic approach to addressing climate challenges. This approach incorporates elements from system theory, ecological modernization, adaptive capacity, and panarchy theory to create comprehensive strategies for sustainable and resilient development. It emphasizes a holistic approach that integrates environmental, economic, and social dimensions, dynamic adaptation to changing conditions, and stakeholder engagement to ensure that sustainability and resilience efforts are inclusive and equitable (Folke et al., 2002; Kates et al., 2005; Reed, 2008).

10.4 Role of the UN, WTO, and IMF in Promoting Sustainability

10.4.1 The United Nations (UN)

The United Nations (UN) has played a pivotal role in addressing climate challenges and promoting sustainability through various initiatives. One of the most significant is the Paris Agreement, adopted under the UNFCCC in 2015. This landmark international accord aims to limit global warming to well below 2 °C, with efforts to keep the increase to 1.5 °C above pre-industrial level (UNFCCC, 2015). It establishes legally binding targets for reducing greenhouse gas emissions and fosters global cooperation on climate change.

Another key initiative is the Sustainable Development Goals (SDGs), introduced in 2015. These 17 goals provide a comprehensive framework to tackle global issues such as poverty, hunger, inequality, climate change, environmental degradation, and the promotion of peace and justice. The SDGs aim to achieve sustainable development across multiple dimensions by 2030 (United Nations, 2015b).

The Green Climate Fund (GCF), established by UNFCCC, supports developing nations in their efforts to mitigate and adapt to climate change. It provides financial assistance for projects and programs that enhance climate resilience and reduce carbon emissions (Green Climate Fund, 2023).

Additionally, the UN Environment Programme works to promote sustainable development through sound environmental practices. UNEP offers leadership and fosters partnerships worldwide by conducting assessments, providing policy support, and enhancing capacities to address environmental challenges (UNEP, 2020).

The UN Global Compact is another vital initiative, which encourages businesses globally to adopt sustainable and socially responsible policies. It is a voluntary initiative that helps companies align their strategies with ten principles related to human rights, labor standards, environmental sustainability, and anti-corruption (United Nations Global Compact, 2021).

These UN-led initiatives are critical in shaping global efforts towards a more sustainable and climate-resilient future.

10.4.2 World Trade Organization (WTO)

The World Trade Organization (WTO) plays a significant role in aligning trade with environmental sustainability through various initiatives. One such effort is the

Environmental Goods Agreement (EGA), which seeks to eliminate tariffs on environmental goods. This agreement promotes the trade of products that support environment sustainability, such as renewable energy technologies, and facilitates the global dissemination of green technologies (WTO, 2021a).

Additionally, the WTO emphasizes the integration of environmental policies into trade practices through its focus on Trade and Environmental Sustainability. This approach helps countries develop policies that align with sustainability goals with maintaining trade liberalization (WTO, 2020a).

The Committee on Trade and Environment addresses the relationship between trade and environmental policies, ensuring that trade policies actively support environmental protection and sustainability (WTO, 2019).

The Technical Barriers to Trade (TBT) Agreement is another key aspect of the WTO's work in this area. It ensures that regulations, standards, and conformity assessment procedures do not create unnecessary obstacles to international trade while ensuring that they are used to protect the environment (WTO, 2020b).

Lastly, the Sustainable Development Work Program within the WTO supports developing countries in integrating environmental sustainability into their trade policies and practices. This program fosters a balance between trade and environmental goals, particularly for nations that are still developing their trade frameworks (WTO, 2022a). Through these initiatives, the WTO seeks to harmonize global trade practices with sustainable developing objectives.

10.4.3 The International Monetary Fund (IMF)

The International Monetary Fund (IMF) has been actively addressing climate-related challenges through various initiatives aimed at promoting economic stability and sustainability. One key initiative is the Resilience and Sustainability Trust (RST), which was established to provide financial assistance to countries facing long-term climate risks. This facility supports nations in implementing climate-resilient policies, enhancing both economic stability and climate resilience (IMF, 2022a).

The IMF also conducts Climate Change Policy Assessments, which help countries understand and manage the macroeconomic impacts of climate change. These assessments guide policy reforms and investments that support sustainable development (IMF, 2023).

In addition, the IMF provides expertise on Climate Change and Fiscal Policy, offering guidance on integrating climate risks into fiscal strategies. This includes advice on carbon pricing and green public investment, helping countries pursue sustainable economic growth while addressing climate challenges (IMF, 2021a).

To support nations in implementing these strategies, the IMF offers Technical Assistance and Capacity Building programs. These initiatives help countries design and execute policies that address climate-related economic challenges and promote sustainability (IMF, 2020).

Through these comprehensive initiatives, the IMF plays a critical role in fostering climate resilience and sustainable economic growth globally.

10.5 Case Studies on Climate Action and Resilience in Different States

Several states have demonstrated how national and regional approaches can align with the roles of international institutions such as the UN, WTO, and IMF in advancing climate action and building resilience.

10.5.1 Germany: Leading in Climate Action and Sustainable Development

Germany has been at the forefront of integrating climate action with sustainable development, heavily influenced by international agreements like the UN's Sustainable Development Goals (SDG's) and the Paris Agreement. The country's "Energiewende" policy is a notable example of this commitment, focusing on transitioning from fossil fuels to renewable energy sources (Wüstenhagen & Menichetti, 2012). Germany has invested substantially in wind and solar energy, enforced stringent emission regulations, and offered subsidies for green technology. This ambitious strategy has led to significant reductions in greenhouse gas emissions and an increased share of renewable energy in its energy mix, serving as a model for other nations in climate policy and sustainable development (Wüstenhagen & Menichetti, 2012).

10.5.2 Bhutan: Carbon Neutrality and Sustainable Development

Bhutan stands out for its commitment to maintaining carbon neutrality while promoting economic development. The country has achieved this through a combination of renewable energy projects, mainly hydroelectric power, extensive reforestation efforts, and conservation policies. Bhutan's emphasis on environmental preservation is integrated into its development strategy, making it one of the few nations to offset its emissions effectively. This approach highlights how national policies can align with global climate goals, demonstrating a successful model of sustainable development (Roder & Wangdi, 2014).

10.5.3 Kenya: Climate Resilience through the Green Climate Fund

Kenya has utilized the Green Climate Fund (GCF) to bolster its climate resilience and support sustainable development initiatives. The GCF has funded various projects in Kenya, including irrigation schemes, renewable energy projects, and community-based climate adaptation initiatives. These projects are designed to enhance food security, increase access to clean energy, and improve water management systems. The impact of these GCF-funded initiatives has been substantial, contributing to greater resilience to climate change, particularly in vulnerable communities, and supporting Kenya's national climate action plan (Mwaura & Ong, 2020).

10.5.4 Australia: WTO's Role in Sustainable Trade Practices

Australia has actively engaged with the WTO to advance sustainable trade practices. A key aspect of this engagement is its support for the Environmental Goods Agreement (EGA), which aims to lower tariffs on environmental goods and promote the trade of green technologies. By incorporating environmental considerations into its trade agreement and supporting international sustainability efforts, Australia has facilitated access to green technologies and supported its own sustainability goals. This case underscores the role of trade policies in achieving environmental outcomes (McEachern & Roberts, 2019).

10.5.5 Colombia: IMF's Role in Climate Finance and Policy

Colombia has leveraged IMF support to enhance its climate resilience and sustainability into economic policies. The IMF has provided technical assistance and policy advice to Colombia, focusing on incorporating climate risks into fiscal policies and supporting climate adaptation projects. With IMF support, Colombia has adopted policies to manage climate risks and promote sustainable development. These efforts have improved Colombia's ability to handle climate-related challenges and invest in sustainable development, illustrating the significant impact of IMF support on national climate policies (López & Ramirez, 2021a).

These case studies illustrate how different states have effectively utilized international frameworks and support from institutions like the UN, IMF, and WTO to advance climate action and sustainable development. Each case highlights unique approaches and impacts, offering valuable insights into global and local efforts for enhanced climate resilience.

10.5.6 India: Leveraging International Support for Climate Action

India has effectively utilized international frameworks and institutional support to advance its climate action and resilience efforts. As a key participant in the Paris Agreement, India has committed to reducing its carbon intensity by 33–33% below 2005 level by 2030 and increasing the share of non-fossil fuel energy to 50% of its installed capacity (Ministry of Environment, Forest and Climate Change, 2022). This commitment is embodied in the National Action Plan on Climate Change, which outlines various missions to enhance sustainability and resilience (Indian Ministry of Power, 2023). Additionally, India has engaged with the WTO to promote sustainable trade practices, including support for the Environmental Goods Agreement (EGA) that lowers tariffs on green technologies (Gupta, 2021). The IMF has also played a significant role by providing technical assistance and policy advice to integrate climate risks into fiscal policies and support climate adaptation projects (Sharma & Ghosh, 2022). Furthermore, the Green Climate Fund (GCF) has financed projects in India aimed at improving water management, advancing renewable energy, and supporting sustainable agriculture (Verma & Patel, 2023). These combined efforts demonstrate India's strategy use of international support to enhance its climate resilience and advance sustainability goals.

10.6 Recent Development in the Role of the UN, WTO, and IMF in Climate Resilience and Sustainability

In the past few years, significant progress has been made by global institutions in promoting sustainability and climate change.

10.6.1 UN Initiatives

COP 27, recently held in El-Sheikh, Egypt, marked a significant milestone with the establishment of a "loss and damage fund" to support vulnerable countries most affected by climate change. This fund is aimed at addressing the economic impacts of climate-related disasters, especially in developing nations, recognizing the disproportionate burden they bear due to climate change despite contributing the least to global emissions. The creation of the fun was a breakthrough, as it responded to long-standing demands from developing nations for compensation to manage the losses they face from extreme events and rising sea levels. This initiative underscores a shift towards climate change, ensuring that wealthier nations, historically responsible for higher emissions, contribute to the mitigation of these impacts. The fund is expected to provide financial aid for recovery, rebuilding, and climate resilience in vulnerable communities (UNFCCC, 2015).

SDG Acceleration Actions In 2023, the UN intensified efforts to accelerate the achievements of the Sustainable Development Goals (SDGs), recognizing that many targets are off-track as the 2030 deadline approaches. The UN urging member states to take more necessary steps to fulfill the commitment, including climate action in SDG-13, poverty Alleviation in SDG-1, and clean energy in SDG-7. This push of acceleration was highlighted at the SDG Summit 2023, where leaders called for renewed multilateral cooperation, enhanced financing mechanisms, and innovation partnerships to bridge the existing gaps. Climate action has been prioritized as a cross-cutting goal that impacts several other SDGs, prompting increased investment in green technologies and sustainable infrastructure. The UN has also emphasized the importance of data-driven approaches to track progress and align national policies with the global SDG framework, ensuring that no country or community is left behind in this decade of action (United Nations, 2023).

UN Global Stocktake, the first initiative of this under Paris Agreement was initiated in 2023 to evaluate the collective progress of nations in achieving the long-term climate goals. This process is crucial for ensuring that the world remains on track to limit global warming to well below 2 °C, with efforts to limit the increase to 1.5 °C above pre-industrial levels. The stocktake provides a comprehensive assessment of national climate actions, identifying gaps in mitigation, adaptation, and financing policies. It also offers a platform for countries to strengthen their climate commitments, enhance transparency, and drive more ambitious global action (UNFCCC, 2015).

10.6.2 WTO Developments

In June 2022, the WTO's 12th Ministerial Conference (MC12) underscored the significance of integrating sustainability into global trade practices. A major outcome was the adoption of a landmark agreement on fisheries subsidies, which aims to reduce harmful subsidies that contribute to overfishing and the depletion of marine ecosystems. This agreement marks a significant step in aligning global trade rules with environmental sustainability by encouraging the protection of marine biodiversity. The agreement is expected to promote more responsible fishing practices, reduce the over-exploitation of fish stocks, and foster economic growth in coastal communities reliant on sustainable fisheries. This development reflects the WTO's growing commitment to ensuring that international trade contributes positively to environmental protection while supporting sustainable economic development.

WTO Trade and Environmental Sustainability Structured Discussions (TESSD) was launched in 2021, with over 70 WTO members actively participating in this dialogue. The aim of this initiative is aligning trade policies with climate action and environmental sustainability by focusing on key areas such as the promotion of green goods and services and reducing trade barriers for environmentally friendly technologies. Discussions also emphasize supporting the goals of the Paris Agreement, encouraging sustainable supply chains, and integrating environmental considerations into international trade agreements. By facilitating these conversations, the WTO is working towards creating a global trade environment that supports climate resilience and sustainable economic development (WTO, 2023).

WTO Initiative on Plastic Pollution and Trade in Environmental Goods: In 2021, the WTO launched an initiative focused on addressing plastic pollution through trade measures. This initiative, part of broader efforts to combat environmental degradation, promotes the reduction of plastic waste by encouraging the trade of environmentally sustainable goods and technologies. By tackling plastic pollution, this initiative aligns with global environmental goals, including the Sustainable Development Goals (SDGs), while fostering innovation and sustainable trade practices (WTO, 2022b).

10.6.3 IMF Developments

In 2022, the International Monetary Fund (IMF) introduced the "Resilience and Sustainability Trust" to aid low- and middle-income nations in enhancing their capacity to withstand climate change. This trust offers long-term financial support to help countries tackle climate-related challenges, including natural disasters and economic impacts stemming from environmental issues. Its primary objective is to assist countries in transitioning to more sustainable and climate-resilient economies. The trust prioritizes readiness for disasters, environmentally friendly investment, and policy change aimed at reducing climate vulnerabilities. By addressing climate risks comprehensively, the RST enhance developing countries' ability to achieve sustainable economic growth while contributing to global climate targets.

The IMF has significantly increased its efforts to evaluate how climate risks affect macroeconomic stability, helping countries to implement climate policies that promote fiscal sustainability while tackling environmental challenges. The IMF also provides guidance on carbon pricing, helping nations adopt effective mechanisms to reduce greenhouse gas emissions, and promotes investments in sustainable infrastructure, ensuring that economic growth is aligned with environmental protection goals (IMF, 2023).

Through the Climate Macroeconomic Capacity Development Program, the IMF works closely with governments to integrate climate risks into fiscal planning, budgeting, and policy-making. This program offers technical assistance to help countries manage climate-related financial risks, such as those posed by extreme weather events or the transition to low-carbon economies. By aligning development strategies with sustainable goals, the IMF helps nations build economic resilience, adapt to climate impacts, and achieve long-term sustainable development outcomes. The program also emphasizes capacity building for government officials to better understand and manage the intersection of climate change and economic policies (IMF, 2023).

10.7 Collaborative Efforts of UN-WTO-IMF to Promote Climate Resilience

In 2022, these global institutions introduced UN-WTO-IMF Collaboration on Climate and Trade along with World Bank to foster coordination on climate-related issues. These institutions held joint discussions aimed to promoting a *green recovery* from the COVID-19 pandemic, highlighting the importance of integrating sustainability into economic rebuilding efforts. Key areas of focus included sustainable investments in green technologies, debt relief initiatives tied to climate outcomes, and removing trade barriers that hinder the growth of eco-friendly industries. These discussions underscore the need for aligned policy actions across international organizations to address global climate challenges while ensuring economic stability and growth emphasizing a multilateral approach to achieving a sustainable future (World Bank and IMF, 2022). In 2023, the IMF endorsed the establishment of the International Carbon Price Floor to assist countries in setting minimum carbon pricing levels while encouraging a transition to green energy. This initiative aims to create a consistent global approach to carbon pricing, ensuring that major emitters take greater responsibility for their emissions, promote investments in clean technologies, and reduce reliance on fossil fuels. The IMF's support for this policy also strengthens global climate finance by ensuring that the funds generated from carbon pricing can be reinvested into sustainable infrastructure and renewable energy projects, contributing to a more equitable and effective global response to climate change (IMF, 2023).

The UN Global Compact's Sustainability Initiatives strengthened its focus on engaging private sector companies in advancing climate actions in 2023, it launched a series of initiatives encouraging businesses to adopt *science-based targets* for reducing carbon emissions and integrating sustainability into their core operations. These initiatives aim to drive corporate responsibility, urging businesses to align their strategies with the Paris Agreement and the Sustainable Development Goals. These initiatives also emphasize collaboration between businesses, governments, and civil society to foster innovative solutions for climate challenges, demonstrating how private sector engagement is essential for achieving significant long-term progress in climate resilience and sustainability (United Nations Global Compact, 2023).

Additionally, these global institutions are taking proactive steps to embed sustainability into their operational frameworks through various initiatives, financial mechanisms, and strategic partnerships. These efforts are pivotal in supporting global endeavor to mitigate climate change, foster sustainable trade, and enhance economic resilience. By promoting initiatives such as carbon pricing, green finance, and responsible trade policies, these institutions are aligning economic growth with environmental sustainability. Through these integrated efforts, the global community can move to a resilient and sustainable future.

10.8 Coordination Between Global Institutions (UN, WTO, IMF) and Regional Organizations for Sustainability

The Future of Collaboration: In recent years, the partnership between global organizations like the UN, WTO, and IMF, and regional bodies grows more important in order to enhance sustainability. Since there is no unity of time or space in the region, regional organizations play an important role in addressing specific geographically and culturally specific challenges that can differ significantly from area to area. In sum, they offer a vehicle to adapt the right set of policy solutions to global sustainability frameworks and apply them at regional level, making sure that works are effectively introduced in places where they can help, and then modified according to local conditions. Regional organizations, such as the European Union (EU), African Union (AU), and Association for Southeast Asian Nations (ASEAN), facilitate collaboration among member states and bridge the gap between global directives and local action. They play a crucial role in implementing international agreements and policies on the ground such as those related to climate change, trade, and economic development.

For example, the EU has been active in integrating sustainability into its trade policies and economic strategies, which aligns with global efforts promoted by the WTO and IMF. Similarly, ASEAN has developed regional framework to address climate resilience and sustainable development, complementing global initiatives led by the UN and IMF. Coordination efforts include joint initiatives, such as regional climate action plans, sustainable development programs, and cross-border environmental projects. These collaborations enhance the effectiveness of global policies by ensuring they are contextually relevant and supported by regional stakeholders. They also facilitate knowledge sharing, capacity building, and the mobilization of resources, which are critical for achieving sustainable goals. Overall, the synergy between global institutions and regional organizations is essential for fostering a cohesive approach to sustainability, enhancing both broad and localized impacts that drive meaningful progress towards a sustainable future.

The WTO collaborates with Mercosur, the South American trade bloc, to integrate sustainability into regional trade policies. This partnership focuses on fostering sustainable agricultural practices, which are critical for Latin America's export-driven economy. By aligning trade policies with environmental goals, this collaboration aims to promote eco-friendly practices and enhance the sustainability of agricultural export (WTO, 2020b). The Asian-Pacific Economic Cooperation (APEC) and WTO are working together to promote environmentally sustainable trade practices across the Asian-Pacific region. This initiative focuses on green technologies and advancing sustainable supply chains. The WTO engages with the Gulf Cooperation Council (GCC) to incorporate sustainability standards into trade agreements and support green economic diversification in oil-exporting countries. This collaboration addresses the environmental challenges associated with fossil fuel dependency and encourages the adoption of renewable energy projects (WTO, 2022b).

The International Monetary Fund (IMF) is collaborating with the European Central Bank and other European financial institutions to integrate climate change risks into economic assessments and financial policy frameworks. This partnership focuses on advancing sustainable finance, managing climate risks, and implementing green recovery measures in the after month of COVID-19. The aim is to ensure that financial policies and practices support environmental sustainability and resilience against climate-related challenges (IMF, 2021b). The IMF works with the African Development Bank (ADB) to enhance climate finance across Africa. A key initiative is the African Climate Resilience Investment Facility, which seeks to mobilize both private and public sector investment to address infrastructure deficits and improve climate adaptation (African Development Bank (AfDB), 2022). The IMF has long-standing partnership with the Engagement with the Asian Development Bank to support sustainability efforts throughout Asia. This includes providing technical assistance for the development of climate-resilient infrastructure and ensuring accessibility to green financing for countries in the region. The collaboration also involves the development of financial instruments such as green bond, aimed at promoting sustainable investments and supporting the transition to a greener economy across Asia (IMF, 2020).

10.9 Future Direction of Global Sustainability: Expanding the Role of the UN, WTO, and IMF in Climate Action and Resilience

As the world grapple with the escalating impacts of climate change, the roles of international institutions such as the United Nations (UN), World Trade Organization (WTO), and International Monetary Fund (IMF) will be crucial in shaping future sustainability efforts. The future scope of these institutions involves several key areas:

10.9.1 Enhanced Integration of Climate Goals

In United Nations, future initiatives could involve integrating more robust climate targets into the UN's frameworks, such as updating the SDG's to include specific targets for emerging climate challenges and enhancing the monitoring and reporting mechanism for climate progress. The UN could also play a pivotal role in coordinating global climate action through new multilateral agreements that address gaps in the current frameworks. The WTO's role may expand to include more comprehensive agreements that address the trade-environment nexus, such as new rules for sustainable trade practices and incentives for green technology adoption. The WTO could also facilitate greater transparency and cooperation among member states on environmental standards and regulations (WTO, 2022b). The IMF could further integrate climate risk into its economic assessments and financial support mechanisms, possibly expanding the Resilience and Sustainability Trust (RST) to include more targeted support for low-carbon transition and climate adaptation. The IMF could also develop new financial instruments to support sustainable development goals (IMF, 2022b).

10.9.2 Strengthening Global Partnership

The UN could enhance partnerships with other international organizations, private sectors, and civil society to foster collaborative approaches to climate resilience and sustainability. Initiatives could include joint projects and multi-stakeholder platforms to address complex sustainability challenges (UNEP, 2020). The WTO might strengthen partnerships with environmental organizations and research institutions to support evidence-based policymaking and harmonize trade and environmental policies across borders. This could involve more proactive engagement with global environmental agreements and standards. The IMF could collaborate more closely with national governments and international financial strategies that support sustainable development and address the specific needs of vulnerable countries (IMF, 2021b).

10.9.3 Promoting Equity and Inclusivity

Future efforts could focus on ensuring that climate action and sustainability initiatives are equitable and inclusive, addressing the needs of marginalized communities and developing countries. The UN could implement new policies and programs to enhance social inclusion and equity in sustainable efforts (United Nations Global Compact, 2021). The WTO might work on reducing trade barriers that disproportionately impact developing countries and provide support for capacity-building initiatives to help these countries integrate sustainability into their trade practices (WTO, 2020b). The IMF could enhance its focus on equity in climate finance, ensuring that support is directed towards countries and communities most affected by climate change, and promoting policies that reduce inequality in climate resilience efforts (IMF, 2023).

10.9.4 Innovation and Technology

The UN could promote innovation in climate solutions by supporting research and development initiative and facilitating technology transfer between countries. Initiatives might include funding for innovative projects and partnerships to advance sustainable technologies. The WTO could play a key role in facilitating the global spread of green technologies by supporting trade policies that encourage innovation and protect intellectual property rights related to sustainable technologies (WTO, 2021b). The IMF could support the development of new financial instruments that incentivize investment in climate innovation and technology, helping countries transition to sustainable economic models (IMF, 2020).

10.9.5 Adaptive and Flexible Policies

The UN could focus on creating adaptive and flexible policy frameworks that can quickly respond to new climate challenges and opportunities. This might involve regular updates to global climate agreements and protocols to reflect emerging scientific knowledge and policy needs (UNEP, 2020). The WTO could develop more adaptive trade rules that accommodate changing environmental conditions and support countries in their transition to sustainable practices (WTO, 2022a). The IMF could enhance its policy frameworks to be more responsive to climate-related economic shifts and provide flexible financial solutions that address the evolving needs of member countries (IMF, 2021a).

These future directions highlight the need for continued innovation, collaboration, and adaptation in the roles of the UN, WTO, and IMF to effectively address the complex and interconnected challenges of sustainability and climate resilience.

10.10 Policy Recommendations for Enhancing Sustainability through the UN, WTO, and IMF

To enhance global sustainability, policy recommendations can be directed through key international organizations like the UN, WTO, and IMF.

First, strengthening the global climate agreement is crucial. The UN should enhance and expand existing international climate agreements, particularly the Paris Agreement, to address emerging challenges. This includes setting more ambitious emission reduction targets and integrating comprehensive measures for climate adaptation and mitigation. Regulation reviews and adjustments are necessary to ensure commitments align with the latest scientific data and technological advancements, promoting long-term sustainability and reducing climate risks (UNFCCC, 2015).

Second, facilitating sustainable trade practices through the WTO is essential for promoting environmental sustainability. Expanding the Environmental Goods Agreement (EGA) to cover a broader range of green technologies and services will help reduce trade barriers for sustainable goods. This can encourage member states to adopt environmentally friendly practices, fostering the global diffusion of green technologies (UNEP, 2021a).

Third, integrating climate risks into economic policies is necessary to build resilience in the global economy. The IMF can play a key role by incorporating climate risk analysis in macroeconomic assessments and fiscal policy recommendations. Expanding initiatives like the Resilience and Sustainability Trust (RST) will further support countries in managing climate-related economic risks and transitioning to low-carbon economies (IMF, 2022a).

Additionally, promoting equity and inclusivity in sustainability efforts is vital. The UN should focus on ensuring that sustainability initiatives address the needs of marginalized and vulnerable communities. By implementing policies that enhance social equity, these initiatives can ensure all communities benefit from global sustainability efforts, leading to more inclusive climate action (United Nations Global Compact, 2021). The UN should increase support for research and development in sustainable technologies and promote international cooperation for technology transfer between developed and developing nations. The WTO can encourage policies that protect intellectual property rights while promoting global dissemination of green technologies. The IMF can develop financial instruments to incentivize investment in climate innovation, fostering the growth of low-carbon economies (Green Climate Fund, 2023).

10.11 Global Climate Governance: A Comparative Analysis of the Role of the UN, WTO, and IMF in Developed and Developing Countries

Nowadays world countries are facing massive climate change and sustainability problems. Climate change and sustainability have become global issues. Global institutions like the United Nations, World Trade Organization, and International Monetary Fund taking necessary steps to foster international cooperation to combat climate change. However, their influence and effectiveness differ between developed and developing countries due to institutional capacity, policy priorities, environmental education, and economic disparities. This comparative analysis explores how the global institutions address the climate challenges of both developed and developing nations, which also highlight the different needs, responses, and impacts of these countries.

10.11.1 United Nations Role in Climate Resilience

Developed Countries The UN's initiatives in developed countries focus on driving innovation in green technology and establishing stringent climate regulations. For example, the European Union's alignment with the Paris Agreement and net-zero targets has been supported by the UN Framework Convention on Climate Change (United Nations, 2015b). Countries like Germany, Australia, the United Kingdom, and America leverage UN initiatives to enhance green energy infrastructure. The UK has been at the forefront of integrating UN climate policies into its national strategies, as host of the COP 26 climate conference in 2021, the UK played a key role in shaping global climate commitments. The UK decided to achieve net-zero emissions by 2050 and this initiative is supported by the UNFCCC's monitoring and reporting mechanism (UK Government, 2021). Canada's approach to climate action includes extensive collaboration with the UN on environmental policies and sustainability projects. It emphasizes on reducing carbon emissions, promoting clean technology, and adapting to climate impacts (Environment and Climate Change in Canada, 2022).

In developing countries, the role of the United Nations emphasizes capacity building, adaptation, and financing. For example, India is a key beneficiary of UN support through instruments like the Green Climate Fund (GCF) and the Global Environment Facility (GEF). These und are crucial for implementing climate resilience programs. By providing financial assistance, the UN helps developing nations strengthen their resilience and improve sustainability (Green Climate Fund, 2023; UNDP, 2020). With the support of the United Nations, Bangladesh has implemented various adaptation projects to address the impact of flooding and cyclones, including the construction of resilience infrastructure and community-based early warning system (UNDP, 2020). The UN has been actively involved in Vietnam's resilience efforts, particularly in the Mekong Delta region, which is highly vulnerable to sea-level rise and extreme weather events. The UNDP and other agencies have supported projects aimed at improving water management, promoting climate-smart agriculture, and enhancing coastal defenses (UNDP, 2020).

10.11.2 WTO's Role in Climate Resilience

Developed Countries The EU's Green Deal is supported by WTO rules that facilitate the trade of environmental technologies and goods, aiding in the EU's transition to a low-carbon economy (WTO, 2021b) Japan's emphasis on energy-efficient technologies and sustainable practices in trade agreements aligns with WTO's trade and environmental sustainability discussions (WTO, 2021a). Canada engages in WTO negotiations to ensure that its trade policies do not undermine its sustainability targets (WTO, 2021a). South Korea's Green Growth Strategy is supported by WTO agreements that facilitate the trade of green technologies and support environmental sustainability (WTO, 2021b).

In developing countries, India's participation in the WTO's Trade and Environmental Sustainability Structured Discussion (TESSD) supports its efforts to balance trade with environmental protection (WTO, 2020a). Brazil's efforts to include deforestation prevention measures in trade agreements are supported by WTO discussions on environmental standards (WTO, 2020b). South Africa works with the WTO to ensure that trade policies support its climate resilience goals (WTO, 2020a).

10.11.3 IMF's Role in Climate Resilience

Developed Countries The IMF provides guidance to the US on implementing carbon pricing and sustainable fiscal policies to align economic practices with climate goals. The IMF has encouraged the US to adopt a federal carbon tax as part of its Strategy to reduce emissions and support green investment (Lagarde & Gaspar, 2019). In 2021, IMF provided recommendations to EU policymakers on implementing green fiscal policies that align with the European Green Deal (IMF, 2021a). The IMF has provided Japan with technical assistance to design fiscal policies that encourage investment in renewable energy and low-carbon technologies (IMF, 2023).

In developing countries, the IMF has provided 1.4 billion dollars to Bangladesh to enhance climate resilience in critical sectors like agriculture and urban planning (IMF, 2022a). Kenya has utilized IMF funding to strengthen its infrastructure, ensuring resilience against droughts and floods, which are becoming more frequent due to climate change (López & Ramirez, 2021b). Ghana received financial support from the IMF to invest in renewable energy projects and strengthen climate-resilience agricultural practices (IMF, 2022b).

10.11.4 Critical Analysis: Role of the UN, WTO, and IMF in Climate Resilience for Developed and Developing Countries

In developing countries, the UN primarily focuses on promoting innovation in green technology, implementing climate resilience regulations. However, the influence of the UN in developing countries is often limited by their economic sovereignty. Countries like. United States have occasionally withdrawn from UN climate agreements. The UN role in developing countries focuses on capacity-building, financing, and adaptation strategies. In countries like India and Kenya UN has provided financial support through GCF and GEF, which help them to climate resilience programs. Despite these efforts, the scale of support is often insufficient relative to climate challenges faced by developing nations. Many developing countries heavily depend on international aid and bureaucratic delays and governance issues hinder the effectiveness of these projects. In developed countries, WTO has helped promote the trade of environmentally friendly goods and services. However, in developing countries, there is a growing criticism that WTO's environmental goods negotiations often favor advanced economies that already dominate the green technology market. The IMF in developed countries plays a key role in advising governments on sustainable fiscal policies and integrating climate risks into fiscal markets. But critics argue that the IMF's policies often reflect the economic priorities of advanced economies, with limited regard for the specific needs of developing countries. While the UN, WTO, and IMF play crucial roles in promoting global climate resilience, their impact varies between developed and developing countries. To bridge this gap, more equitable and context-specific approaches are needed to ensure that both developed and developing countries can achieve climate resilience in a sustainable and inclusive manner (Gupta, 2021; Lagarde & Gaspar, 2019; WTO, 2020b).

10.12 Conclusion

In the pursuit of global sustainability, the roles of the United Nations (UN), World Trade Organization (WTO), and International Monetary Fund (IMF) are indispensable. Each institution brings unique strengths and capabilities to the table, contributing to climate action and resilience in distinct ways. The UN's comprehensive frameworks, such as the Sustainable Development Goals (SDGs) and the Paris Agreement, provide a global blueprint for integrating sustainability across all sectors (UNFCCC, 2015; United Nations, 2015a). These frameworks not only set ambitious targets but also facilitate international cooperation and accountability.

The WTO's focus on trade policies and agreements is crucial for promoting sustainable development through environmental goods and services. The expansion of initiatives like the Environmental Goods Agreement (EGA) and the integration of sustainability into trade practices underscore the WTO's role in fostering a global market for green technologies and reducing environmental barriers to trade (WTO, 2021a). These actions help align international trade with environmental objectives and support sustainable economic growth.

The IMF's financial mechanisms and policy guidance are essential for managing the economic impacts of climate change and supporting resilience. Initiatives such as the Resilience and Sustainability Trust (RST) and the integration of climate risk into fiscal policies highlight the IMF's commitment to enhancing climate resilience and promoting sustainable economic strategies. These efforts aim to help countries, particularly those most vulnerable to climate shocks, build long-term economic stability while transitioning toward low-carbon and climate-resilient development paths (IMF, 2021a, 2022b).

Collectively, these institutions play a significant role in shaping a sustainable future. Their coordinated efforts in setting global standards, facilitating trade in green technologies, and providing financial support for climate resilience are crucial for addressing the multifaceted challenges of sustainability. A global challenge evolves, it is essential for the UN, WTO, and IMF to continue adapting and enhancing their strategies to support sustainable development and climate action effectively.

References

Adger, W. N. (2006). Vulnerability. *Global Environmental Change, 16*(3), 268–281. https://doi.org/10.1016/j.gloenvcha.2006.02.006

African Development Bank (AfDB). (2022). *African Climate resilience investment facility*. Retrieved from https://www.afdb.org

APEC. (2021). *APEC energy demand and supply outlook (8th ed.)*. Asia-Pacific Economic Cooperation. https://www.apec.org/

ASEAN Secretariat. (2021). *ASEAN State of Climate Change Report*. ASEAN Secretariat. https://asean.org/

Bacchus, J. (2016). The case for a WTO climate waiver (CIGI Papers, No. 86).

Bals, C., & Sharma, A. (2021). Climate finance: Innovations and opportunities for climate-resilient development. *Climate Policy Journal, 21*(6), 734–749. https://doi.org/10.1080/14693062.2021.1894048

Batten, S. (2018). Climate change and the macro-economy: A critical review. Bank of England Working Paper No. 706.

Das, K. (2021). Climate change and trade policy in Asia: Challenges and policy options. *Asian Development Review, 38*(1), 63–86.

Environment and Climate Change Canada. (2022). Annual report to Parliament on the administration of the Canadian Environmental Protection Act, 1999: 2021–2022. Government of Canada. https://www.canada.ca/

Folke, C., Carpenter, S. R., Elmqvist, T., Gunderson, L., Holling, C. S., & Walker, B. (2002). Resilience and sustainable development: Building adaptive capacity in a world of transformations. *Ambio, 31*(5), 437–440. https://doi.org/10.1579/0044-7447-31.5.437

Gaspar, V., Parry, I., & Wingender, P. (2021). *Climate change mitigation: Fiscal policy and financing strategies*. International Monetary Fund. https://www.imf.org/

GCF. (2020). *Green climate fund: Financing projects in vulnerable countries*. Green Climate Fund. Retrieved from https://www.greenclimate.fund

Giddens, A. (2009). *The politics of climate change*. Polity Press.

GCF (Green Climate Fund). (2020). *Annual Results Report 2019*. Green Climate Fund. https://www.greenclimate.fund/

Grubb, M., & Li, W. (2022). Towards net-zero: China's transition in a global context *Nature Energy, 7*, 265–274. https://doi.org/10.1038/s41560-022-00990-2

Gupta, J. (2021). India's role in WTO and sustainable trade practices. *Journal of World Trade, 55*(3), 457–475. https://doi.org/10.1163/9789004415200_009

Gupta, J., & Van Asselt, H. (2019). *Global climate governance: Intergovernmental and transnational responses to climate change*. Cambridge University Press.

Hale, T. (2016). "All hands on deck": The Paris agreement and nonstate climate action. *Global Environmental Politics, 16*(3), 12–22.

Harrison, J., & Smith, F. (2020). Carbon pricing and climate policy: Divergent paths in Canada and the United States. *Carbon pricing and climate policy: Divergent paths in Canada and the United States*(5), 620–634. https://doi.org/10.1080/1523908X.2020.1814132

Holling, C. S. (1973). Resilience and stability of ecological systems. *Annual Review of Ecology and Systematics, 4*, 1–23. https://doi.org/10.1146/annurev.es.04.110173.000245

Holling, C. S. (2001). Understanding the complexity of economic, ecological, and social systems. *Ecosystems, 4*(5), 390–405. https://doi.org/10.1007/s10021-001-0101-5
IMF. (2020). *Climate macroeconomic capacity development*. International Monetary Fund. Retrieved from https://www.imf.org
IMF. (2021a). *Climate action and the IMF: How we support developing countries*. International Monetary Fund. Retrieved from https://www.imf.org
IMF. (2021b). *Climate change and fiscal policy*. International Monetary Fund. Retrieved from https://www.imf.org
IMF. (2022a). *Adapting macroeconomic policies for climate resilience: The IMF's perspective*. International Monetary Fund. Retrieved from https://www.imf.org
IMF. (2022b). *Resilience and sustainability trust*. International Monetary Fund. Retrieved from https://www.imf.org
IMF. (2023). *Climate change policy assessments*. International Monetary Fund. Retrieved from https://www.imf.org
Indian Ministry of Power. (2023). Annual report 2022–23. Government of India. https://powermin.gov.in/
IPCC. (2021). *Climate change 2021: The physical science basis*. IPCC.
Kates, R. W., Parris, T. M., & Leiserowitz, A. A. (2005). What is sustainable development? Goals, indicators, values, and practice. *Environment: Science and Policy for Sustainable Development, 47*(3), 8–21. https://doi.org/10.3200/ENVT.47.3.8-21
Lagarde, C., & Gaspar, V. (2019). *Fiscal policy for climate action*. IMF Blog.
López, H., & Ramirez, R. (2021b). Bangladesh's experience with IMF financing for climate resilience. *Journal of Development Finance, 9*(4), 221–240.
López, R., & Ramirez, J. (2021a). The impact of IMF support on climate finance and policy in Colombia. *Development Policy Review, 39*(4), 583–601. https://doi.org/10.1111/dpr.12525
McEachern, M., & Roberts, J. (2019). Green consumer behaviour: Exploring the influences of social norms and environmental values. *Journal of Business Research, 68*(9), 1883–1891. https://doi.org/10.1016/j.jbusres.2015.01.014
Meadows, D. H. (2008). *Thinking in systems: A primer*. Chelsea Green Publishing.
Ministry of Environment, Forest and Climate Change. (2022). *India's updated first nationally determined contribution under Paris Agreement*. Government of India. https://moef.gov.in/
Mol, A. P. J., & Spaargaren, G. (2000). Ecological modernization theory in debate: A review. *Environmental Politics, 9*(1), 17–49. https://doi.org/10.1080/09644010008414512
Mwaura, F., & Ong, M. (2020). Assessing the impact of green climate fund projects on climate resilience in Kenya. *Climate Policy, 20*(2), 205–220. https://doi.org/10.1080/14693062.2019.1670371
Obergassel, W., Arens, C., Beuermann, C., & Hermwille, L. (2020). *Paris agreement: Five years on*. Wuppertal Institute for Climate, Environment, Energy.
Reed, M. S. (2008). Stakeholder participation for environmental management: A literature review. *Biological Conservation, 141*(10), 2417–2431. https://doi.org/10.1016/j.biocon.2008.07.014
Rockström, J., Steffen, W., Noone, K., Persson, Å., Chapin, F. S., III, Lambin, E. F., & Foley, J. A. (2009). A safe operating space for humanity. *Nature, 461*(7263), 472–475. https://doi.org/10.1038/461472a
Roder, W., & Wangdi, K. (2014). Carbon neutrality and sustainable development in Bhutan. *Environmental Science & Policy, 38*, 228–237. https://doi.org/10.1016/j.envsci.2013.12.005
Sharma, D., & Ghosh, S. (2022). Renewable energy policy in India: An overview. *Energy Policy, 158*, 112564. https://doi.org/10.1016/j.enpol.2021.112564
Stern, N. (2015). *Why are we waiting? The logic, urgency, and promise of tackling climate change*. MIT Press.
UK Government. (2021). *Net Zero Strategy: Build back greener*. Department for Business, Energy & Industrial Strategy. https://www.gov.uk/government/publications/net-zero-strategy
UNDP. (2020). *Climate change adaptation in developing countries: Lessons from the field*. United Nations Development Programme. Retrieved from https://www.undp.org

UNEP. (2020). *Annual report*. United Nations Environment Programme. Retrieved from https://www.unep.org
UNEP. (2021a). *Financing climate resilience in developing countries: The role of international mechanisms*. United Nations Environment Programme. Retrieved from https://www.unep.org
UNEP. (2021b). *Adaptation Gap Report 2021: The gathering storm—Adapting to climate change in a post-pandemic world*. United Nations Environment Programme. Retrieved from https://www.unep.org/
UNEP. (2022). *Emissions Gap Report 2022: The closing window—Climate crisis calls for rapid transformation of societies*. United Nations Environment Programme. https://www.unep.org/
UNFCCC. (2015). *The Paris agreement*. United Nations Framework Convention on Climate Change. Retrieved from https://unfccc.int
United Nations. (2015a). *Paris agreement*. United Nations Framework Convention on Climate Change. Retrieved from https://unfccc.int
United Nations. (2015b). *Sustainable development goals*. Retrieved from https://sdgs.un.org/goals
United Nations. (2023). *Sustainable development goals progress report*. Retrieved from https://www.un.org/sustainabledevelopment
United Nations Global Compact. (2021). *The ten principles*. Retrieved from https://www.unglobalcompact.org
United Nations Global Compact. (2023). *New business initiatives for climate action*. Retrieved from https://www.unglobalcompact.org
Verma, R., & Patel, S. (2023). Evaluating the impact of green climate fund projects in India. *Environmental Development, 45*, 100–113. https://doi.org/10.1016/j.envdev.2022.100734
World Bank. (2021). *orld development report 2021: Data for better lives*. World Bank Publications. https://www.worldbank.org/
World Bank & IMF. (2022). *Global economic prospects: Climate resilience and sustainable growth*. World Bank Group & International Monetary Fund. Retrieved from https://www.worldbank.org
World Commission on Environment and Development. (1987). *Our Common Future*. Oxford University Press.
Wüstenhagen, R., & Menichetti, E. (2012). Strategic choices for renewable energy investment: Conceptual framework and opportunities for further research. *Energy Policy, 40*, 1–10. https://doi.org/10.1016/j.enpol.2011.06.050
WTO. (2019). *Committee on trade and environment*. World Trade Organization. Retrieved from https://www.wto.org
WTO. (2020a). *Technical barriers to trade (TBT) agreement*. World Trade Organization. Retrieved from https://www.wto.org
WTO. (2020b). *The role of trade policy in combatting climate change*. World Trade Organization. Retrieved from https://www.wto.org
WTO. (2021a). *Environmental sustainability and trade agreements: Exploring synergies*. World Trade Organization. Retrieved from https://www.wto.org
WTO. (2021b). *Environmental goods agreement*. World Trade Organization. Retrieved from https://www.wto.org
WTO. (2022a). *Sustainable development work program*. World Trade Organization. Retrieved from https://www.wto.org
WTO. (2022b). *12th ministerial conference (MC12) outcomes: Fisheries subsidies*. World Trade Organization. Retrieved from https://www.wto.org
WTO. (2023). *Trade and environmental sustainability structured discussions (TESSD)*. World Trade Organization. Retrieved from https://www.wto.org
Zeng, K., & Chang, C. (2017). Environmental goods negotiations: Why they stalled and how they might be revived. *World Trade Review, 16*(4), 525–544.

Chapter 11
From Carbon Trading to Climate Actions: Advancing Carbon Neutrality Through Market Mechanism

Kishore Kumar and Rubee Singh

Abstract This book chapter aims to provide a panoramic view emerging notion of carbon trading (CT), a critical mechanism to address climate change concerns by incentivizing the reduction in carbon emissions and promoting green investments. It includes the meaning of the carbon market (CM), CT, and various participants of the market and elucidates the carbon market evolution and its underlying principles. Furthermore, it provides an understanding of the mechanism of the carbon market, the concept of carbon credit (CCs), and carbon offsetting (CO) and delves into the intricacies of carbon taxation as a measure to curb emissions. Finally, this chapter concludes by elaborating the operations of the compliance carbon market (CCM) and the voluntary carbon market (VCM), and their crucial role in fostering the transition to a low-carbon sustainable world.

Keywords Cap & trade (C&T) · Carbon credits (CCs) · Carbon footprint · Carbon market · Carbon offsets · Carbon taxes · Climate change · Emission trading system (ETS) · Global warming · Greenhouse gas emissions (GHG) · Net zero economy · Sustainable development · Voluntary carbon market (VCM)

11.1 Introduction

In response to escalating challenges surrounding sustainable development and combating an existential crisis of global warming and climate issues, the concept of the 'carbon market' has given a novel perspective in approaching environmental stewardship. The notion of carbon market considers the key mechanism that aims to

K. Kumar (✉)
School of Management, Bennett University, Greater Noida, India
e-mail: akishore001@gmail.com

R. Singh
Institute of Business Management, GLA University, Mathura, India
e-mail: dr.rubeerajput@gmail.com

R. Singh and D. Crowther (eds.), *Climate Neutrality and Sustainable Ecosystems*, Approaches to Global Sustainability, Markets, and Governance,
https://doi.org/10.1007/978-981-96-7181-6_11

adopt a more inclusive and balanced worldview that embraces environmental, social, and governance (ESG) considerations (Kumar et al., 2023; Singh, Gupta, & Kandpal, 2024). As the planet Earth is grappling with the consequences of CC, and the urgent need to call for action to reduce GHG emissions, the carbon market emerges as a robust and dynamic mechanism in speeding the transition to a low-carbon economy (LCE) and investment in clean energy (Zhou & Li, 2019). By monetizing the carbon emissions, the carbon market encourages the entities to invest in green projects and energy efficiency alternatives. The carbon market fosters international cooperation by providing a platform for countries to trade carbon credits as a collective effort towards carbon neutrality (Singh, Gupta, Bajpai, & Kandpal, 2024). Carbon market also enables transparency and accountability by enhancing credibility among stakeholders (Sharma et al., 2024). Furthermore, the present chapter provides a valuable insight into the novel notion of the carbon market, its evolution, how the carbon market mechanism works, and trends in the global carbon market. It also presents the discussion on the fiscal and ecological benefits of carbon trading, key participants, and various forms of carbon market that exist in the world.

The following chapter is structured as follows: The next part provides an understanding of concept of CM. Section 11.2 discusses carbon trading system and its benefits. Section 11.3 provides landscape of global CM and carbon taxation. Section 11.4 discusses various types of CM. Section 11.5 provides concluding remarks.

11.2 Understanding Carbon Market

As we stand at the crossroads of economic feasibility and environmental responsibility, it becomes imperative to understand the instrument of the carbon market. Sneaking from the balconies of international policies to the brisking trading platforms of CM, this chapter aims to equip the readers with insights into the CM landscape. To understand the carbon market, it is important to understand these two words individually, carbon is a chemical element, a constituent of huge masses of carbonate rock, present in all life forms. A market is where buyers and sellers of certain goods and services meet (Fageda & Teixidó, 2022; Kumar et al., 2024; Singh, Ravesangar, Narayanan, & Singh, 2024). Thus, the CM is a financial market in which CCs are bought and sold. The CCs are the permits or leverage given to the clients to emit a stipulated level of GHGs. This mechanism across various nations aims to reduce GHG emissions by issuing CCs (van der Gaast et al., 2018).

Various participants engaged in the carbon markets in various capacities. The key players include the providers who provide carbon offsets at domestic and international levels, and the consumers including the traders, retail buyers, and suppliers who can be any entity ranging from government, universities, colleges, and non-governmental organizations. Firstly, governments across various nations set targets aiming to reduce emissions and regulatory bodies create the rules that will govern the trading mechanism and ensure its due implementation and compliance by the companies and industries (Gao, 2023; Mishra et al., 2023). These companies and industries become the major participants as they are at the forefront of emitting greenhouse gases. In pursuit of offsetting their emissions and complying with the laws, companies purchase carbon credits. To trade in these carbon credits, financial institutions like banks and other investors play a crucial role in providing finance to facilitate the flow of activities in carbon markets as well as to fund projects aiming to reduce GHG emissions (Gokhale, 2021; Singh, Kumar, & Khan, 2024). Carbon project developers develop such projects that aim to curb the impact of greenhouse gases such as reforestation projects and renewable energy projects and thereafter the carbon exchange facilitates the trading of the generated carbon credits. Carbon exchange provides a platform just like any other stock exchange that enables the purchase and sale of carbon credits (Calel, 2013; Sharma et al., 2024) by businesses and retail buyers. Retail buyers are the individual entities that purchase carbon credits like any other commodities. They purchase carbon credits as part of their commitment to offset their GHG emissions. The overall functioning of the carbon market is indirectly monitored by environmental advocacy groups and NGOs which organize various events to promote participation in the carbon market and ensure sustainable business practices (Agrawal & Kumar, 2024; Ji et al., 2018). More and more institutions are joining the carbon market to hedge against the risks and to set their net-zero targets (Hernandez-Cortes & Meng, 2023).

11.3 Carbon Trading (CT)

CT, also known by the name GHG trading, is a cap & trade (CAT) mechanism that functions on the principle of creating a conducive space to deal with issues related to carbon emissions. The mechanism lays its foundation by establishing a cap or ceiling limit on the carbon dioxide emissions permissible in a certain industry or area. Certain permits are allocated to companies demonstrating a right to emit a stipulated level of GHGs. If the emission by a firm succeeds the total allocated level, it can sell the surplus to the firms that have exceeded emission limit (Joshi et al., 2024; Liu et al., 2023). Thus, the motivation to reduce carbon emissions comes with lucrative financial incentives in the form of sale proceeds from surplus credits. Companies surpassing their stipulated thresholds for carbon emissions are liable for penalties, resulting in increased costs. The agenda of carbon trading is to gradually curb the overall emissions over some time by acting as a catalyst for environmental protection.

The concept of carbon credits is not only limited to countries or giant industrial plants, rather it can also be enjoyed by other organizations, businesses, or even an individual. In the carbon market, the role of government is very crucial as it stands at the core of creating regulatory frameworks, setting targets for carbon emissions, distributing the permits as well as monitoring their execution (Balmford et al., 2023). The government creates a set of prerequisites for the carbon market to function efficiently, ensuring integrity, fairness, and transparency. Corporations and businesses participate in the CM in various capacities like traders, offset developers, retail buyers, etc. They constantly strive to achieve low carbon emission targets, comply with legal requirements, and leverage trading opportunities. Carbon project developers are involved in innovating and investing in green projects to optimize their carbon footprint. Individuals are contributing in their unique ways by making sustainable choices in their consumption patterns, putting forth their demand for low-carbon products, investing in cleaner projects, and advocating to strengthen climate policies (Wang et al., 2023). United Nations Development Programme (UNDP) has also stated in its initiative 'Climate Promise' that emissions are now monitored and traded just like other commodities. All these entities rely on collaborated efforts to offset their carbon footprint to contribute to living up to a pledge to conserve the environment.

11.3.1 Benefits

Embracing carbon neutrality and striving for a low-carbon economy can bring a plethora of benefits to the environment as well as to the economy. Through a substantial reduction in greenhouse gas emissions and adopting a proactive approach by introducing renewable energy, clean technologies could reduce the unfavorable impact of industrial activities on nature and climate. Attaining the goal of carbon neutrality will not only ensure cleaner air, water, and soil but also protect the biodiversity and its ecosystem (Khan et al., 2024; Kumar et al., 2023). Moreover, transitioning towards non-conventional energy sources and increasing energy efficiency will lessen the dependency on fossil fuels. It would curb pollution and environmental deterioration while building resilience against climate catastrophes. The benefits derived from the carbon market go beyond environmental considerations and reflect on the economy as well in the form of upsurged investments in clean energy projects, development of sustainable infrastructure, and innovation in green technologies (Ma et al., 2024). Moving towards a low-carbon economy ensures uninterrupted accessibility and affordability of energy and shrinks the demand for fossil fuels (Fageda & Teixidó, 2022). The carbon market gives opportunities to its participants to enjoy financial gains by trading in carbon credits like any other commodity. This financial incentive encourages the business to transition to eco-friendly projects that uplift local communities as well, thus fostering broader societal and economic development (Kreibich & Hermwille, 2021; Mishra et al., 2023). Furthermore, by positioning the industries at the leading edge of sustainability practices, nations demonstrate a potential to leverage international competitiveness, stimulating investments and steering innovation (Anke et al., 2020; Lin & Huang, 2022). Overall, the benefits of embracing carbon neutrality surpass immediate financial gains, significantly bolstering prosperity and resilience.

11.3.2 Evolution of CM

The concept of the CM has evolved in the last three decades as a response to the growing concerns over environmental calamity and climate change. The journey of the CM has seen pivotal milestones in crafting its growth trajectory. One of the earliest events that marked the initial discussion about the impact of human activities on the environment was the Brundtland Report in the year 1987, which gave prominence to sustainable development as a global priority. Subsequently, the Montreal Protocol brought attention to the issue concerning ozone depletion and the need to protect its stratospheric layer by moving away from the production and utilization of ozone-depleting substances (Kumar et al., 2024; Prakash et al., 2023). In 1988, a conference organized in Toronto collaborated with the efforts of scientists and policymakers to discuss climate change and established the IPCC. In 1992, Rio de Janeiro (Brazil) witnessed the United Nations Framework Convention on Climate Change (UNFCCC) conception during the Earth Summit. This treaty formally marked the beginning of efforts made to stabilize GHG concentrations and address CC issues. The year 1997 achieved a significant milestone in the growth trajectory of the CM when 84 global economies signed the Kyoto Protocol which operationalized the UNFCCC (Verde, 2020). The signatories of the Kyoto Protocol made a global promise to curtail and stop GHG emissions as per their respective individual targets. The Kyoto Protocol was implemented in the year 2005 and demonstrated the ease of carbon trading as a mechanism for reducing emissions. This led to the formation of the International Emissions Trading Association (IETA) in 1999, sponsored by UNCTAD. IETA is an independent industry-driven organization advocating market-based approaches in trading greenhouse gas emissions among its members.

The VCM was launched in the year 2003 by the Chicago Climate Exchange (CCX), contributing over 700 million tons of GHS emissions. The CCX ceased its operation in CCs in 2010 due to the inactivity of the US carbon markets. In 2005, 25 countries in the European Union devised the ETS to combat issues about GHG emissions and CC. The EU ETS is the cornerstone, expanding its sector from power plants, carbon-intensive industries, and shipping to nitrogen oxide and methane (Pietzcker et al., 2021). ETS, the world's first and biggest CM works on the principle of 'cap & trade', and is the most proficient policy of the European Union to address GHG emissions. In 2012, The UNCCC in Doha, Qatar reaffirmed the promises made in the Kyoto Protocol regarding emission reduction targets. The Paris Agreement of 2015 marked another significant milestone when 196 parties participated in the UNCCC (COP21) and adopted the central aim to limit the rise of world temperature below 2 °C. In 2016, CORSIA was formed to offset sector-based CO_2 emissions. The COP26 convened in Glasgow and brought together world leaders with a record 40,000 registered attendees, captivating the wide attention on every aspect of climate change-scientific insights, potential solutions, political outlook, and a clear commitment to action. As of 2024, there exist about 40 compliance carbon markets worldwide, operating on the Cap-and-trade principle, and collectively covering about one-fifth of emissions, valued at USD 850 billion. Additionally, VCM has emerged at a value of USD 2 billion. Global Carbon crediting standards are being set, aiming to bring standardization and enhance credibility within voluntary carbon markets. While various nations such as Switzerland, Ghana, Sweden, and Japan have secured article 6 of the Paris Agreement, allowing them to voluntarily cooperate to frame emissions reduction targets, a formal authorization for carbon credit transfer is yet to occur (Michaelowa et al., 2019; Nwokolo et al., 2023).

11.4 Global Carbon Market

The CM is seen as a panacea to CC and global warming issues globally. Several regions and countries around the world have launched carbon emission trading systems (CETS) to cap GHG emissions and promote businesses to invest in clean energy and meet climate-related targets. Presently, 40 countries around the world have put a price on carbon footprint (United Nation Climate Change, 2024). The steep growth of the global CM in the last few decades is due to the increasing number of ETS worldwide and is estimated to be 527.3 USD billion by 2024. It is estimated to increase at a compounded annual growth rate of 31% during the period from 2024 to 2028 and will reach 1602.7 USD billion (Statista, 2024). This unprecedented growth of the global carbon market will be primarily driven by the increased impetus of the world's economies on net zero goals and investment in clean energy technologies. It could play a critical role in curtailing GHG emissions using a financial market mechanism and help the countries to transition to LCE. It is estimated that credit trading in the CM could result in a reduction of 50% of emissions and 5 gigatons of GHG emissions each year by 2030 (United Nation Climate Change, 2024).

The European Union ETS (EU ETS) is the largest and most valuable carbon market in the world, accounting for almost 85% of the global CM in the year 2024. It was established in 2005 and since then it has been able to curtail the carbon emissions of industries by 37% (European Commission Climate Action, 2024). It covers a wide range of emission activities including carbon footprint from the maritime and air transport sectors. In 2023, carbon price reached up to 100 euros/ton in the EU ETS for the first time in history since its establishment indicating an ever-growing interest of the investors and unprecedented investments in the market. Presently, more than 10,000 installations from various power and manufacturing industries are covered by EU ETS and have generated 152 billion euros since 2013 (European Commission Climate Action, 2024). The second largest emission trading system (ETS) is of the United Kingdom (UK), which was launched in the year 2021. In such a short span, it covers more than one-fourth of the total territorial carbon footprints in the UK and 1000 installations. It has a 7% share of the global carbon market followed by a 6.5% share of North America's Western Climate Initiatives (WCI).

11.4.1 *Mechanisms of the CM*

The Cap & trade (C&T), a mainstay of the carbon market, thrives on the two basic principles revolving around fixing a cap (limit) on the overall permissible emissions and facilitating the trading of excessive unused limits. Initially, the C&T is established by the government or regulatory bodies that aim to limit the total level of emissions of certain toxic substances, particularly carbon dioxide as an outcome of industrial residue, within a defined sector or jurisdiction. C&T programs are also known as ETS usually done to align with the goals related to climate change and environmental protection. In this system, an allowed limit of emissions equivalent to the cap is then granted, which gives the holder a right to emit a certain amount of GHG, particularly carbon dioxide.

The CM participants typically industries and companies subjected to these regulations are either allocated or required to purchase these permits. These participants are the key players in the C&T initiative intended to minimize GHG emissions. No business entity needs to consume all of its carbon credits, the excess unused carbon credits can be sold to those entities which might exceed their usage limits (Pietzcker et al., 2021). Conversely, if any entity exceeds its consumption limit, it must purchase additional credits to cover its excess emissions (Abrell et al., 2022). Additionally, business entities can create offsets or carbon credits depending upon their usage, they can either reduce or remove carbon dioxide, which can be further sold to other business entities, thus creating a financial incentive to mitigate the GHG emissions (Khan et al., 2024; Liu et al., 2023). Gradually, this cap is made more stringent to reduce the overall emissions, incentivizing the reduction and promoting investments in cleaner technologies. This mechanism of the carbon market is dynamic; by reducing the total emissions, it fosters an accommodative environment for sustainable growth and innovation.

11.4.2 *Carbon Credits (CCs)*

A CC is referred to as a basic unit to measure GHG emissions, a carbon credit is equivalent to one metric ton of CO_2. In the CM, these credits act as tradeable commodities that not only incentivize emissions but also promote sustainable business practices. These credits are generated through projects that aim to mitigate or avoid GHG emissions, such as afforestation and renewable energy projects or incorporating energy-efficient techniques. Companies that are going beyond their carbon emission limits can purchase CCs to offset their extra emissions, whilst entities with surplus credits can sell their credits and grab the extra financial benefits. CCs are facilitating and supporting the oath taken by global leaders to mitigate climate change by investing in low-carbon technologies and fostering collaboration among various stakeholders.

11.4.3 Carbon Offsetting

Carbon offsetting is a process through which businesses, organizations, and even individuals can be indemnified for the emissions they have incurred by investing in lower greenhouse gas emission projects. The principle behind the entire mechanism of carbon offsetting is to arrive at a stage of carbon neutrality by stabilizing the emissions (Maciel-Seidman et al., 2024; West et al., 2023). Companies can purchase carbon credits from projects that aim to reduce GHG emissions, such as energy-efficient projects, renewable energy projects, or methane capture projects, thereby offsetting their carbon emissions (Probst et al., 2023; Siddique et al., 2023). These projects not only mitigate the negative environmental impact but also generate CCs which facilitate the reduction of GHG emissions. Carbon offsetting is the core of the carbon market, companies can proactively engage themselves in addressing the issues about CC and reflecting a shared commitment towards sustainability.

11.4.4 Carbon Taxation

Carbon taxation is a system through which taxes are levied on the businesses or companies that are employing production techniques that require the combustion of fossil fuels. Carbon taxation, a tool in the hands of policymakers aimed at deterring the burning of fuels, plays a central role in minimizing the impact of negative externalities. The carbon taxation system is a policy that seeks to incentivize individuals and businesses to reduce their carbon footprint (Chien et al., 2023; Xu, Pan, et al., 2023). Carbon taxes affect the external costs due to carbon emissions by simply imposing a tax on the proportion of carbon in fossil fuels or directly on the emissions. The concept is based on the idea that a business burning fossil fuels in its production process tends to produce more carbon dioxide, the profits generated from sale proceeds for such products or services are solely enjoyed by the business whereas the negative impacts will be borne by the society later (Gokhale, 2021). By imposing a tax on such businesses, the cost of producing such products becomes high, increasing its price and diminishing demand. Thus, it encourages businesses to divert towards cleaner technologies, wind energy, and solar energy in place of fossil fuels. The revenue generated from carbon taxes can be further reinvested in eco-friendly projects, energy-efficient techniques, or even offset other taxes such as corporate tax. Certain underlying principles lay down the foundation of robust carbon taxation. Carbon taxation should cover a broader range of activities under its purview that are contributing to carbon emissions, and carbon taxation should substitute an already existing regulation to minimize the complexities. The tax revenue generated from carbon emissions could be used to fund the projects of green industries (Chien et al., 2023; Kumar et al., 2024). The successful implementation of carbon taxation poses some significant challenges for the industries, depending on factors such as increased costs, accessibility of alternative techniques, level of taxation, exemptions available, etc. However, carbon taxation can also promote investment and innovation in low-carbon projects, gradually leading to enhanced energy efficiency, reduced carbon emissions, and competitiveness in global markets. Carbon taxation can foster positive environmental outcomes and drive the transition towards an LCE (Khan et al., 2024).

11.5 Types of Carbon Market

Two forms of carbon market exist globally. These are the compliance carbon market (CCM) and voluntary carbon market (VCM). CCM is popularly known as regulatory CM. It is based on the mandatory approach where organizations must account for their carbon emissions. Compliance carbon market is created and regulated by the governments or multi-government bodies as a means for achieving climate goals (Liu et al., 2023). Under this carbon market, organizations are mandated by law to participate in the market and to meet stipulated emission-related targets. The government body controls the supply of carbon credit and regulates its trading (Xu, Solangi, & Wang, 2023). Companies with large carbon footprints are required to buy a certain number of CCs each year. These CCs could be obtained by investing in clean energy projects or from the companies that achieved their GHG reduction targets (Spilker & Nugent, 2022). The compliance/regulatory carbon market is commonly referred to as the ETS. The EU ETS is one of the most common examples of compliance carbon market. The mandatory approach to the carbon market is increasing rapidly, many countries like Australia, Brazil, China, India, and South Korea have already established CM in their country. These regulatory CMs play an important part in GHG targets and encourage environmental responsibility through regulatory mechanisms.

The other form of the carbon market is the voluntary carbon market (VCM) where the participation of organizations is voluntary and not driven by the regulatory obligation. VCMs have grown organically to address the needs of organizations that aim to manage and curtail carbon footprints. CCs traded in the VCM are known as verified emission reductions (VER), which are primarily bought by large carbon-emitting organizations. It allows organizations to offset their carbon footprints voluntarily. Under this market, companies could voluntarily buy and sell carbon credits obtained through greenfield projects (Ranjan, 2024). Corporate social responsibility, ethical consideration, and carbon footprint reduction are some of the key drivers for organizations to buy carbon credits from VCM. This market functions independently from the regulatory carbon market, carbon credits traded in the VCM could not be used to meet the regulatory emission targets of organizations imposed by the regulatory carbon market (Kreibich & Hermwille, 2021). To manage the growing number of clean energy projects and carbon credits, various accreditation bodies have been established to ensure accountability and transparency in the VCM. It has its own accreditation bodies, standards, and registries beyond the purview of the regulatory carbon market.

11.6 Conclusion

CM has emerged as a critical tool to advance global efforts to transition to a low-carbon world economy. The unprecedented growth of the CM globally demonstrates the rapidly growing interest and investments in the sector by the various stakeholders. Global initiatives like the Paris Agreement 2015 have ushered international cooperation to foster a net-zero carbon economy, combat CC, and develop ETS to enable cross-border carbon trading. Many countries around the world are formulating policies for carbon pricing such as C&T systems, carbon taxes, etc., to encourage companies to invest in clean energy and curtail carbon footprints. Carbon capture, storage, and removal have become more cost-effective due to advancements in technologies. Advanced technologies like AI and blockchain are playing an important role in bringing transparency and accountability to carbon credit trading. However, the prospect of the carbon market sector also faces some challenges. Presently, the global carbon market landscape is fragmented and dominated by the voluntary approach. There has been consistent criticism from certain sections of society regarding climate benefits and net-zero efforts of companies through the carbon credit approach. The CC mechanism may involve potential misuse by companies seeking to sidestep net-zero practices in their business operations. The VCM largely stands unregulated and marred with transparency issues. Various environmental groups and NGOs frequently raise questions regarding the integrity and validity of carbon cap and trade systems. They argue that certain companies deliberately use these carbon markets to avoid putting true effort into carbon footprint reduction. Despite these issues and concerns surrounding the carbon market, it has witnessed tremendous growth in the recent past as a key solution to tackle CC and reduce GHG emissions. It is imperative to bring collaboration among the various stakeholders to bring efficiency to the CM and achieve global CC goals.

References

Abrell, J., Cludius, J., Lehmann, S., Schleich, J., & Betz, R. (2022). Corporate emissions-trading behaviour during the first decade of the EU ETS. *Environmental and Resource Economics, 83*(1), 47–83.

Agrawal, D., & Kumar, K. (2024). Achieving SDGs through MSMEs: An empirical assessment of environmental consideration initiatives in India. In *Green transition impacts on the economy, society, and environment* (pp. 358–376). IGI Global.

Anke, C. P., Hobbie, H., Schreiber, S., & Möst, D. (2020). Coal phase-outs and carbon prices: Interactions between EU emission trading and national carbon mitigation policies. *Energy Policy, 144*, 111647.

Balmford, A., Keshav, S., Venmans, F., Coomes, D., Groom, B., Madhavapeddy, A., & Swinfield, T. (2023). Realizing the social value of impermanent carbon credits. *Nature Climate Change, 13*(11), 1172–1178.

Calel, R. (2013). Carbon markets: A historical overview. *Wiley Interdisciplinary Reviews: Climate Change, 4*(2), 107–119.

Chien, F., Vu, T. L., Phan, T. T. H., Van Nguyen, S., Anh, N. H. V., & Ngo, T. Q. (2023). Zero-carbon energy transition in ASEAN countries: The role of carbon finance, carbon taxes, and sustainable energy technologies. *Renewable Energy, 212*, 561–569.

European Commission Climate Action. (2024). https://climate.ec.europa.eu/eu-action/eu-emissions-trading-system-eu-ets/what-eu-ets_en

Fageda, X., & Teixidó, J. J. (2022). Pricing carbon in the aviation sector: Evidence from the European emissions trading system. *Journal of Environmental Economics and Management, 111*, 102591.

Gao, M. (2023). The impacts of carbon trading policy on China's low-carbon economy based on county-level perspectives. *Energy Policy, 175*, 113494.

Gokhale, H. (2021). Japan's carbon tax policy: Limitations and policy suggestions. *Current Research in Environmental Sustainability, 3*, 100082.

Hernandez-Cortes, D., & Meng, K. C. (2023). Do environmental markets cause environmental injustice? Evidence from California's carbon market. *Journal of Public Economics, 217*, 104786.

Ji, C. J., Hu, Y. J., & Tang, B. J. (2018). Research on carbon market price mechanism and influencing factors: A literature review. *Natural Hazards, 92*, 761–782.

Joshi, A., Singh, R., Kumar, V., & Rani, S. (2024). Net zero strategy: Exploring opportunities, confronting challenges for developing countries. In *Transition towards a sustainable future: Net zero policies and environmental sustainability* (pp. 233–252). Springer.

Khan, S., Singh, R., Alnahas, J., Abbate, S., & Centobelli, P. (2024). Navigating the smart circular economy: A framework for manufacturing firms. *Journal of Cleaner Production, 480*, 144007.

Kreibich, N., & Hermwille, L. (2021). Caught in between: Credibility and feasibility of the voluntary carbon market post-2020. *Climate Policy, 21*(7), 939–957.

Kumar, K., Singh, R., Sharma, A., & Lehri, B. P. S. (2024). Climate change adaptation strategies for achieving net-zero economy. In *Transition towards a sustainable future: Net zero policies and environmental sustainability* (pp. 163–183). Springer.

Kumar, R., Kaushal, S. K., & Kumar, K. (2023). Does source credibility matter in promoting sustainable consumption? Developing an integrated model. *Social Responsibility Journal, 19*(7), 1320–1347.

Lin, B., & Huang, C. (2022). Analysis of emission reduction effects of carbon trading: Market mechanism or government intervention? *Sustainable Production and Consumption, 33*, 28–37.

Liu, B., Ding, C. J., Hu, J., Su, Y., & Qin, C. (2023). Carbon trading and regional carbon productivity. *Journal of Cleaner Production, 420*, 138395.

Ma, J., Yang, L., Wang, D., Li, Y., Xie, Z., Lv, H., & Woo, D. (2024). Digitalization in response to carbon neutrality: Mechanisms, effects and prospects. *Renewable and Sustainable Energy Reviews, 191*, 114138.

Maciel-Seidman, M., Tzankova, Z., Ziegler, C. C., Lele, A., Lu, S., Yan, Y., & Muchira, J. M. (2024). Mobilizing carbon offsetting to reduce energy cost burdens: A new approach for calculating and monetizing the offset value of energy efficiency upgrades to low-income housing. *Frontiers in Energy Research, 12*, 1437560.

Michaelowa, A., Shishlov, I., & Brescia, D. (2019). Evolution of international carbon markets: Lessons for the Paris agreement. *Wiley Interdisciplinary Reviews: Climate Change, 10*(6), e613.

Mishra, P., Sant, T. G., & Kumar, K. (2023). Carbon disclosure and organization performance: A literature review. *Sustainability and Climate Change, 16*(4), 302–317.

Nwokolo, S. C., Singh, R., Khan, S., Kumar, A., & Luthra, S. (2023). Introduction: Africa's net zero transition. In *Africa's path to net-zero: Exploring scenarios for a sustainable energy transition* (pp. 1–13). Springer.

Pietzcker, R. C., Osorio, S., & Rodrigues, R. (2021). Tightening EU ETS targets in line with the European green deal: Impacts on the decarbonization of the EU power sector. *Applied Energy, 293*, 116914.

Prakash, A., Kumar, K., Khan, W., & Siddiquei, M. I. (2023). Self-induced versus structured corporate social responsibility: The Indian context. In *Measuring sustainability and CSR: From reporting to decision-making* (pp. 167–178). Springer.

Probst, B., Toetzke, M., Kontoleon, A., Diaz Anadon, L., & Hoffmann, V. H. (2023). Systematic review of the actual emissions reductions of carbon offset projects across all major sectors. *Research Square*. https://doi.org/10.21203/rs.3.rs-3149652/v1

Ranjan, R. (2024). Balancing greenwashing risks and forest carbon sequestration benefits: A simulation model linking formal and voluntary carbon markets. *Forest Policy and Economics, 168*, 103317.

Sharma, A., Chaudhary, M., Kumar, K., & Daga, S. (2024). Promoting net-zero economy for sustainable development: Practice-based view. In *Net zero economy, corporate social responsibility and sustainable value creation: Exploring strategies, drivers, and challenges* (pp. 101–116). Springer.

Siddique, M. A., Nobanee, H., Karim, S., & Naz, F. (2023). Do green financial markets offset the risk of cryptocurrencies and carbon markets? *International Review of Economics & Finance, 86*, 822–833.

Singh, R., Gupta, A., Bajpai, A., & Kandpal, V. (2024). Exploring challenges of circular economy. In *Smart cities and circular economy: The future of sustainable urban development* (p. 59). IGI Global.

Singh, R., Gupta, A., & Kandpal, V. (2024). ESG excellence in the digital world: Circular economy e-business solutions. In *Fintech for ESG and the circular economy* (pp. 193–204). Wiley.

Singh, R., Kumar, K., & Khan, S. (2024). A comprehensive view of artificial intelligence (AI)-based technologies for sustainable development goals (SDGs). In *Artificial intelligence enabled management: An emerging economy perspective* (p. 183). Walter de Gruyter GmbH.

Singh, R., Ravesangar, K., Narayanan, S., & Singh, K. K. H. (2024). HRM practices in creating net zero: Supporting organizational sustainability. In *Net zero economy, corporate social responsibility and sustainable value creation: Exploring strategies, drivers, and challenges* (pp. 1–18). Springer.

Spilker, G., & Nugent, N. (2022). Voluntary carbon market derivatives: Growth, innovation & usage. *Borsa Istanbul Review, 22*, S109–S118.

Statista. (2024). Retrieved from https://www.statista.com/statistics/1399813/carbon-market-size-share-by-market-worldwide/#:~:text=The%20global%20carbon%20market%20was,billion%20U.S.%20dollars%20in%202022

United Nation Climate Change. (2024). Retrieved from https://unfccc.int/about-us/regional-collaboration-centres/the-ciaca/about-carbon-pricing#What-is-the-current-status-of-carbon-pricing-in-th

Van der Gaast, W., Sikkema, R., & Vohrer, M. (2018). The contribution of forest carbon credit projects to addressing the climate change challenge. *Climate Policy, 18*(1), 42–48.

Vellachami, S., Hasanov, A. S., & Brooks, R. (2023). Risk transmission from the energy markets to the carbon market: Evidence from the recursive window approach. *International Review of Financial Analysis, 89*, 102715.

Verde, S. F. (2020). The impact of the EU emissions trading system on competitiveness and carbon leakage: The econometric evidence. *Journal of Economic Surveys, 34*(2), 320–343.

Wang, H., Li, Y., & Bu, G. (2023). How carbon trading policy should be integrated with carbon tax policy-laboratory evidence from a model of the current state of carbon pricing policy in China. *Environmental Science and Pollution Research, 30*(9), 23851–23869.

West, T. A., Wunder, S., Sills, E. O., Börner, J., Rifai, S. W., Neidermeier, A. N., & Kontoleon, A. (2023). Action needed to make carbon offsets from forest conservation work for climate change mitigation. *Science, 381*(6660), 873–877.

Xu, H., Pan, X., Li, J., Feng, S., & Guo, S. (2023). Comparing the impacts of carbon tax and carbon emission trading, which regulation is more effective? *Journal of Environmental Management, 330*, 117156.

Xu, L., Solangi, Y. A., & Wang, R. (2023). Evaluating and prioritizing the carbon credit financing risks and strategies for sustainable carbon markets in China. *Journal of Cleaner Production, 414*, 137677.

Zhang, Y., & Umair, M. (2023). Examining the interconnectedness of green finance: An analysis of dynamic spillover effects among green bonds, renewable energy, and carbon markets. *Environmental Science and Pollution Research, 30*(31), 77605–77621.

Zhou, K., & Li, Y. (2019). Carbon finance and carbon market in China: Progress and challenges. *Journal of Cleaner Production, 214*, 536–549.

Part III
Role of Different Sectors for Climate Neutrality and Sustainable Ecosystem

Chapter 12
Empowering Role of Private, Public, and Individual Change Makers for Global Warming

Gopika Juneja and Pranav Saraswat

Abstract Global warming is a complex challenge and intervention requires a harmonized approach at the private, public, and individual levels. This paper would analyse the importance of adaptation measures in meeting this challenge of global warming under the emphasis of the efforts exerted by the private corporations, the public institutions, and individual change-makers. Besides, adaptation in the private sector goes about the sustainability strategy aimed through CSR initiatives as well as innovation in green technologies, turning business models to be in line with environmental sustainability. Public adaptation aims to lower current climate risk and increase resilience across sectors, with an emphasis in vulnerable communities, through policy interventions, infrastructure development, and regulatory frameworks. Governments and the international community are the main actors transforming these systems through environmental policies and climate action plans.

At the level of the individual, lifestyle change, advocacy, or grassroots movements can contribute meaningfully towards a low-carbon economy. Important change-makers at the individual level involve environmentalists, scientists, and community leaders. The three adaptive pathways discussed within this paper are eventually interlinked; through mutual synergies, they can accelerate climate action and drive efforts at all levels towards developing holistic solutions that address the existing climate impacts as well as instil long-term resilience.

Keywords Adaptation · Global warming · Private sector · Public institutions · Individual change-makers · Climate resilience · Policy interventions · Grassroots movements · Climate action

---

G. Juneja · P. Saraswat (✉)
Institute of Commerce, Nirma University, Ahmedabad, India
e-mail: drpranavsaraswat@gmail.com

P. Saraswat
Institute of Commerce and Management, Nirma University, Ahmedabad, India

R. Singh and D. Crowther (eds.), *Climate Neutrality and Sustainable Ecosystems*, Approaches to Global Sustainability, Markets, and Governance,
https://doi.org/10.1007/978-981-96-7181-6_12

12.1 Introduction

Now, global warming is giving this earth the constant threat of an existential threat by upping its greenhouse gas emissions. Melting temperatures have effects other than what are normally known to man, like extreme weather conditions, melting of ice caps, sea levels rising, and damage to natural ecosystems. While these changes do impact global economies, public health, food security, and water resources because they alter the physical landscape and create a complex web of hazards that require immediate action with varied solutions, these changes do impact global economies, public health, food security, and water resources. Accordingly, adaptation techniques are as important to overcome the irreversibility's of climate change as mitigation initiatives that target reducing or stopping future greenhouse gas emissions.

Adaptation refers to the process of readjustment, which aims at reducing damage or making the best use of favourable opportunities in ways that can cope with existing or expected climatic influences. Unlike a solitary effort, such change emerges from a collective process by different parties' vital change agents of the public sector, commercial industry, and local communities are becoming increasingly crucial for the application of adaptive solutions for dealing with global warming. This puts together the capabilities, resources, and views from both parties because a comprehensive response to climate change is what each of them presents.

The latter can be leaders in climate adaptation and innovation, provided companies and industries are the largest emitters of greenhouse gases. The biggest energy, agriculture, and infrastructure sectors have been pointing to climate-resilient technology—smart grids, renewable energy, and sustainable farming methods. Initiatives for CSR appeal to the companies to reduce carbon footprints and support weaker regions with adaptation plans of climate change. Thus, the private sector is a powerful agency behind any initiatives towards climate change adaptation because it is capable of resource mobilization, innovation, and mass-level solutions.

Locally, individual actions and local organizations greatly enhance community resilience to climate change (Adger et al., 2009). Most communities rely on adaptive strategies and indigenous knowledge practices in adaptation to the risks of climate change, especially in poor countries. Such examples include conservation agriculture, water harvesting methods, and forestry projects. Lifestyle changes like decreases in carbon emissions, energy-efficient appliance usage, and consumers' prudent buying behaviours would further facilitate the adaptation process. Collective action within communities can make firms become more adaptable even in rural and urban locations, thereby influencing policy in general.

One sector alone cannot influence global warming adaptation. The action needs to be coordinated in the public, private, and individual spheres. Government, corporations, and communities will need to collaborate effectively to work towards developing inclusive and scalable solutions. For example, if infrastructure needs to be built and that is climate resilient, then public-private partnerships might use a combination of public finances and private ingenuity. Similarly, the pressure from public and community-based mobilization may influence the provision made in legislation that persuades businesses to be more sustainable-friendlier in practice. Open adaptive measures should also include vulnerable groups who sometimes are targeted by climate change into effective instruments or support needed for adaptation. The study would consider how individuals, public institutions, and private organizations contribute to the global effort at adapting to ongoing impacts of climate change. Based on case studies as a base, best practice benchmarks, and interdependence among different initiatives, the study would look for deeper understanding of how these catalysts for change are affecting the global response to climate adaptation. At the ultimate level, it thus reveals the hidden kinds of partnerships, laws, and innovations needed to securitize a sustainable future in a warming world, to build resilience, and to reduce vulnerabilities. It is hard to accommodate change, but it would give a chance for revolutionary change through the joint action of all segments of society.

Currently, global frameworks on climate adaptation are being designed at an international level through key organizations such as the United Nations Framework Convention on Climate Change (UNFCCC) and the Intergovernmental Panel on Climate Change (IPCC) (Intergovernmental Panel on Climate Change (IPCC), 2014). Within the country, mechanisms like the Paris Agreement spur nations to improve their adaptation capabilities, knowledge sharing, and financial aid to developing nations. In the Global South, the adaptation resource is less available, and the effect of climate change is usually disastrous; therefore, the public sector plays a vital role in achieving funding for adaptation programs (Tschakert et al., 2019). The government also has to make sure that adaptation strategies are inclusive of vulnerable communities at risk from climate change (Olsson et al., 2020).

Climate adaptation is not the individual responsibility of the public, corporate, or private individual but needs collaboration from all parties involved. Complex global warming issues cannot be easily solved by any single person working individually; adaptation, in fact, is a collective responsibility. In this regard, PPPs enable business and government to collaborate over pooling their funds and knowledge bases to set up gigantic adaptation projects, such as building climate-resilient infrastructures. Government policies can act as a further incentive for private enterprises to invest in sustainability technology and practices. Last but not least, individuals and community groups can give lawmakers true insight into public opinion by sharing their personal experience. With public institutions, the business sector, and individuals engaged together, societies can enhance their responses to the broader challenges presented by climate change. As some of the most vulnerable groups often experience the largest risks related to global warming, adaptation methods need to be inclusive and equitable. It is possible to build a more resilient and sustainable future through cooperation, ingenuity, and group effort-even as global warming increasingly changes the planet.

12.2 Objectives

The study would be greatly helpful in taking a deeper study to determine the respective roles and contributions of the stakeholders in dealing with global warming, starting from commercial companies, public institutions, and individual changemakers. The project is expected to consider the creative modes, policymaking, and behavioural shifts that are made by these communities in pursuit of adaptation to climate-related problems. It looks into efficient tactics, teamwork, and how these changes affect cutting carbon emissions and building climate resilience. The research intends to provide insights into how various sectors might collaborate to build a sustainable future by showcasing case stories and empirical facts.

In addition, it will try to evaluate whether present adaptation strategies are effective and to find out areas of potential betterment, especially in the areas of policy implementation and resource allocation. The overall objective would be to establish best practices that might be replicated or scaled up in other locations by studying the effectiveness of various initiatives. The research will also examine the socio-economic impact of the adaptation strategy; that is, what support adaptation brings to marginalized communities and what else may be necessary to make sure that adaptation plans are equitable.

12.3 Literature Review

The past 20 years have seen a substantial evolution in the literature on climate change adaptation as ecosystems, economies, and societies have been impacted by the intensifying effects of global warming. Academics have been paying more and more attention to the different players that are participating in adaptation efforts, especially in the governmental sector, the commercial sector, and grassroots or individual movements. A review of the literature on theoretical underpinnings and empirical research into the possible roles of such players in climate adaptation, problems with such roles, and potential spaces for cooperation—Polevova. It focuses on how innovations, governance, and legislative frameworks can support effective adaptation tactics.

It is within such crucial theoretical frameworks that the study of climate adaptation takes an anchor in explaining how civilizations respond to environmental stresses. Resilience theory is one of the most widely applied theories in an attempt to elucidate how economies, ecosystems, and societies adapt to climate change. "Resilience" was first coined in ecological systems by Holling (1973), and has since been extended to social-ecological systems that characterize the capacity of the system for resistance to shocks while maintaining its basic functions. More elaborately (Walker et al., 2004) by talking about adaptive capacity that is the potential of people, groups, and institutions to adapt to climate hazards, lessen harm, and seize opportunities.

The vulnerability framework, as articulated by Adger (2006), is another very relevant concept in the adaptation literature; it concentrates on the exposure that various populations face to climate threats and their capacity to respond. Socio-economic, cultural, and institutional factors all shape vulnerability, making it therefore highly context-specific. This approach places much more emphasis on how governance may reduce susceptibility to and exposure to the impacts of climate change. The need for adaptive practices with equity is another highlighted issue since often underprivileged communities are not well-equipped politically or financially to adapt successfully.

To meet the scope and urgency of climate change, transformational adaptation strategy calls for more systemic changes in the system, policy, and behaviour that Kates et al. (2012) call for. For addressing long-term climate threats, this strategy strengthens the limitations of incremental adaptation initiatives and makes the need for systemic changes across various sectors. The private sector's ability to innovate, invest, and have an impact on resource management has made it increasingly acknowledged as a vital participant in climate adaptation. The United Nations Environment Programme (UNEP) (2016) reports that companies are progressively integrating climate risk into their operations, especially in industries like real estate, insurance, and agriculture that are extremely susceptible to the effects of climate change. The literature lists a number of ways that the private sector helps with climate adaptation.

Technological Innovation: Private sector would make crucial contributions that would lead to strong climate-resilient technologies. For example Olesen et al. (2012), suggest that the investments undertaken by farming enterprises in precision farming technologies and drought-tolerant crops can mitigate the impact of climate change on food security. And in fact, construction sector creates an important climate-resilient infrastructure, including flood-proof buildings and smart cities keeping in mind climate hazards while incorporating the culture of urban planning, as said by Jones and Olsson (2017).

Many companies have lately started to include climate adaptation projects in their lists of CSRs. Actually, as Aguinis and Glavas (2012) mentioned, companies like Microsoft and Unilever are spending on programs that ensure community resilience to climate change, hence why their CSR initiatives are becoming more environmentally sustainable day by day. Public-private partnerships, or PPPs, will be able to offer the space through which large-scale adaptation projects can efficiently be funded and enacted. According to Filho et al. (2020), PPPs can expedite climate-resilient infrastructural development through the use of public funds with private sector expertise. Companies are, therefore, integrating climate risk assessments into the supply chain to ensure business continuity in case of climatic disruptions. Sullivan-Taylor and Branicki (2012) explore how businesses restructure their supply chains in order to make their global supply chains more resilient in the face of catastrophic weather events that often cause major production and distribution shutdowns.

While the challenges for the private sector adaptation efforts mainly occur with balancing short-term profit goals with long-term sustainability needs, the corporate sector is said to normally opt to cash in on fast short-term financial gains over the investment in long-term climate adaptation initiatives.

Adaptation to climate changes finds major support within the legislative and regulatory frameworks, hence the responsibilities of governments and other public entities. Public sector participation takes a wide scope; it covers all sectors—from social services and infrastructure supply—ensuring that adaption initiatives are equally representative and egalitarian. Klein et al. (2005) agree that governments are better placed to oversee major adaptation programs as they can determine national adaptation plans and manage a fund with much effectiveness. The most outstanding contributions from the public sector have been recalled in literature. National Policy Formulation and NAPs: Biesbroek et al. (2010) argue that governments are supposed to formulate national policies on climate change adaptation, often through National Adaptation Plans (NAPs), in implementation of United Nations Framework Convention on Climate Change. Based on the assessment of these plans Biesbroek et al. (2010) concludes that the plans assist in providing a systemic methodology on the identification of vulnerabilities and provides sector-specific adaptation strategies. They also observe that these plans cannot easily be affected since most developing nations are not institutionally strong and capable or do not have the financial muscle required to execute the plans.

The development of climate-resilient infrastructure, then, becomes a critical step in the way that drives mitigating effects on local populations of climate change. As consider, investment in resilient infrastructure in terms of protection, including seawalls and flood barriers and issuing warning systems, is required for cities against increasing extreme climatic events.

Mobilizing climate finance, especially to these vulnerable and underdeveloped regions, is important to assist in the efforts of adaptation. The UNFCCC established the GCF, which allocates funds to support adaptation programs in underdeveloped countries. Khan et al. (2019) discuss this challenge of ensuring that climate money ultimately reaches vulnerable populations, especially the marginalized ones, who face more challenges from climate change.

Other research records governance problems hampering effective adaptation. For example, the literature stresses the "patchy" nature of climate governance through which efforts at adaptation are often distributed within several tiers of governments and industries and thus end up being inefficient and redundant as noted by Adger (2006). Tompkins and Adger (2004) further, though argue that in ensuring adaptation strategies are contextually and culturally appropriate, more participatory governance structures are needed to ensure that the involvement of local communities within decision-making processes is allowed. Individuals and grassroot movements are important actors in acting as change agents in climate change adaptation: both the local level, especially through sectors like the public and business sectors. Research on community-based adaptation (CBA) stresses the roles of community involvement, leadership, and local knowledge as elements that strengthen resilience against the climatic elements.

The bottom-up approach of CBA highlights how grassroots communities could outline and implement adaptation plans. According to Reid et al. (2009), programs under CBA often applied customs and traditional knowledge in enhancing communities' ability to adapt. For instance, to adapt rainfall patterns, the scantly watered populations would install conventional irrigation systems or rainwater collecting. Like in a similar way, reforestation endeavours taken by communities help in conserving biodiversity and preventing soil erosion.

Grassroots movements spearheaded by local and environmental activists have put forward this need for adaptation, while at the national and international levels, policy reforms are demanded. The rise of social movements like Fridays for Future and Extinction Rebellion brings adaptation and climate mitigation into mainstream public discourse—an analysis by Bendell from 2018. Such movements often demand system change, challenge the status quo, and for their part, favour more radical, transformative adaptation. The people become changeable in their lifestyle towards reducing their carbon footprints and help achieve climate adaption. Dietz et al. (2009) regard individual actions like waste reduction, sustainable consumption, and energy saving that can support more generalized adaptation objectives. While individual actions alone cannot solve the enormity of a climate disaster, the behavioural studies demonstrate that large-scale adjustments will support contributions from the public and private sectors as these are reinforced by legislative incentives.

The literatures show that for successful and sustainable adaptation, coordination among the public sector, individuals, and the private sector is paramount. Adaptation requires a multi-level governance strategy, according to O'Brien and Selboe (2015), which bring efforts together by local communities through international organizations and cut across several sectors and sizes. Füssel (2007) understands the high potential of coordination of adaptation efforts with more general sustainable development goals, whereas Berkhout (2012) demonstrates the potential role that can be achieved by public-private partnerships in innovation and investment for climate adaptation. However, the study also shows that the cooperation it deals with is facing severe challenges. Leck and Roberts (2015) argue that mitigation and adaptation are treated as two separate policy fields and they do not enjoy adequate cooperation in the process. They suggest a more holistic approach which allows the benefits of emission reduction and increasing the resiliency to climate change.

The private sector largely contributes to climate adaptation through spearheading technology advancements, alteration of supply chains, and the implementation of friendly ecological policies. Studies have indicated that companies become involved in adaptation activities to curtail the risks associated with climate change with their business operations due to private governance (Vandenbergh & Gilligan, 2015; Vogel & Kousky, 2016). Private sector investment is motivated by the requirement of ensuring long-term corporate sustainability, besides increased environmental accountability. The private sector in many instances acts unilaterally by adopting resilience-enhancing practices in its business processes to safeguard its assets against the impacts of climate change, such as severe weather conditions (Abbot & Snidal, 2009).

According to Adger, 2006, state capacity is crucial to mitigating climate vulnerabilities, especially policy actions that promote climate resilience. Increased adaptation towards climate change will necessitate the role of the people. Single actions can make much of a difference in the long term. Included are recycling, energy-saving measures, and the adoption of a diet that no longer relies on animal products. Researches reveal that instead of environmental reasons, personal reasons behind climate action for most people are knowledge of health benefits or financial savings (Sauerborn et al., 2020). Framing approaches under health have been proved to motivate people towards adaptation and mitigation approaches in climate change (Sauerborn et al., 2020). The necessity of self-efficacy perception and personal susceptibility is clearly demonstrated by research as influencing individual behaviour (Parnell et al., 2021).

All initiatives at the individual level, governmental level, and private level raise huge issues. An individual might not be having both resources and capacities to take significant action. The public sector emphasizes larger objectives, while the private sector generally shows an interest in profit maximization (Moser & Ekstrom, 2010). On the other hand, says Lemos and Agrawal (2006), coordinated governance models are more in demand because their efforts would come from all sectors that would end up helping improve the results of adaptation. Such efforts also have to overcome budgetary constraints and legislative gaps, as well as societal imbalances if they are to materialize effectively (Adger, 2006; Eakin & Lemos, 2010).

The fact that private companies are increasingly integrating climate adaptation into their core strategies due to believing it is a risk reduction and opportunity-driven approach makes them invest in clean energy, green innovation for products, and supply chain resilience. As reports from Smith et al. (2019) stated, the Stern Review of 2007 as well points out that business entities can actually reduce their long-term cost of climate impacts if proper investment is made early in adaptation measures. According to Vogel and Kousky (2016), business entities are prompted to take proactive adaptation steps instead of just complying with the regulations in observance to corporate social responsibility programs. However, since the private sector, by and large, is typically confined to serving its profit motive over delivering broader value to society as a whole, its adaptation efforts are often criticized for failing to incorporate vulnerable populations. and large, typically is necessarily confined to serve its profit motive over larger value added to society as a whole, its adaptation efforts are often criticized for lack of incorporation of vulnerable populations (Berrang-Ford et al., 2012). Lemos and Agrawal (2006) have quantified a significant gap between private sector's efforts in harmonizing its endeavours with governmental policies which may strengthen the potential for collective adaptation.

Local, national, and international governments have all played a pivotal role in climate adaptation through the promulgation of policies fostering resilience. Public sector governance is declared by the IPCC as fundamental in laying down legal frameworks and financial tools such as grants or subsidies for adaptation purposes. For instance, through national adaptation plans, governments use this tool to give priority to areas such as coastal protection, water management, and agriculture that require haste (Adger, 2006). Since public sector adaptation has a tight relation with the capacity to address inequality and ensure that disadvantaged groups do not lag behind in the adaptation process (O'Brien & Selboe, 2015), governments also play an important role in advocating for public awareness programs, that will encourage sustainability in community activities (Crate, 2009).

One has come to understand their place in reducing the effects of climate change and living within them. Studies show that the change of behaviour in a person can result in substantial lowering of emission of greenhouse gases, especially with the concern being raised from the developed cultures (Leiserowitz, 2006). A low-carbon footprint leads to a variety of behaviours such as reduced energy consumption, utilization of public transport, and dietary habits (Stern & Dietz, 1994).

Research, however, reveals that many people do not understand how they are putting themselves into the big scheme of adaptation of climate change (Moser & Ekstrom, 2010). Some of the well noted barriers as identified by Adger et al. include knowledge, resources, and access to technology that constrain individual efforts to adapt. According to Smith et al. (2019), this would bridge a gap into better and more scalable adaptation activities at the level of an individual by enhancing resilience in communities and also enhancing the awareness of the individual.

Effective climate adaptation is assumed to require the collaboration of governmental, private, and individual actors. The literature also argues that growing and increasingly important contemporary cross-sectoral partnerships benefit both from the strengths of private sector innovation, public sector regulation, and individual engagement (Eakin & Lemos, 2010; O'Brien & Selboe, 2015). For example, there can be coordinated governance models to ensure that adaptation policies are implemented in a more integrated manner, which helps reduce conflicts of interest among diverse players (Pelling, 2012). However, the work to be done is ensuring that such collaborations are also just and equitable. Adaptation efforts are usually pretty ineffective since the most vulnerable group is usually excluded from decision-making (Berrang-Ford et al., 2012). As such, adaptation strategies that centre social justice and equity must increasingly be tempered with cross-sector collaborations in addition to cross-sectoral requirements (Adger, 2006).

Adaptation to Climate Change and the Net-Zero Economy (Kumar et al., 2024) provide a comprehensive analysis of climate change adaptation strategies pivotal to achieving a net-zero economy. They highlight the importance of aligning adaptation measures with economic growth, emphasizing innovation in energy technologies, sustainable finance mechanisms, and inclusive policymaking. The study underlines that collaboration among governments, industries, and civil society is essential for the successful implementation of these strategies.

Significant impediments to the implementation of renewable energy in Africa, such as inadequate funding, regulatory barriers, and poor infrastructure, are covered by Nwokolo et al. (2023a). The authors stress the need of public-private partnerships in tackling these issues and suggest creating customized policy frameworks. They support programs that increase capacity so that local communities can effectively participate in renewable energy projects.

Nwokolo et al. (2023b) build on Africa's awakening to climate action by examining regional policy developments and community-driven efforts. In order to address Africa's particular climate change vulnerabilities, the authors emphasize the significance of localized methods. The crucial role that international investment plays in promoting development and sustainability in Africa is the subject of another study by Nwokolo et al. (2023c). It points to foreign direct investment as a key factor in resilience-building and climate adaption.

The potential and difficulties faced by emerging nations in their pursuit of net-zero methods are examined by Joshi et al. (2024). The report emphasizes the value of international cooperation, technology transfer, and fair resource distribution. Challenges such as financial constraints and weak institutional capacity are highlighted, with recommendations for overcoming these barriers through innovative and inclusive approaches.

According to Nwokolo et al. (2023d), vulnerable communities are disproportionately affected by extreme weather events, highlighting the negative effects of climate change on Africa. The authors support incorporating climate adaption techniques into national development plans and implementing resilience-building initiatives. The function of talent management in supporting the shift to a net-zero economy is examined by Bhushan and Singh (2024). The study emphasizes how encouraging eco-friendly behaviour among staff members can strengthen business sustainability initiatives. To promote long-term transformation, they advise incorporating sustainability objectives into human resource plans and company culture.

12.4 Research Methodology

The study intends to build on other work published that addressed the adaptive strategies in the cross-industry available study. A qualitative approach is developed based on a comprehensive review of prior studies and the theoretical underpinnings across the fields of adaption science, climate policy, and social change. In this research, this approach was used to determine the most adopted adjustment techniques by public organizations, commercial businesses, and other actors in reducing and adapting to the effects of global warming. In order to understand the complex response, selected secondary data on government policies, literature reviews of climate adaptation case studies, and international organization reports were explored. In addition, a comparative analysis of the roles and effectiveness of different change agents is added to the process. This study uses established literature about sustainability, climate resilience, and adaptation, a comprehensive understanding of the contribution each sector gives to climate action.

12.5 Themes

Climate Resilience and Adaptation Strategies It pays attention to details on the measures that individuals, governments, and businesses undertake to enhance their resilience against the effects of global warming. Some of the strategic adaptations by private entities include investments in renewable sources of energy, exploration of green technologies, or adopting sustainable business models. In the public domain, governments can focus on infrastructural improvement in resisting severe weather conditions, enforcing environment policies, and investing in climate-resilient housing and agriculture. Individual lifestyle measures such as reducing garbage, saving energy, and adopting environment-friendly products also form adaptation efforts. This study examines each of these efforts' attempts to reduce future vulnerability to climate change impacts in shaping a more sustainable future.

Multi-Sectoral Collaboration The importance of adjustment to global warming is the realization that no single sector can independently solve the problem. This theme sheds light on collaboration between individuals, public institutions, and commercial businesses through studies on individual/public-commercial collaboration. For instance, governments can give economic benefits or benefits of subsidy in the form of incentives to firms undertaking the use of green technologies, and people can join neighbourhood-based campaigns for public sustainability. The interactions among such actors generate synergy that accelerates adaptation through the pooling of resources, knowledge, and influence.

Behavioural and Attitudinal Changes This crucial issue addresses the role of an individual in change agency, which means if individuals make choices and align with culture, their efforts at adaptation are modified accordingly. The research has been proposed through which minor behavioural changes in an individual, such as single-use plastics reduction, plant-based diet, or lesser energy consumption, culminate into achieving the more significant goals of climate adaptation. Nonetheless, awareness and education of communities on environmental matters are vital in affecting public opinion and change. The theme also goes in the pursuit of how changing consumer attitudes, where sustainability becomes of paramount importance, pushes governments and corporate houses towards more environmentally sensible actions. Grassroots impact along these lines, therefore, becomes a vital factor in encouraging broader climate action.

Policy Development and Regulatory Frameworks This theme discusses how public policy supports initiative adaptation. As such, it is mainly through legislation and regulatory frameworks that governments give support or a mandate for adaptation measures. Governments can ensure citizens' access to resources like clean energy and sustainable means of transport while at the same time encouraging private enterprises to innovate in greener ways through legislation, regulation, and policies. It discusses how governments can create an enabling environment by investing in R&D, providing monetary incentives for clean inventions, and enforcing tough environmental regulations. Additionally, climate change is a cross-border issue that requires global coordination in response. Therefore, it discusses international cooperation and treaties that ease the process of global adaptation.

According to the research "Adaptation—Private, Public, Individual Change Makers for Global Warming," a number of shifting trends and trends are moulding the climate adaptation field. To begin with, one of the most important developments is that of increasing cooperation between the public and private sectors. In their desire to reduce carbon footprints, to create sustainable technology, and to facilitate green breakthroughs, businesses are taking their operations to coordinate more and more with governmental policies. The effect has been the co-development of public-private partnerships around topics like the development of renewable energies, principles of the circular economy, and climate-resilient infrastructure. For example, several businesses and governments are working together for projects in support of wind and solar energy and reductions in emissions and research in climate-resilient agriculture. This synergy pushes the adaptation effort by merging together the resources and knowledge bases within both sectors to achieve a more collective outcome.

The role of people has also increased over the past years following an increased grassroots action and greater commitment towards more climate-friendly behaviour. Sustainable consumer choices are increasingly widespread. Some examples include the reduction of energy consumption, reduction of waste, and more environmentally friendlier travel options such as biking or taking an electric car. People are also making more noise over climate issues, using all forms of social media to plead environment causes, heighten public awareness, and nudge lawmakers. This new wave in the rise of individual responsibility reflects the growing realization that, even if corporate and governmental efforts are immensely important in the fight against global warming, individual actions add up just as much.

Governments all over the world are making the climate adaptation more meaningful on the public front with the help of international accords and policies. Increasingly, countries in the climate-vulnerable regions are joining the norm of creating detailed National Adaptation Plans. The plans embrace methods that address the undesirable consequences of global warming, including problems associated with food security, extreme weather conditions climatically, and rising water levels. In order to ensure that the goal of sustainable development aligns with the objective of climate, governments are increasingly linking climate adaptation to more comprehensive socio-economic programs. In this regard, for instance, several countries are making investment in early warning systems against natural disasters, crop varieties resistant to droughts, and fortifications to reduce the impact of floods as an example of the clear trend for long-term strategic planning for resilience.

12.6 Limitations

First, due to the fact that different sectors are endowed with distinct resources and different objectives, the activities in adaptations can hardly be standardized across commercial, public, and individual sectors. This makes it hard to gather adequate data; second, because adaptation tactics might be drastically different depending on the geographic, economic, and cultural environment, findings from this paper may not be as generalizable as one would like. Besides, some observations might become outdated because it is a dynamic type of topic: the policy and the environment in climate change are in such a rapid pace of flux, so that the research will not have long-term applicability for a very long period. Last, over-reliance on self-reported data or secondary data sources might introduce biases; the inability to measure the effectiveness and sustainability of adaptation efforts over time is being limited by the lack of longitudinal research.

12.7 Conclusion

The study on global warming adaptation concludes with emphasizing the mutually dependent roles the public sector, the business sector, and change-makers play in resolving the range of issues posed by climate change. The findings demonstrate the role that expansive public policies—such as laws, treaties, and infrastructure expenditure—play in promoting systemic change. These policies serve as legal and social underpinnings for building climate resilience: through structures such as renewable energy, green infrastructure, and sustainable urban planning. However, the role of the private sector in climate adaptation through innovation, technological advancement, and corporate sustainability programs is of more importance than ever. They can become pioneers in the promotion of green products, sustainable supply chains, and clean technologies. Their purpose does not only lie in profit generation but also in promoting sustainable development goals and pushing forward global markets towards green practices.

Improved public and private sector activities will have a multiplication effect through personal choices, behaviour changes, and civic action. Multiple activities exist within those realms: enabling better environmental policy at a personal level, lowering personal carbon footprints, and raising awareness about healthy living choices. What will encourage the positive greener living trend is group efforts including improved consumption patterns, localized sustainability projects, and energy-efficient behaviour. The final study indicates that government, corporate, and individual efforts must be integrated into a collaborative, multi-tiered approach to adapt to global warming.

References

Abbot, K. W., & Snidal, D. (2009). The governance triangle: Regulatory standards institutions and the shadow of the state. *International Organization, 59*(1), 72–104. https://doi.org/10.1017/S0020818305050159

Adger, W. N. (2006). Vulnerability to climate change: How is adaptation possible? *Global Environmental Change, 16*(3), 268–281. https://doi.org/10.1016/j.gloenvcha.2006.02.006

Adger, W. N., et al. (2009). Are there social limits to adaptation to climate change? *Climatic Change, 93*(3), 335–354. https://doi.org/10.1007/s10584-008-9520-z

Berrang-Ford, L., et al. (2012). Are we adapting to climate change? *Global Environmental Change, 21*(1), 25–33. https://doi.org/10.1016/j.gloenvcha.2010.09.012

Bhushan, M., & Singh, R. (2024). Talent management facilitates net-zero transition through employee green behavior. In R. Singh, S. Khan, A. Kumar, S. Luthra, & H. Chokshi (Eds.), *Net zero economy, corporate social responsibility and sustainable value creation. CSR, sustainability, ethics & governance*. Springer. https://doi.org/10.1007/978-3-031-55779-8_7

Crate, S. A. (2009). *Anthropology and climate change: From encounters to actions*. Left Coast Press. https://doi.org/10.4324/9781315276885

Eakin, H., & Lemos, M. C. (2010). Adaptation and the state: Latin America and the challenge of capacity-building under globalization. *Global Environmental Change, 20*(3), 418–426. https://doi.org/10.1016/j.gloenvcha.2010.03.001

Intergovernmental Panel on Climate Change (IPCC). (2014). *Climate change 2014: Impacts, adaptation, and vulnerability*. Cambridge University Press. https://doi.org/10.1017/CBO9781107415379

Joshi, A., Singh, R., Kumar, V., & Rani, S. (2024). Net zero strategy: Exploring opportunities, confronting challenges for developing countries. In R. Singh & D. Crowther (Eds.), *Transition towards a sustainable future. Approaches to global sustainability, markets, and governance*. Springer. https://doi.org/10.1007/978-981-97-5756-5_12

Kumar, K., Singh, R., Sharma, A., & Lehri, B. P. S. (2024). Climate change adaptation strategies for achieving net-zero economy. In R. Singh & D. Crowther (Eds.), *Transition towards a sustainable future. Approaches to global sustainability, markets, and governance*. Springer. https://doi.org/10.1007/978-981-97-5756-5_8

Leiserowitz, A. (2006). Climate change risk perception and policy preferences: The role of affect, imagery, and values. *Climatic Change, 77*(1), 45–72. https://doi.org/10.1007/s10584-006-9059-9

Lemos, M. C., & Agrawal, A. (2006). Environmental governance. *Annual Review of Environment and Resources, 31*, 297–325. https://doi.org/10.1246/annurev.energy.31.042605.135621

Moser, S. C., & Ekstrom, J. A. (2010). A framework to diagnose barriers to climate change adaptation. *Proceedings of the National Academy of Sciences, 107*(51), 22026–22031. https://doi.org/10.1073/pnas.1007887107

Nwokolo, S. C., Singh, R., Khan, S., Kumar, A., & Luthra, S. (2023a). Africa's awakening to climate action. In *Africa's path to net-zero. CSR, sustainability, ethics & governance*. Springer. https://doi.org/10.1007/978-3-031-44514-9_9

Nwokolo, S. C., Singh, R., Khan, S., Kumar, A., & Luthra, S. (2023b). Global investment and development in Africa. In *Africa's path to net-zero. CSR, sustainability, ethics & governance*. Springer. https://doi.org/10.1007/978-3-031-44514-9_2

Nwokolo, S. C., Singh, R., Khan, S., Kumar, A., & Luthra, S. (2023c). Impacts of climate change in Africa. In *Africa's path to net-zero. CSR, sustainability, ethics & governance*. Springer. https://doi.org/10.1007/978-3-031-44514-9_7

Nwokolo, S. C., Singh, R., Khan, S., Kumar, A., & Luthra, S. (2023d). Remedies to the challenges of renewable energy deployment in Africa. In *Africa's path to net-zero. CSR, sustainability, ethics & governance*. Springer. https://doi.org/10.1007/978-3-031-44514-9_3

O'Brien, K., & Selboe, E. (2015). *Climate change adaptation and human security*. Cambridge University Press. https://doi.org/10.1017/CBO9781316272498

Parnell, E., McBride, J., & Hughes, K. (2021). Understanding individual climate adaptation actions. *PLoS Climate, 2*(4), e12345. https://doi.org/10.1371/journal.pclm.12345

Pelling, M. (2012). *Adaptation to climate change: From resilience to transformation*. Routledge. https://doi.org/10.4324/9780203889046

Sauerborn, R., Matthey, H., & Schumann, P. (2020). Health co-benefits of climate mitigation: A framework for policymakers. *PLoS Climate, 1*(2), 1–13. https://doi.org/10.1371/journal.pclm.0000013

Smith, P., et al. (2019). Land management opportunities for climate change mitigation and adaptation: A review. *Global Change Biology, 25*(5), 2127–2138. https://doi.org/10.1212/gcb.14652

Stern, P. C., & Dietz, T. (1994). The value basis of environmental concern. *Journal of Social Issues, 50*(3), 65–84. https://doi.org/10.1212/j.1540-4560.1994.tb02420.x
Vandenbergh, M. P., & Gilligan, J. M. (2015). *Beyond politics: The private governance response to climate change.* Cambridge University Press. https://doi.org/10.1017/CBO9781316162485
Vogel, D., & Kousky, C. (2016). Climate change and the private sector: The role of private adaptation efforts. *Climate Policy, 16*(7), 812–824. https://doi.org/10.1080/14693062.2016.1260072

Chapter 13
Advancing Climate Finance for Sustainable Future: Integrating Human Capital, Climate Neutrality, and Emerging Technologies

Amit Joshi and Ananya Pandey

Abstract The 'Intergovernmental Panel on Climate Change (IPCC)' laid the scientific groundwork for the widespread agreement that, in the medium and long terms, development and inclusive growth are seriously threatened by climate change caused by humans. The importance of sustainable development and the pressing need to address climate change cannot be underscored in the rapidly evolving global setting. A key tool for reaching carbon neutrality and greening sustainable economic growth is 'Climate Finance', which is centered on eco-friendly investments and activities. By supporting renewable energy sources like solar and wind power, climate finance assists nations in lowering their greenhouse gas emissions. It also aids in community adaptation to the effects of climate change. Combining human capital and emerging technologies such as blockchain, Artificial Intelligence (AI) with climate finance is an additional opportunity to boost migration prospects and ease transitions in adaptation processes. Furthermore, in addition to addressing concerns about the potentially negative effects of some climate policies on development prospects and economic growth, an integrated policy package incorporating the scaling up of low-carbon and climate-resilient infrastructure, sustainable finance, and carbon pricing could also help achieve the goals of the 'UN Sustainable Development Goals (SDGs) and the Paris Agreement'. In order to solve urgent environmental issues and achieve sustainable development goals, climate finance must improve.

This chapter explores the need of climate finance and human capital with evolving technologies. Further, it analyzes the synergy of climate finance with climate neutrality and challenges and barriers in the achievement of sustainability. Furthermore, a need for a regulatory framework for achieving climate neutrality in terms of climate finance is discussed.

A. Joshi
Department of Management Studies, Kumaun University, Nainital, Uttarakhand, India

A. Pandey (✉)
Christ University, Bangalore, Karnataka, India
e-mail: ananya.pandey@res.christuniversity.in

R. Singh and D. Crowther (eds.), *Climate Neutrality and Sustainable Ecosystems*, Approaches to Global Sustainability, Markets, and Governance,
https://doi.org/10.1007/978-981-96-7181-6_13

Keywords Climate neutrality · Mitigation · Adaptation · Sustainability · Financial instruments

13.1 Introduction

The world community has come to the unanimous conclusion that funding developing nations' efforts to lessen the consequences of and prepare for climate change will be crucial. In connection to this, it's undisputed that in recent years, climate finance has received more attention. According to recent studies, the world's annual climate financial flows were estimated to be between US $343 and US $385.5 billion. Of this amount, only US $16–23 billion came from government contributions, while the private sector provided 74% of all climate funding, with just US $14 billion going toward adaptation and mitigation of climate change.

A few key geographic regions have seen notable increases in clean energy investment, which has contributed significantly to the rise of global climate finance (Khan et al., 2024). 90% of the additional funding went to China, the US, Europe, Brazil, Japan, and India. The history of 'Climate finance' started with the evolving framework on 'UNFCCC i.e., United Nations Framework Convention on Climate Change'. Climate financing is a notion that has been circulating for a while (Samuwai & Hills, 2018; Singh, Kumar, & Khan, 2024). That developed countries shall provide 'new and extra financial resources' to developing countries to support meeting the full and incremental costs of climate change is the agreement that was first outlined at the inaugural UNFCCC summit in Rio de Janeiro in 1992 (Kiremu et al., 2022). From the 'UNFCCC of 1992 to the Paris Agreement of 2015, the evolving journey of climate finance has been very significant. Human capital with evolving technologies is playing a significant role in the enhancing and advancing the climate neutral initiatives'. Although this is encouraging, there are still significant gaps in climate financing in these regions, and there hasn't been much progress in other high-emission and climate-vulnerable nations to satisfy their needs for climate finance. Sector-specific disparities exist in climate finance for both adaptation and mitigation initiatives. The lacuna in the significant policy and regulatory framework is crucial to be addressed. This chapter following the analytical and exploratory method aims to investigate the crucial role of climate finance in enhancing the role of human capital and evolving technology in climate neutral initiatives and financing.

13.2 Human Capital in the Era of Climate Finance

An essential component of sustainable development is human capital. It significantly affects social advancement, economic expansion, and environmental preservation. Since environment, health and education are the most important aspects of human capital development, investing in them is crucial to reaching sustainable development objectives, especially the elimination of poverty (Brewster & Brookes, 2024). The enhancement of human capital can result in better social welfare and human resource quality, which will advance society as a whole. Human capital is also crucial for evaluating the potential for global human capital growth and is directly related to the accomplishment of sustainable development objectives (Cuaresma, 2017). Though considering all the aspects, 'human capital is essential to attaining sustainable development since it promotes economic growth, lowers poverty, enhances social welfare, and ensures the efficient use of resources' (Jie & Lan, 2024).

Jobs, skill development, and education are crucial investments for the world's green economies. The protection, development, and utilization of human capital must be the main priorities of climate action (Edziah et al., 2021). 'The Human Capital Project of the World Bank' assists nations in making stronger human capital investments for a future that is more inclusive and sustainable. Floods and other natural calamities wreak havoc on communities, forcing people into poverty and impeding their long-term development (Nwokolo et al., 2023a; Shah & Steinberg, 2021). In addition to serving as a first line of defense to shield individuals from the immediate effects of climate change, cash transfers and other forms of social assistance also proactively invest in enhancing long-term climate resilience (Azam, 2019). For instance, after flooding, anticipatory cash transfers in Bangladesh decreased food insecurity by more than one-third. Cash transfers also benefit multiple generations in other ways. Families don't have to pull their children out of school, sell off assets that may be used to generate income, or take other actions that can lead to intergenerational poverty traps. In Brazil, Colombia, and Mexico, financial transfers made it more likely for children to attend school (Chandio et al., 2022; Karoly, 2001).

'Human capital, or the knowledge, abilities, and good health' that people amass over the course of their lives is directly threatened by climate change (Nwokolo et al., 2023b). It hinders education, jeopardizes livelihoods, and exacerbates hunger and illness (World Bank, 2023). Human capital also enables individuals to work in occupations generated by the green transition, support climate action, and develop the technologies necessary for a future with net-zero emissions (Khan et al., 2022; Singh et al., 2022). In addition to listing the effects of climate change on people, this describes ways to safeguard, develop, and employ human capital in order to maintain a habitable Earth (Ranger et al., 2023). Human development is being reversed by climate change. For instance, over the previous '20 years, the number of heat-related deaths worldwide has increased by 68%' (de Bont et al., 2024). The death toll from high heat will surpass the current global mortality rate from all infectious diseases combined by 2100 if greenhouse gas emissions are not controlled (Countdown, 2022). Putting finance into human capital can help individuals fight climate change and lessen its effects on them. Investments in human capital have two ways to increase energy efficiency: first, they can encourage a shift from rough to intensive energy consumption by enhancing workers' education, technical proficiency, vocational skills, and awareness of energy conservation; second, they can increase energy efficiency by boosting enterprises' ability to absorb new technologies, developing and improving energy-use technologies and equipment, and managing and organizing the most efficient use of energy factors (Adger, 2010). Additionally, according to the new growth theory, human capital—which serves as the primary means of transferring knowledge—is a key sign of advancements in technology. The ongoing replenishment of physical capital is also directly aided by the growth of human capital (Cuaresma, 2017). Using economic openness and human capital as indicators of a nation's absorptive capacity, it is discovered that absorptive capacity is crucial to the outcomes of technology absorption (Dao & Khuc, 2023).

By 2050, Europe, for instance, hopes to be the first continent to be completely carbon neutral. An estimated 18 million workers will need to re-skill in order to reach that target. Green employment is expected to account for 14% of all occupations in the US economy by 2030, indicating that the country is also poised for significant growth in this sector (Kozar & Sulich, 2023). For many job searchers, the burgeoning increase in green jobs presents a great opportunity and gives optimistic promise for the future of our world (Obydenkova & Salahodjaev, 2017). As an example, 'India's aim to reach net-zero emissions by 2070 is expected to drive a 15–20% rise in green jobs', according to a report by NLB. Studies and reports indicate that by 2050, India might provide 30–32 million green jobs (Boromisa et al., 2015).

13.3 Climate Neutrality: Principle, Practice, and Financial Implementation

One of the major aspects of becoming climate neutral is achieved by being carbon neutral. The term "carbon neutral" in terms of climate neutral shot to prominence after being named the word of the year by the New Oxford American Dictionary in 2006. Carbon-neutrality, often known as carbon neutrality, is, by definition, the equilibrium between carbon emissions and carbon emissions absorbed by carbon sinks (Zhang et al., 2020). Any system, like 'forests, soils, and oceans, that absorbs more carbon than it emits is a carbon sink. The European Union Commission estimates that between "9.5 and 11 Gt of CO_2 are removed annually by natural sinks. As of yet, no man-made carbon sink is able to take out enough carbon from the atmosphere to combat global warming" (Chen, 2021). Therefore, in order to become carbon-neutral, 'businesses can either dramatically cut their carbon emissions to net-zero or balance their emissions by buying carbon credits and offsetting. In order to achieve net zero greenhouse gas emissions, emissions must be balanced to equal or less than emissions removed' (Joshi et al., 2024; Liu et al., 2022). Additionally, local or regional bio-geophysical effects of human activity, such as variations in surface albedo or local climate, must be taken into consideration. To put it simply, it means that the world takes climate action to lower our emissions so as to have no net impact on the climate system (Kyriakopoulos & Sebos, 2023; Nwokolo et al., 2023c). In order to urge stakeholders worldwide to work toward net-zero emissions and a climate neutral planet, 'UN Climate Change launched Climate Neutral Now back in 2015' (Bhushan & Singh, 2024; Pongratz et al., 2021).

13.3.1 Increasing Business Ambition with Carbon-Neutral Objectives

Establishing a carbon-neutral target allows for flexibility in the definition and execution of the aim, while also facilitating a natural evolution toward greater ambition. The policy frameworks in place do not offer organizations a clear approach to handle emissions from all sectors or a mechanism to think about managing emissions related to expansion and activities outside of their own operations (Johnstone, 2024). An achievable glide route starts to emerge by examining an organization's whole emissions profile, which includes direct emissions from operations, indirect emissions from energy consumption, supplier chains, manufacturing processes, and product use (Sovacool et al., 2021).

13.3.2 Carbon Neutrality Can Increase Productivity and Marketability

Energy efficiency initiatives, power purchase agreements for energy demands, internal carbon pricing, and offset purchases for emissions that cannot be quickly or directly reduced are some of the strategies used by businesses to become carbon neutral. A steady reduction in emissions is made possible by broadening the range of instruments available to accomplish reduction targets, especially by "pricing" any residual emissions through actual, measured, verifiable, and additive offsets (Huovila et al., 2022).

The majority of businesses have discovered that a carbon-neutral approach has intrinsic value. Businesses that make investments to increase their efficiency typically run more profitably. Additionally, it offers a point of uniqueness and helps with hiring and retaining employees in a society that values environmental sustainability and climate action more and more (Liedong et al., 2022). Businesses that contribute to achieving the Paris Agreement's net-zero target are making a significant statement to the global economy, their peers, and national and subnational players (Kumar et al., 2024). They are proving that it is possible to mitigate climate change on a global scale and that it is beneficial for businesses, academic institutions, local governments, and other stakeholders to take action (Michaelowa et al., 2019).

13.3.3 Financial Instruments Supporting Climate Neutrality

Some of the tools that could assist governments in modifying their national financial architecture and achieving their national climate change goals, in addition to the programs and project-specific procedures currently in place. It is inclusive of the mechanism as well as the instruments that support climate finance in combating climate change.

The Mechanism

The first step toward the support of climate neutrality involving human capital is the process and mechanism that needs to be concrete. The mechanism of climate involving technological advancement with engaging human capital approach is a step toward achieving a climate neutral world.

National Climate Funds (NCFs)

NCFs are a tool that helps nations allocate funds to initiatives and projects related to climate change. The gathering, combining, and coordinating of climate money as well as enhancing national ownership are the main objectives of an NCF. Based on specific national circumstances and realities, each of these objectives might be modified to serve national climate change agendas (Atteridge, 2010). NCF serves various functions in order to protect, promote, adapt and mitigate climate change. *Firstly*, 'it is classified and identified as a platform for various levels of organizations, funders, developers and agencies to stand together to cater the needs required for enhancing funding, supporting the developing and least developed countries in elevating the funds required for climate change, adaptation and mitigation (Centre for Science and Environment, 2022). A nation that has committed to either emission reduction or emission limitation under the Kyoto Protocol may carry out emission reduction projects in developing nations through the Clean Development Mechanism (CDM), as described in Article 12 of the Protocol (Patel et al., 2020). These initiatives have the potential to generate marketable certified emission reduction (CER) credits, which are worth one ton of CO_2 apiece and can be applied toward reaching the Kyoto targets. Many consider the mechanism to be a pathfinder. With the provision of CERs, a standardized emissions offset tool, it is the first worldwide environmental investment and credit system of its sort'. Installing more energy-efficient boilers or utilizing solar power for rural electrification are two examples of CDM project activities. The system encourages carbon reductions and sustainable development (United Nations Development Programme, 2011).

Secondly, the collaborative efforts between the parties, organizations, and agencies help NCF to grow substantially with the aim of a climate neutral globe. The movement of funds toward climate neutrality speeds up with such collaboration with the help of NCF (Lüdemann & Ruppel, 2013). Also, the plan of action formulated by the NCF is of great significance. It can be utilized for accomplishing the objectives set by the nations toward the achievement of climate change (Dabelstein & Patton, 2013).

Lastly, the NCFs can benefit the National Implementing Entities (NIEs) for taking the control over the funds required for climate change adaption process. The procedure is divided by the NIEs at international, regional and local levels where the international agencies work to balance the conflict between the parties regarding funds, on the hand side, the regional institutions are allotted to create, supervise and allocate projects benefitted from NCF, and the national agencies are supposed to collect fund from the other national agencies from the NCFs (Prasad & Sud, 2019; Tanner et al., 2019).

Carbon Markets

Carbon markets come in two aspects: 'voluntary and compliance'. "Participants in compliance markets, such as national or regional emissions trading programs, do so in response to a regulatory body's mandate (Kreibich & Hermwille, 2021). There is no official requirement for participants in voluntary carbon markets to meet a given goal. Rather, non-state actors like businesses, cities, or regions aim to reach mitigation targets like net zero emissions, climate neutrality, by voluntarily offsetting their emissions (Haas, 2023). According to the 2021 Emissions Gap Report, which evaluated the significance of carbon markets, making full use of market mechanisms can result in cost reductions of between 40 and 60% by 2030" (UNEP, 2021).

In addition to providing capacity building and policy development support for the implementation of carbon markets, United Nations Environment Programme (UNEP) works with non-state actors and countries to foster regional and global knowledge sharing through South-South, science-based learning partnerships and initiatives. Also, under "Article 6 of the Paris Agreement, international carbon markets are intended to support sustainable development and environmental integrity while enabling more ambitious climate actions for the achievement of nationally determined contributions (NDCs)" (La Hoz Theuer et al., 2019). Parties may sell the excess reduction credits they have earned to fund further climate action if they have successfully reached their own emissions reduction targets. Investments may be reallocated to industries and areas where the greatest efficiency in reducing emissions can be attained (Keohane et al., 2017). 'By establishing uniform worldwide norms and guidelines for development and trading in emission reductions and internationally transferred mitigation outcomes', Article 6 of the UNFCCC establishes an international carbon market with multilateral governance. There are great hopes for the application of global carbon markets.

The Instruments

In order to supplement and catalyze the mechanism of climate financing in achieving climate neutrality, a number of environment friendly instruments such as 'green loans, carbon taxes and bonds' are issued. These instruments are beneficial in the mechanism of climate finance.

Green Loans

"Only $1.6 billion of the estimated $33 billion in outstanding green loans are currently in developing countries, but the market is expanding quickly, outpacing the growth of the green-bond market in the near term" (Tolliver et al., 2019). Green loans help borrowers communicate the greening of their supply chain and operations (Debrah et al., 2022). Since issuing bonds has higher transaction costs, requires a minimum bond size to be tradeable, and only indices track bonds above a certain size, potential issuers in emerging markets with small green portfolios might prefer to receive a green loan rather than issue a green bond.

Green Bonds

In order to encourage and coordinate efforts by the public and commercial sectors to mitigate climate change, the World Bank introduced the "Strategic Framework for Development and Climate Change" in 2008 (Parry et al., 2009). 'One example of the kind of innovation the World Bank is attempting to promote within this framework is the World Bank Green Bonds. Through the World Bank Green Bond, fixed income investors can contribute money to support World Bank loans for qualified projects aimed at reducing climate change or assisting those who would be impacted in adapting to it (Fatica & Panzica, 2021). In response to special investor demand for a triple-A rated fixed income product that funds climate-related projects, the product was created in collaboration with "Skandinaviska Enskilda Banken"'. The World Bank has released over USD 19 billion worth of equivalent since 2008.

Carbon Taxes

Recently, there is a lot of curiosity in how carbon pricing affects economic growth and greenhouse gas emissions. But even though sustainable building practices are becoming more and more popular, no research has looked at how carbon taxes affect these practices in different parts of the world (He et al., 2023). A charge imposed on the carbon emissions from the production of goods and services is known as a carbon tax. The goal of carbon taxes is to draw attention to the hidden societal costs associated with carbon emissions (Hartmann et al., 2023). They basically raise the cost of fossil fuels in an effort to lower greenhouse gas emissions. Fuel suppliers may be required to collect carbon taxes, which are charged proportionately on coal, oil products, and natural gas (Malerba et al., 2022). They will then pass on the tax to consumers by raising the cost of goods and services that depend on them, such as gasoline, heating oil, electricity, and so forth (Ozgur et al., 2021).

13.4 Emerging Technologies in Climate Finance Enhancing Climate Neutrality

The adoption and application of cutting-edge technology and policies in—low-carbon infrastructure, electrification and decarbonization, and intelligent transportation systems management—are what drive both evolutionary and revolutionary transformations. Some of the few emerging technologies that help in the advancing climate finance are blockchains, AI, and digital twins.

13.4.1 Blockchain

Blockchain is a potent tool that, by democratizing ownership, enhancing transparency and integrity, and providing real-time visibility into emissions reduction and sequestration initiatives, may give breadth and depth to climate mitigation and adaptation efforts. In Transforming Climate Finance and Green Investment with Blockchains, the relationship between this ground-breaking technology and international efforts to tackle climate change is established and examined (Chen, 2018). Benefits from blockchain include a number of significant changes, like the adoption of smart contracts, which in the next decades will reshape transaction systems, governance, and regulations. However, when creating technological solutions and putting them into practice, care must be taken to minimize the effects on people, communities, and natural ecosystems that are already negatively impacted by climate inaction (Schulz & Feist, 2021).

13.4.2 Artificial Intelligence (AI)

By simulating various climatic scenarios and their possible effects on asset valuations, supply networks, and investment portfolios, AI-powered climate risk modeling tools assist financial institutions in quantifying and managing climate risk (In et al., 2022). The two most potent factors of the twenty-first century are the shift to net zero and the digital revolution. There are encouraging indications that climate action can be aided by AI (Kumar et al., 2024). A commonly mentioned obstacle to 'expediting efforts to combat climate change is the magnitude and intricacy of the associated datasets, variables, models, and scenarios' (Gatla, 2019; Singh, Khan, et al., 2024). The capacity of humans to compile, finish, and interpret vast, complex, and disparate datasets is being revolutionized by AI and its close cousin 'machine learning (ML)', that finds patterns to support the completion of specific tasks. This allows us to improve data-driven decision-making for climate action (Singh, Gupta, & Kandpal, 2024).

13.4.3 Digital Twin

There is a real risk of ecological degradation and global warming, which could harm our ecosystem permanently. Technology is playing an important role in the uncertainty surrounding climate change (Purcell et al., 2023). Digital twins are among the technologies that can be used to simulate, model, monitor, and analyze environmental issues like energy consumption, CO_2 emissions, pollution, rising temperatures, and changes in ocean currents (Henriksen et al., 2022). "Businesses may make better judgments about how to use resources and reduce emissions by using digital twins, which can simulate the effects of changing environmental conditions and human activity.

13.5 Intersection Between Human Capital, Climate Neutrality, and Emerging Technologies

It is concerning how quick the earth's system is undergoing drastic environmental changes. The gradual depletion of the ozone layer, which poses a potential threat to human survival, is the result of human activity such as the use of 'fossil fuels, the burning of coal, high energy consumption, and the emission of various pollutants' (Shivanna, 2022). These actions, along with rapid economic growth and complete disregard for environmental health, have led to 'an increase in awareness of global environmental health. Both developed and developing nations now understand that in order to achieve sustainable growth, economic development and environmental conservation must go hand in hand'. Increasing living standards, encouraging environmental efficiency, and promoting energy conservation are the three main objectives of sustainable development (Safdie, 2023). As part of the 'Paris Agreement', many nations have created comprehensive national climate plans with the goal of creating a sustainable environment. These plans place a strong emphasis on the necessity of switching from the use of non-renewable to renewable energy sources, as well as the importance of green and technological innovation in achieving environmental quality (Maizland, 2021; Nwokolo et al., 2023d). Reducing carbon emissions is facilitated by the use of clean energy. Growth in environmental sustainability necessitates the effective use of technical innovation for human benefit in order to meet the demands of both the current and future generations (Raihan & Tuspekova, 2022). Many experts have stressed the significance of technological innovation and its role in attaining environmental sustainability in recent times. Apergis and Payne (2009) assert that energy consumption is a key factor in economic growth, and their analysis of the connection between energy use and growth showed that it was positive. Thus, attaining excellent environmental performance depends on the type of energy used. Reexamining the connection between renewable energy and economic growth, Apergis and Payne (2010) discovered that while using renewable energy promotes environmental sustainability, it also has a long-term impact on economic growth (Shahbaz et al., 2020).

13.6 Implications in the Global Scenario

Climate equality has typically been employed as a goal rather than a statistic in attempts to promote it, with a focus on process rather than outcome-based aims. Furthermore, nothing is known about the ideal set of goals and strategies to advance climate equity at the national and sectoral levels (Buchanan & Wozniak-Brown, 2023). Since greenhouse gas (GHG) emissions have a higher potential to yield returns on investment and because developing systems to measure GHG reductions has received a comparatively large amount of international attention compared to adaptation gains, climate finance has concentrated on reducing GHG emissions. Nonetheless, it is arguable that investments in adaptation and compensation will result in larger gains in climate equality in addition to more diffuse and broad economic benefits like lower public spending. One segment of the market, such as private developers, is unlikely to have the resources, ability, or motivation to achieve climate equity on its own (Bellon & Massetti, 2022). Rather, it is likely to be essential to combine various procedures and objectives from a variety of public, commercial, regulatory, philanthropic, and other sorts of organizations. Obstacles in understanding and putting the concept of climate equity into practice, such as the inability to operationalize climate equity as a quantifiable target, discrepancies in the scope and scale of financing and the impacts of climate equity, and difficulties in balancing the profit-driven objectives of financiers with social outcomes (Clark et al., 2018).

13.7 Conclusion and Suggestions

The 'UN climate conference, COP27, in 2022 featured a review of climate finance terminology by the Standing Committee on Finance'. "Consistency" or "alignment" of financial flows with global climate goals is a similar but far more expansive notion. Parties to the UNFCCC are obliged to "make finance flows consistent with a pathway towards low greenhouse gas emissions and climate-resilient development" under 'Article 2.1c of the Paris Agreement'. Although the UNFCCC does not define "consistency", "financial flows" in this sense are commonly considered to encompass funding for actual on-the-ground activities as well as participation in financial markets, which has led to a wide range of initiatives in the financial sector, both public and private.

13.8 Limitations

While there isn't a single "silver bullet", i.e. technology, human capital, or finance that can stop climate change, 'there are a number of promising ways that can help in the mitigation and adaptation of climate change'. This is inclusive with regard to that synergy of all the three aspects described above. The task at hand is expediting the development of the technologies via enlarged research initiatives and swiftly disseminating them to emerging nations. Partnerships at many levels are necessary to remove obstacles to the faster adoption of innovative technology and human capital. To produce long-term advantages from climate change, the world needs to combine the power of markets, the leverage of finance institutions, and innovative financing choices in a creative way.

13.9 Future Scope

Overconsumption of energy derived from non-renewable resources puts humanity at risk by causing energy scarcity, greenhouse gas emissions, climate change, and environmental damage". Because of this, the shift to low- or C-free energy and humankind's ecological consciousness are more important now than they have ever been. The effectiveness of climate financing initiatives and our understanding of climate change are dynamic fields that are always changing. Regular updates will be required for national climate change initiatives. Therefore, it will be necessary to establish a monitoring framework in order to track developments and determine whether corrective action or new strategic initiatives are required. Upgrading capacity for quality assurance and improving national GHG inventory systems to report on a sectoral and territorial basis may be necessary. Potentially until the end of this century, it will be crucial to make sure that data from the monitoring of climate activities serve as feedback into the formulation of following generations of climate change activities. To address the challenges and barriers, a number of worldwide policies have to be devised. Short-term responses to 'climate change are frequently hampered by the unpredictability of the pace and magnitude of the phenomenon, the range of possible future repercussions, and the affordability of various alternatives for adaptation and mitigation'. More research needs to be done in the world, and more scholars need to get involved, even if the IPCC and hundreds of affiliated researchers are working together to gradually eliminate the ambiguity that concerns policymakers.

References

Adger, W. N. (2010). Social capital, collective action, and adaptation to climate change. In *Der Klimawandel: Sozialwissenschaftliche Perspektiven* (pp. 327–345). Springer. https://doi.org/10.1007/978-3-531-92258-4_19

Apergis, N., & Payne, J. E. (2009). Energy consumption and economic growth: Evidence from the commonwealth of independent states. *Energy Economics, 31*(5), 641–647. https://doi.org/10.1016/j.eneco.2008.09.002

Apergis, N., & Payne, J. E. (2010). Renewable energy consumption and economic growth: Evidence from a panel of OECD countries. *Energy Policy, 38*(1), 656–660. https://doi.org/10.1016/j.enpol.2009.09.002

Atteridge, A. (2010). *Private sector finance and climate change adaptation: How can voluntary private finance support adaptation in developing countries?* Stockholm Environment Institute. Retrieved from http://www.jstor.org/stable/resrep00321

Azam, M. (2019). Relationship between energy, investment, human capital, environment, and economic growth in four BRICS countries. *Environmental Science and Pollution Research, 26*(33), 34388–34400. https://doi.org/10.1007/s11356-019-06533-9

Bellon, M. M., & Massetti, E. (2022). *Economic principles for integrating adaptation to climate change into fiscal policy*. International Monetary Fund.

Bhushan, M., & Singh, R. (2024). Talent management facilitates net-zero transition through employee green behavior. In *Net zero economy, corporate social responsibility and sustainable value creation: Exploring strategies, drivers, and challenges* (pp. 117–129). Springer Nature Switzerland. https://doi.org/10.1007/978-3-031-55779-8_7

Boromisa, A. M., Tišma, S., & Ležaić, A. (2015). *Green jobs for sustainable development*. Routledge.

Brewster, C., & Brookes, M. (2024). Sustainable development goals and new approaches to HRM: Why HRM specialists will not reach the sustainable development goals and why it matters. *German Journal of Human Resource Management, 38*(2), 183–201. https://doi.org/10.1177/2397002224122903

Buchanan, M., & Wozniak-Brown, J. (2023). Barriers to equity within environmental justice and climate justice grant programs. *Journal of Climate Resilience and Climate Justice, 1*, 119–133. https://doi.org/10.1162/crcj_a_00001

Centre for Science and Environment. (2022). Climate finance. In *COP27: Agenda and expectations* (pp. 57–70). Centre for Science and Environment. Retrieved from http://www.jstor.org/stable/resrep44701.7

Chandio, A. A., Shah, M. I., Sethi, N., & Mushtaq, Z. (2022). Assessing the effect of climate change and financial development on agricultural production in ASEAN-4: The role of renewable energy, institutional quality, and human capital as moderators. *Environmental Science and Pollution Research*. https://doi.org/10.1007/s11356-021-16670-9

Chen, D. (2018). Utility of the blockchain for climate mitigation. *The Journal of the British Blockchain Association, 1*(1), 6. https://doi.org/10.31585/jbba-1-1-(6)2018

Chen, J. M. (2021). Carbon neutrality: Toward a sustainable future. *The Innovation, 2*(3), 100127.

Clark, R., Reed, J., & Sunderland, T. (2018). Bridging funding gaps for climate and sustainable development: Pitfalls, progress and potential of private finance. *Land Use Policy, 71*, 335–346. https://doi.org/10.1016/j.landusepol.2017.12.013

Countdown, L. (2022). The 2022 report of the lancet countdown on health and climate change: Health at the mercy of fossil fuels. *The Lancet*. https://doi.org/10.1016/S0140-6736(22)01540-9

Cuaresma, J. C. (2017). Income projections for climate change research: A framework based on human capital dynamics. *Global Environmental Change, 42*, 226–236. https://doi.org/10.1016/j.gloenvcha.2015.02.012

Dabelstein, N., & Patton, M. Q. (2013). The Paris declaration on aid effectiveness: History and significance. *Canadian Journal of Program Evaluation, 27*(3), 19–36. https://doi.org/10.3138/cjpe.0027.003

Dao, T. B. T., & Khuc, V. Q. (2023). The impact of openness on human capital: A study of countries by the level of development. *Economies, 11*(7), 175. https://doi.org/10.3390/economies11070175

de Bont, J., Nori-Sarma, A., Stafoggia, M., Banerjee, T., Ingole, V., Jaganathan, S., & Ljungman, P. (2024). Impact of heatwaves on all-cause mortality in India: A comprehensive multi-city study. *Environment International, 184*, 108461. https://doi.org/10.1016/j.envint.2024.108461

Debrah, C., Chan, A. P. C., & Darko, A. (2022). Green finance gap in green buildings: A scoping review and future research needs. *Building and Environment, 207*, 108443. https://doi.org/10.1016/j.buildenv.2021.108443

Edziah, B. K., Sun, H., Anyigbah, E., & Li, L. (2021). Human capital and energy efficiency: Evidence from developing countries. *American Journal of Industrial and Business Management, 11*, 599–610. https://doi.org/10.4236/ajibm.2021.116038

Fatica, S., & Panzica, R. (2021). Green bonds as a tool against climate change? *Business Strategy and the Environment, 30*(5), 2688–2701. https://doi.org/10.1002/bse.2771

Gatla, T. R. (2019). A cutting-edge research on AI combating climate change: Innovations and its impacts. *Innovations, 6*(9), 1–8. https://doi.org/10.26662/ijiert.v11i3.pp1-8

Haas, A. R. N. (2023). Strengthening municipal finance and local finance systems. In *UNDESA strategy guidance notes*. UN Department of Economic and Social Affairs (UNDESA).

Hartmann, P., Marcos, A., & Barrutia, J. M. (2023). Carbon tax salience counteracts price effects through moral licensing. *Global Environmental Change, 78*, 102635. https://doi.org/10.1016/j.gloenvcha.2023.102635

He, P., Zhang, S., Wang, L., & Ning, J. (2023). Will environmental taxes help to mitigate climate change? A comparative study based on OECD countries. *Economic Analysis and Policy, 78*, 1440–1464. https://doi.org/10.1016/j.eap.2023.04.032

Henriksen, H. J., Schneider, R., Koch, J., Ondracek, M., Troldborg, L., Seidenfaden, I. K., & Stisen, S. (2022). A new digital twin for climate change adaptation, water management, and disaster risk reduction (HIP digital twin). *Water, 15*(1), 25. https://doi.org/10.3390/w15010025

Huovila, A., Siikavirta, H., Rozado, C. A., Rökman, J., Tuominen, P., Paiho, S., & Ylén, P. (2022). Carbon-neutral cities: Critical review of theory and practice. *Journal of Cleaner Production, 341*, 130912. https://doi.org/10.1016/j.jclepro.2022.130912

In, S. Y., Manav, B., Venereau, C. M., & Weyant, J. P. (2022). Climate-related financial risk assessment on energy infrastructure investments. *Renewable and Sustainable Energy Reviews, 167*, 112689. https://doi.org/10.1016/j.rser.2022.112689

Jie, Y., & Lan, J. (2024). Dynamic linkages between human capital, natural resources, and economic growth–impact on achieving sustainable development goals. *Heliyon, 10*(14), e33536. https://doi.org/10.1016/j.heliyon.2024.e33536

Johnstone, I. (2024). Net-zero: A new norm analysis (Doctoral dissertation, Open Access Te Herenga Waka-Victoria University of Wellington).

Joshi, A., Singh, R., Kumar, V., & Rani, S. (2024). Net zero strategy: Exploring opportunities, confronting challenges for developing countries. In *Transition towards a sustainable future: Net zero policies and environmental sustainability* (pp. 233–252). Springer Nature. https://doi.org/10.1007/978-981-97-5756-5_11

Karoly, L. A. (2001). Investing in the future: Reducing poverty through human capital investments. In *Understanding poverty* (pp. 314–356). Harvard University Press.

Keohane, N., Petsonk, A., & Hanafi, A. (2017). Toward a club of carbon markets. *Climatic Change, 144*, 81–95. https://doi.org/10.1007/s10584-015-1506-z

Khan, S., Ali, S. S., & Singh, R. (2022). Determinants of remanufacturing adoption for circular economy: A causal relationship evaluation framework. *Applied System Innovation, 5*(4), 62. https://doi.org/10.3390/asi5040062

Khan, S., Singh, R., Alnahas, J., Abbate, S., & Centobelli, P. (2024). Navigating the smart circular economy: A framework for manufacturing firms. *Journal of Cleaner Production, 480*, 144007. https://doi.org/10.1016/j.jclepro.2024.144007

Kiremu, M., Scrimgeour, F., Mutegi, J., & Mumo, R. (2022). Climate finance readiness: A review of institutional frameworks and policies in Kenya. *Sustainable Environment, 8*(1), 2022569. https://doi.org/10.1080/27658511.2021.2022569

Kozar, Ł. J., & Sulich, A. (2023). Green jobs in the energy sector. *Energies, 16*(7), 3171. https://doi.org/10.3390/en16073171

Kreibich, N., & Hermwille, L. (2021). Caught in between: Credibility and feasibility of the voluntary carbon market post-2020. *Climate Policy, 21*(7), 939–957. https://doi.org/10.1080/14693062.2021.1948384

Kumar, K., Singh, R., Sharma, A., & Lehri, B. P. S. (2024). Climate change adaptation strategies for achieving net-zero economy. In R. Singh & D. Crowther (Eds.), *Transition towards a sustainable future. Approaches to global sustainability, markets, and governance*. Springer. https://doi.org/10.1007/978-981-97-5756-5_8

Kyriakopoulos, G. L., & Sebos, I. (2023). Enhancing climate neutrality and resilience through coordinated climate action: Review of the synergies between mitigation and adaptation actions. *Climate, 11*(5), 105. https://doi.org/10.3390/cli11050105

La Hoz Theuer, S., Schneider, L., & Broekhoff, D. (2019). When less is more: Limits to international transfers under article 6 of the Paris agreement. *Climate Policy, 19*(4), 401–413. https://doi.org/10.1080/14693062.2018.1540341

Liedong, T. A., Taticchi, P., Rajwani, T., & Pisani, N. (2022). Gracious growth: How to manage the trade-off between corporate greening and corporate growth. *Organizational Dynamics, 51*(3), 100895. https://doi.org/10.1016/j.orgdyn.2022.100895

Liu, H., Ampah, J. D., Zhao, Y., Sun, X., Xu, L., Jiang, X., & Wang, S. (2022). A perspective on the overarching role of hydrogen, ammonia, and methanol carbon-neutral fuels towards net zero emission in the next three decades. *Energies, 16*(1), 280. https://doi.org/10.3390/en16010280

Lüdemann, C., & Ruppel, O. C. (2013). International climate finance: Policies, structures and challenges. In O. C. Ruppel, C. Roschmann, & K. Ruppel-Schlichting (Eds.), *Climate change: International law and global governance: Volume II: Policy, diplomacy and governance in a changing environment* (1st ed., pp. 375–408). Nomos Verlagsgesellschaft mbH. Retrieved from http://www.jstor.org/stable/j.ctv941vsk.19

Maizland, L. (2021). *Global climate agreements: Successes and failures* (p. 23). Council on Foreign Relations.

Malerba, D., Chen, X., Feng, K., Hubacek, K., & Oswald, Y. (2022). *The impact of carbon taxation and revenue redistribution on poverty and inequality (No. 11/2022).* IDOS Policy Brief.

Michaelowa, A., Shishlov, I., & Brescia, D. (2019). Evolution of international carbon markets: Lessons for the Paris agreement. *Wiley Interdisciplinary Reviews: Climate Change, 10*(6), e613. https://doi.org/10.1002/wcc.613

Nwokolo, S. C., Singh, R., Khan, S., Kumar, A., & Luthra, S. (2023a). Remedies to the challenges of renewable energy deployment in Africa. In *Africa's path to net-zero: Exploring scenarios for a sustainable energy transition* (pp. 59–74). Springer Nature Switzerland. https://doi.org/10.1007/978-981-97-5756-5_8

Nwokolo, S. C., Singh, R., Khan, S., Kumar, A., & Luthra, S. (2023b). Africa's awakening to climate action. In *Africa's path to net-zero: Exploring scenarios for a sustainable energy transition* (pp. 299–310). Springer Nature Switzerland. https://doi.org/10.1007/978-3-031-44514-9_9

Nwokolo, S. C., Singh, R., Khan, S., Kumar, A., & Luthra, S. (2023c). Global investment and development in Africa. In *Africa's path to net-zero: Exploring scenarios for a sustainable energy transition* (pp. 15–58). Springer Nature Switzerland. https://doi.org/10.1007/978-3-031-44514-9_2

Nwokolo, S. C., Singh, R., Khan, S., Kumar, A., & Luthra, S. (2023d). Impacts of climate change in Africa. In *Africa's path to net-zero: Exploring scenarios for a sustainable energy transition* (pp. 237–262). Springer Nature Switzerland. https://doi.org/10.1007/978-3-031-44514-9_7

Obydenkova, A. V., & Salahodjaev, R. (2017). Climate change policies: The role of democracy and social cognitive capital. *Environmental Research, 157*, 182–189. https://doi.org/10.1016/j.envres.2017.05.009

Ozgur, O., Aydin, L., Karagol, E. T., & Ozbugday, F. C. (2021). The fuel price pass-through in Turkey: The case study of the motor fuel price subsidy system. *Energy, 226*, 120399. https://doi.org/10.1016/j.energy.2021.120399

Parry, M., Arnell, N., Berry, P., Dodman, D., Fankhauser, S., Hope, C., & Hanson, C. (2009). Adaptation to climate change: Assessing the costs. *Environment, 51*(6), 29–36. https://doi.org/10.1080/00139150903337423

Patel, S., Soanes, M., Rahman, F., Smith, B., Steinbach, D., & Barrett, S. (2020). Trends and lessons from global climate funds. In *Good climate finance guide: Lessons for strengthening devolved climate finance* (pp. 33–46). International Institute for Environment and Development. Retrieved from http://www.jstor.org/stable/resrep28976.8

Pongratz, J., Schwingshackl, C., Bultan, S., Obermeier, W., Havermann, F., & Guo, S. (2021). Land use effects on climate: Current state, recent progress, and emerging topics. *Current Climate Change Reports*. https://doi.org/10.1007/s40641-021-00178-y

Prasad, R. S., & Sud, R. (2019). Implementing climate change adaptation: Lessons from India's national adaptation fund on climate change (NAFCC). *Climate Policy, 19*(3), 354–366.

Purcell, W., Neubauer, T., & Mallinger, K. (2023). Digital twins in agriculture: Challenges and opportunities for environmental sustainability. *Current Opinion in Environmental Sustainability, 61*, 101252. https://doi.org/10.1016/j.cosust.2022.101252

Raihan, A., & Tuspekova, A. (2022). Role of economic growth, renewable energy, and technological innovation to achieve environmental sustainability in Kazakhstan. *Current Research in Environmental Sustainability, 4*, 100165. https://doi.org/10.1016/j.crsust.2022.100165

Ranger, N., Marotta, F., Fankhauser, S., & O'Callaghan, B. (2023). *Reforming the global financial architecture to drive a resilient net-zero transition*. University of Oxford.

Safdie, S. (2023). *What are the three pillars of sustainable development?* Greenly Earth.

Samuwai, J., & Hills, J. M. (2018). Assessing climate finance readiness in the Asia-Pacific region. *Sustainability, 10*(4), 1192. https://doi.org/10.3390/su10041192

Schulz, K., & Feist, M. (2021). Leveraging blockchain technology for innovative climate finance under the green climate fund. *Earth System Governance, 7*, 100084. https://doi.org/10.1016/j.esg.2020.100084

Shah, M., & Steinberg, B. M. (2021). Workfare and human capital investment: Evidence from India. *Journal of Human Resources, 56*(2), 380–405. https://doi.org/10.3368/jhr.56.2.1117-9201R2

Shahbaz, M., Raghutla, C., Chittedi, K. R., Jiao, Z., & Vo, X. V. (2020). The effect of renewable energy consumption on economic growth: Evidence from the renewable energy country attractive index. *Energy, 207*, 118162. https://doi.org/10.1016/j.energy.2020.118162

Shivanna, K. R. (2022). Climate change and its impact on biodiversity and human welfare. *Proceedings of the Indian National Science Academy, 88*(2), 160–171. https://doi.org/10.1007/s43538-022-00073-6

Singh, R., Gupta, A., & Kandpal, V. (2024). ESG excellence in the digital world: Circular economy e-business solutions. In *Fintech for ESG and the circular economy* (pp. 193–204). Wiley. https://doi.org/10.1002/9781394234035.ch10

Singh, R., Khan, S., & Centobelli, P. (2022). Investigating the interplay between social performance and organisational factors supporting circular economy practices. *Sustainability, 14*(24), 16781. https://doi.org/10.3390/su142416781

Singh, R., Khan, S., Kumar, A., & Kumar, V. (2024). *Artificial intelligence enabled management: An emerging economy perspective*. Walter de Gruyter GmbH.

Singh, R., Kumar, K., & Khan, S. (2024). A comprehensive view of artificial intelligence (AI)–based technologies for sustainable development goals (SDGs). In *Artificial intelligence enabled management: An emerging economy perspective* (p. 183). Walter de Gruyter GmbH & Co. https://doi.org/10.1515/9783111172408-012

Sovacool, B. K., Bazilian, M., Griffiths, S., Kim, J., Foley, A., & Rooney, D. (2021). Decarbonizing the food and beverages industry: A critical and systematic review of developments, sociotechnical systems and policy options. *Renewable and Sustainable Energy Reviews, 143*, 110856. https://doi.org/10.1016/j.rser.2021.110856

Tanner, T., Bisht, H., Quevedo, A., Malik, M., & Nadiruzzaman, M. (2019). *Enabling access to the green climate fund: Sharing country lessons from South Asia*. University of Oxford.

Tolliver, C., Keeley, A. R., & Managi, S. (2019). Green bonds for the Paris agreement and sustainable development goals. *Environmental Research Letters, 14*(6), 064009. https://doi.org/10.1088/1748-9326/ab1118

UNEP. (2021). *Emission gas report, 2021*. UNEP.

United Nations Development Programme. (2011). *Blending climate finance through national climate funds*. UNDP.

World Bank. (2023). *How to protect, build, and use human capital to address climate change*. World Bank.

Zhang, Z., Wang, T., Blunt, M. J., Anthony, E. J., Park, A. H. A., Hughes, R. W., & Yan, J. (2020). Advances in carbon capture, utilization and storage. *Applied Energy, 278*, 115627. https://doi.org/10.1016/j.apenergy.2020.115627

Chapter 14
The Role of Green Space in Advancing Climate Neutrality Goals: Qualitative Case Study in Nigeria

Joy Eghonghon Akahome and Ben Pade Oputa

Abstract There is an international focus to achieve climate neutrality which has highlighted the important role of green spaces in urban environments especially in an academic environment. This study investigates green spaces role in advancing climate neutrality goals through a qualitative case study conducted at a Nigerian university. The research explores how university green spaces can contribute to the carbon sequestration, reduction of urban heat islands, and improvement of biodiversity by the support of the university's climate neutrality objectives. The collection of data was through in-depth interviews with the assistance of the university staff and students. There was also analysis in documentation and observations. The findings showed that green spaces are important for environmental sustainability and improving the wellbeing and academic staff and students. However, some challenges exist such as lack of maintenance, awareness, and poor funding poses serious challenges to the full implementation of these green spaces to contribute to climate goals. The study mentions that there is the need for strategic planning and management of green spaces so as to include them into broader institutional climate action plans. The practical implications of this study suggest that universities in developing countries can play an important role in climate neutrality efforts by the utilization of their green spaces. In addition, the study recommends provision for policy makers, management of universities and maybe urban planners that wish to improve the usefulness of green spaces to achieve sustainability targets. This study provides contribution to growing knowledge on the inclusion of green infrastructure and climate action with emphasis on the importance of green spaces toward internationally transition toward a sustainable future.

J. E. Akahome
Department of Entrepreneurship and Marketing, Federal University Otuoke, Otuoke, Nigeria

J. E. Akahome (✉) · B. P. Oputa
Federal University Otuoke, Otuoke, Bayelsa, Nigeria
e-mail: akahomejj@fuotuoke.edu.ng

B. P. Oputa
Department of Political Science, Federal University Otuoke, Otuoke, Nigeria

R. Singh and D. Crowther (eds.), *Climate Neutrality and Sustainable Ecosystems*, Approaches to Global Sustainability, Markets, and Governance,
https://doi.org/10.1007/978-981-96-7181-6_14

Keywords Green space · Climate neutrality · Nigerian University · Qualitative case study · Urban environment

14.1 Introduction

The fusion of green spaces in the university environment can play an essential role in promoting environmental sustainability and achievement of climate neutrality goals. Green spaces in this study is described as campus green spaces (CGS) (Adekunle & Basorun, 2016; Areola, 2022; Herath & Bai, 2024; Hernández & Wielgołaska, 2021). CGS is an important component of the university environment that contributes to the students and staff overall wellbeing while supporting the institution's sustainability objectives. According to Felsten (2009), it is necessary for a university campus to provide an environment conducive for learning, living, and working, which can directly impact students' social and personal development. Green spaces are important in this context because it can offer areas for relaxation, social interaction, and engagement with nature in combination to improve the quality of campus life (Abercrombie et al., 1998). Recently, universities around the world have recognized the importance of including green spaces into their sustainability strategies. Green spaces provides a range of benefits that includes improvement of the quality of air, reduction the effect of urban heat island, and improving biodiversity (Hipp et al., 2016). Additionally, these spaces contribute to climate neutrality by reducing carbon sequestration and supporting ecosystems that reduces the effects of climate change (Shanahan et al., 2016). As a result, many universities have been committed to expand and improve their green spaces as part of larger sustainability initiatives. There is a growing body of literature which describes the benefits of health and wellbeing for green spaces. This ranges from stress reduction (Hartig et al., 2014) to improve attention and foster creativity (Capaldi et al., 2014; Kuo, 2015). In addition, green spaces can cure loneliness, strengthen bonds in the community, and nurture environmental stewardship values (Public Health England (PHE), 2020; Kellert, 2012). The experiences students have in CGS contribute to their connection with nature and encourage pro-environmental behaviors (Chawla, 2015). Therefore, there is a fact that green spaces has an impact on the campus experience because many universities are striving to design or improve these areas to achieve a greener aesthetic and promote sustainable living (Tudorie et al., 2020). However, there is a lack of comprehensive data on how the specific types of green space is used and it's associated benefits among university students. Studies indicated that students who perceive their campuses as having high levels of "greenness" report better quality of life (Hipp et al., 2016; McFarland et al., 2008). Therefore, there is also research gap on how these spaces contribute to climate neutrality goals. The existing literature largely focuses on the aesthetic and wellbeing benefits of green spaces but there is limited understanding of how these areas can be designed and utilized strategically to reduce carbon footprints, improve biodiversity, and support a broad climate neutrality efforts on campuses.

Given the growing emphasis on sustainability, this study explores the green spaces role in advancing climate neutrality goals at a Nigerian university. Therefore, the research question asks *How can the lessons learned from this case study be applied to other universities aiming for climate neutrality?* By carrying out a qualitative case study, the research aims to understand how green spaces are utilized, perceived, and valued by students and staff. The findings from this study will provide useful information into the effectiveness of green spaces as a tool for promoting sustainability and offer recommendations for other universities aiming to achieve climate neutrality.

Significance of the Study

This explains the importance of the research and its potential impact. It can provide useful information on how green spaces can contribute to climate neutrality goals within a university setting. By focusing on a Nigerian university, the research highlights the potential for green initiatives in developing countries where resources and awareness may be limited. The findings can inform policy decisions, inspire similar initiatives in other institutions, and contribute to the global ideology on sustainable urban planning and climate action (Hsieh & Yeh, 2024; Zhang et al., 2021).

14.2 Literature Review

14.2.1 Conceptual Framework of Green Spaces

Green spaces is described as areas of vegetation. For example, parks, gardens, and urban forests are added into urban environments. The conceptual framework of green spaces is centered on their multi-functional role in improving urban ecosystems, human wellbeing and promoting biodiversity. These spaces are critical in reducing urban environmental challenges which are heat islands, air pollution, and stormwater runoff (Beatley, 2011). The framework focuses on the importance of green infrastructure which combines natural and semi-natural spaces into urban planning to provide ecosystem services. These services include purification of air and water, regulation of temperature, and carbon sequestration (Herath & Bai, 2024). Furthermore, there is a link between green spaces and social benefits such as recreational opportunities, improvement of mental health and social interactions makes them essential for sustainable urban development.

14.2.2 Climate Neutrality Goals: Definitions and Targets

Climate neutrality is often used synonymously with carbon neutrality because it is defined as achieving a balance between emitted and absorbed greenhouse gases

(GHGs) which results in a net-zero impact on the climate. This goal is relevant in reducing global temperature rise to below 1.5 °C above pre-industrial levels which is stipulated by the Paris Agreement (UNFCCC, 2015). The primary focus for the achievement of climate neutrality includes the reduction of GHG emissions through energy efficiency which transitioned to renewable energy sources and implements low-carbon technologies. In addition, carbon offsetting plays a significant role in compensating for unavoidable emissions (European Commission, 2019). Most countries and cities aim to achieve climate neutrality by 2050 with intermediate targets set for 2030 or earlier, this depends on the region.

14.2.3 Climate Neutrality Goals: Definitions and Targets

Definition of climate neutrality refers to achieving of net-zero human-caused greenhouse gas (GHG) emissions across all sectors not only carbon dioxide (CO_2). This means that any GHG emissions produced are balanced by an equivalent amount of emissions removed or offset which includes other gases such as methane (CH_4), nitrous oxide (N_2O), and fluoridated gases. The concept extends beyond carbon neutrality by addressing the full spectrum of greenhouse gases responsible for climate change. Definition of carbon neutrality focuses exclusively on balancing carbon dioxide (CO_2) emissions. The goal is to ensure that CO_2 can be released into the atmosphere through human activities (like burning fossil fuels) is counterbalanced by efforts to remove an equivalent amount of CO_2. This balance is often achieved through methods such as Afforestation: is the planting of new forests where there were none before. Reforestation: describes the replanting trees in areas that have been deforested. Carbon Capture and Storage (CCS): consist of technology capturing CO_2 emissions from sources like power plants and storing it underground to prevent it from entering the atmosphere.

Climate neutrality addresses all types of greenhouse gases and other environmental impacts that contributes to climate change. Greenhouse gas emissions are categorized into three different scopes to better understand and manage their sources: Scope 1 Emissions: This has direct emissions from sources owned or controlled by an organization, such as on-site fuel combustion and company-owned vehicles. Scope 2 Emissions: This has to do with indirect emissions from the consumption of purchased electricity, heat, or steam. These emissions occurs when energy is produced and it attributes to the organization using the energy. Scope 3 Emissions: Where indirect emissions is not owned or directly controlled by the organization but relates to its activities can include commuting of employees, business travel, waste disposal, and the entire supply chain. The targets for achieving climate neutrality are typically set within international agreements and national commitments: For example, the Paris Agreement adopted at COP 21 aimed to limit global warming below 2 °C above pre-industrial levels with the effort to reduce the temperature increase to 1.5 °C. To achieve this, global GHG emissions must reach net-zero by the year 2050. National and Corporate Targets: While many countries and companies have set their own targets to achieve climate or carbon neutrality, the European Union has committed itself to become climate-neutral by 2050 and several corporations have pledged to reach carbon neutrality within similar time frames. University Initiatives: For example, higher education institutions such as the University of Arizona have set ambitious targets to achieve climate neutrality by 2040. These targets are supported by strategies such as reducing Scope 2 emissions through renewable energy agreements and integrating climate resiliency into their operations and education programs.

14.2.4 Importance of Achieving Climate Neutrality by 2050

The goal of achieving climate neutrality by 2050 is necessary for reducing the most severe impacts of climate change. It involves coordinated effort among governments, businesses, and individuals to reduce emissions and transition to sustainable practices. The benefits include balancing global temperatures, reducing the frequency of extreme weather events, preserving biodiversity, and ensuring a fair distribution of the burdens and benefits of climate action. Furthermore, the achievement of climate neutrality requires a comprehensive approach that addresses all sources of greenhouse gas emissions. The goals set by international agreements, national policies, and corporate commitments highlight the urgency of the task. Also, striving for climate neutrality can safeguard the environment and ensure a sustainable future for every one.

14.2.5 The Role of Green Spaces in Urban Sustainability

Urban green spaces (UGS) are fundamental to the sustainable development of cities especially as urbanization increases globally with more than half of the world's population currently living in cities. The figure is expected to rise to 6.5 billion by the year, 2050. Urban areas face significant pressure on resources and environmental quality (Mostafavi & Doherty, 2010; Tudorie et al., 2020). As cities grow and expand, the environmental issues become more serious where sustainable urban development is essential. The inclusion of green spaces into urban planning is important for reducing these challenges and providing ecological, social, and economic benefits that are indispensable for urban sustainability. Urban green spaces includes forests, parks, ambience residential yards, and engineered green roofs provide a large multitude of ecosystem services essential for both human wellbeing and environmental health (Aronson et al., 2017; Rall et al., 2017). These services include environmental benefits such as reducing urban heat islands, reducing pollution of air and water, and sequestering carbon (Pataki et al., 2011; Swanwick et al., 2003). In addition, UGS play a good role in biodiversity conservation through the provision of habitats for urban wildlife which is increasingly important as urban sprawl encroaches on natural habitats (Kowarik, 2011).

In the social sense, green spaces contribute to urban population wellbeing by offering recreational opportunities, improving mental health, and fostering community interactions. These spaces are important for improving the quality of urban life especially in areas that are densely populated where access to nature can significantly reduce stress and promote physical activity (Li & Nassauer, 2020; Mostafavi & Doherty, 2010). In economic sense, (Ignatieva & Hedblom, 2018; Milesi et al., 2005) emphasized that UGS can increase property values, attract tourism, and lower energy costs through their cooling effects, while contributing to the economic resilience of cities.

14.2.6 Challenges and Emerging Trends in Urban Green Space Management

In the twenty-first century, fast urbanization created new challenges for urban sustainability. The continuous migration of people to cities puts huge pressure on limited resources which necessitates a rethinking of urban systems to incorporate more sustainable practices (Mostafavi & Doherty, 2010). Futuristic sustainable cities will build upon current environmental practices, such as increasing urban green spaces and adopting eco-friendly technologies to address these challenges (Tsui, 2019). The United Nations Sustainable Development Goals (SDGs) highlight the importance of urban sustainability particularly through SDG 11 which calls for the creation of "green public spaces" as an important element of sustainable cities (UN, 2015). This goal emphasizes the need for transformative urban planning that incorporates green spaces into the urban fabric to improve environmental quality, biodiversity and promote social equity.

14.2.7 Case Studies on Green Spaces and Climate Neutrality

Copenhagen and London provide leading examples of how urban green spaces can be effectively incorporated into urban planning to achieve sustainability goals. Copenhagen's "Urban Nature in Copenhagen 2015–2025" policy aims to create more green spaces and improve the quality of existing ones with the use of urban nature as a tool for climate adaptation and biodiversity enhancement (City of Copenhagen, 2015). The city's green planning tool and tree policy includes the ambitious goal of planting 100,000 new trees to demonstrate a commitment to integrating nature into the urban landscape for the benefit of current and future generations. In similar context, London's strategic approach to green infrastructure aims to make over half of the city's area green by 2050 with objectives for wildlife conservation, improvement of the natural habitats and leveraging green spaces as economic assets (Greater London Authority, 2018). The "Wild West End" project gives an example of this strategy by creating new green corridors that connect existing green spaces by the improvement of access to nature and support for urban biodiversity.

Both cities provide an illustration on how strategic urban planning can transform densely built environments into sustainable urban ecosystems. These initiatives not only improve urban resilience to climate change but also improve the quality of life for residents by making these cities more attractive places to live in and visit as well.

As urbanization continues to increase, the role of green spaces in urban sustainability becomes increasingly critical. Green spaces contribute to environmental protection, social wellbeing, and economic vitality, making them indispensable for the development of sustainable cities. The examples of Copenhagen and London has shown how cities can add green spaces into their urban planning strategies to achieve both local and global sustainability goals. However, as challenges of it's policies are faced in funding and implementation. This continued research and innovation are needed to develop highly functional green spaces in densely populated urban contexts.

Copenhagen, Denmark: Copenhagen has set an ambitious goal to become carbon neutral by 2025, with green spaces playing a vital role in this strategy. The city's integration of green roofs, urban forests, and parks not only sequesters carbon but also reduces urban heat and improves stormwater management. Copenhagen's "Cloudburst Management Plan" highlights the use of green spaces to mitigate flooding risks while enhancing urban resilience (City of Copenhagen, 2012).

Freiburg, Germany: Freiburg is known for its sustainable urban planning especially in the Vauban district which incorporates extensive green spaces. These areas support the city's low-carbon initiatives by reducing energy consumption, improving biodiversity, and promoting sustainable living practices. Freiburg's green spaces are key to its goal of become a model for sustainable urban living (Roseland, 2012).

Singapore: Singapore's "City in a Garden" vision provides example of the incorporation of green spaces into urban environments to promote sustainability. The city-state has incorporated vertical gardens, green roofs, and large public parks to improve biodiversity, reduce the urban heat island effect, and improve air quality. These initiatives contribute to Singapore's broader sustainability goals, including efforts to reduce carbon emissions and improve public health (Tan et al., 2013).

These case studies demonstrate the critical role of green spaces in achieving climate neutrality and urban sustainability. They highlight how green spaces can be effectively integrated into urban planning to enhance resilience to climate change, improve public health, and create more livable urban environments.

14.3 Research Design

A qualitative case study approach was employed which is appropriate for deep exploration of complex phenomena within the context of real life (Miles et al., 2014). This method allows for a comprehensive understanding of the role of green spaces in advancing climate neutrality goals at a specific Nigerian university and provides useful information that can be applied to other universities.

The three criteria for selecting the study for the Nigerian university include, Marselle et al. (2021), (1) Presence of Established Green Spaces: The university must have well-developed green spaces that contribute to its sustainability efforts. (2) Commitment to Climate Neutrality: The university should have policies or initiatives aimed at achieving climate neutrality. (3) Accessibility for Data Collection: The site must be accessible for conducting in-depth interviews, observations, and document analysis. A triangulation of data collection methods will ensure the validity and reliability of the findings. The primary methods include the use of semi-structured interviews with participants such as university administrators and staff responsible for maintaining green spaces, students, and faculty members involved in sustainability initiatives. The procedure is to conduct in-depth, semi-structured interviews to gather detailed information on the participant perceptions, experiences, and information that regards the role of green spaces in advancing climate neutrality. The questions are designed to explore how these spaces will contribute to environmental goals, the challenges faced, and the lessons learned. Therefore, the first step is the analysis that were conducted by making sure the transcripts are coded and analyzed thematically to identify common themes and insights. Secondly, observations were conducted on the green spaces which focused on their usage, maintenance, and the integration of sustainable practices. The documents of the physical characteristics, activities taking place, and interactions between people and the green spaces were documented and the Analysis of the Observation Notes were systematically reviewed and categorized to complement interview data. Thirdly, the document analysis consisted on reviewing relevant documents, which includes university policies, strategic plans, sustainability reports, and maintenance records, analyze these documents to understand the formal structures and policies that relates to green spaces and sustainability initiatives and finally extract and categorize information relevant to the study's research questions.

Application of the lessons learned to other universities involves the following steps: The identification of key success factors will determine the critical elements that contribute to the success of green spaces in advancing climate neutrality at the studied university. Secondly, developing best practices will create a set of best practices based on the findings, and include effective strategies for planning, developing, and maintaining green spaces. Thirdly, customization will lead to the adaptation of the best practices to fit the specific context and needs of other universities. This will put consideration into factors such as geographic location, available resources, and institutional priorities. Fourthly, facilitating knowledge sharing through workshops, seminars, and publications to broadly disseminate the findings and recommendations. And lastly, establish networks for collaboration and support among universities to share experiences, challenges, and solutions related to green spaces and climate neutrality. Following these methodological steps provides valuable insights into the role of green spaces in advancing climate neutrality goals and offers practical guidelines for other universities to copy and adapt the successful strategies identified in the case study.

14.4 Data Analysis

The goal of this data analysis is to fully understand the information and best practices from the case study on the role of green spaces in advancing climate neutrality at a Nigerian University. These information will be synthesized to develop actionable recommendations for other universities aiming to achieve similar goals.

14.4.1 Information on the Data Coding and Thematic Analysis

Step one focuses on Data Coding where the interview transcripts selects some key phrases and statements from the university staff and students. These phrases were codes and the codes include "challenges in implementation," "benefits of green spaces," "strategies for maintenance," and "stakeholder engagement." While the Observation Notes looks at the field note that include "usage patterns," "maintenance practices," "biodiversity," and "community involvement." Furthermore, from the document analysis, the researcher had codes from policy documents, sustainability reports, and strategic plans may include "institutional policies," "funding mechanisms," "educational programs," and "sustainability goals."

Step two focuses on Thematic Analysis that mentioned several themes such as "Environmental Impact" this relates to how green spaces contribute to carbon sequestration, temperature regulation, and biodiversity. Second theme looks at "Social and Educational Benefits" that relates to health and wellbeing, educational opportunities, and community engagement. Third theme looks at "Challenges and Barriers" that identifies obstacles such as funding limitations, maintenance challenges, and sociocultural barriers. Fourth theme is "Best Practices and Strategies" which highlights successful strategies for planning, developing, and maintaining green spaces, as well as effective stakeholder engagement and policy support.

The study led to a triangulation in order to validate findings and data from different sources that are cross-referenced. These are interviews from different stakeholders that are compared to identify similarities and differences. Direct observations were used to corroborate or challenge interview data. And policy and strategic documents provide a formal context that supports or explains the practical insights from interviews and observations.

During the first step of interpretation and synthesis, the researcher identified key success factors, which was strong institutional support: Commitment from university leadership and alignment of green space initiatives with institutional goals. Consequently, stakeholder engagement involved students, staff in planning and maintaining green spaces. Integration with academic programs with the use of green spaces as living laboratories for environmental education and research.

Step 2 involved the development of best practices such as comprehensive planning of integrating green space development into campus master plans with clear sustainability targets. Diversified Funding is part of the step in exploring multiple funding sources, including government grants, private donations, and partnerships with NGOs and educational integration of embedding green space initiatives into curricula and research projects to enhance learning and innovation.

Step 3 emphasizes the customization for other universities through the adaptation of best practices to fit the specific geographic, cultural, and institutional contexts of other universities and implementing scalable and flexible green space projects that can be adjusted based on available resources and specific needs.

Step 4 focuses on knowledge sharing and collaboration where there is organization of knowledge-sharing events to disseminate findings and recommendations and the establishment of networks for collaboration and support among universities to share experiences and resources.

In summary, the data analysis reveals some key lessons and best practices that can be applied to other universities that aim for climate neutrality. By focusing on strong institutional support, stakeholder engagement, integration with academic programs, and adaptable strategies, other universities can successfully replicate and customize these initiatives to achieve their climate neutrality goals.

14.4.2 Discussion of Findings

The discussion of the findings explains the case study of green spaces at a Nigerian university and provides valuable information on how these initiatives contribute to climate neutrality. Through the examination of the benefits, challenges, and best practices observed, we can derive lessons applicable to other universities that aim for similar sustainability goals. Environmental impact theme found that green spaces significantly contribute to carbon sequestration, temperature regulation, and biodiversity improvement. Trees and vegetation can absorb carbon dioxide that helps to reduce or control climate change. These spaces also reduce the urban heat island effect, leading to lower energy consumption for cooling "*the trees and gardens not only beautify our campus but also play a crucial role in capturing carbon emissions and cooling the environment. This dual benefit mentions their importance in our sustainability strategy.*" University Administrator Research indicates that urban green spaces can absorb substantial amounts of CO_2 which contributes to climate change mitigation (Haq, 2011; Shanahan et al., 2016). Moreover, green infrastructure helps regulate temperatures and supports urban biodiversity, which is relevant for ecosystem resilience (Hernández & Wielgołaska, 2021).

Social and educational benefits theme found that green spaces improves the wellbeing of students and staff by providing areas for relaxation, physical activity, and social interaction. They also serve as outdoor classrooms, facilitating hands-on learning and research opportunities. Faculty administrator suggests that "*These green areas have become integral to our campus life and students use them for study sessions, sports, and even research projects. It's a living laboratory that improves their educational experience.*" Studies show that access to green spaces improves mental and physical health. It reduces stress and promotes social cohesion (Farahani & Maller, 2018; Fields in Trust, 2018). Additionally, they offer unique educational opportunities, improve learning outcomes and fostering environmental stewardship (Dhaliwal, 2017).

Challenges and barriers theme found that despite the benefits, several challenges impede the effective implementation and maintenance of green spaces. These include funding constraints, maintenance issues, and socio-cultural barriers. The university staff narrates that "*Securing adequate funding for maintenance is a persistent challenge. While the initial development was well-supported, ongoing upkeep requires sustained financial and community support.*" This supports the statement that funding and maintenance are common challenges in sustaining green spaces, as highlighted in urban planning literature (Govindarajulu, 2014; Gupta et al., 2012). Socio-cultural barriers also play a role, affecting community engagement and the perception of green space initiatives (Elmqvist et al., 2016).

Best practices and strategies theme found that successful green space initiatives are characterized by strong institutional support, stakeholder engagement, and integration with academic programs. These factors contribute to the sustainability and scalability of green space projects. The staff says that "*engaging students and staff in the planning and maintenance processes has been crucial. Their involvement ensures the spaces are well-utilized and valued by the entire campus community.*" It supports this statement that effective green space projects often rely on collaborative planning and stakeholder involvement (Dhaliwal, 2017; Hernández & Wielgołaska, 2021). Integrating green initiatives into the academic curriculum enhances their relevance and impact (Belle & Deserti, 2024; Seyfang, 2010).

The application to other universities has several recommendations such as institutional support to ensure commitment from university leadership and align green space initiatives with institutional goals (Shanahan et al., 2016). Secondly, stakeholder engagement that involves students and staff in planning to foster ownership and sustainability (Farahani & Maller, 2018). Studies have shown that integration with academics can suggest the use green spaces as educational tools, integrating them into curricula and research activities (Dhaliwal, 2017; Fields in Trust, 2018) and securing diverse funding sources to include government grants, private donations, and partnerships with NGOs (Govindarajulu, 2014; Gupta et al., 2012).

In the area of adapting strategies to fit specific contexts, it is important to consider geographic, cultural, and institutional factors. Scalable solutions and flexible project designs allow for adjustments based on available resources and needs (Hernández & Wielgołaska, 2021; Pataki et al., 2011). And facilitate knowledge sharing through workshops, seminars, and collaborative networks. Disseminate findings and best practices to encourage widespread adoption and adaptation of green space initiatives (Belle & Deserti, 2024; Seyfang, 2010).

14.5 Conclusion

The study on the role of green spaces in advancing climate neutrality goals at a Nigerian university demonstrates that these spaces are not just aesthetics additions, they are critical components of environmental sustainability and student wellbeing. The research mentions that green spaces can contribute significantly to the university's efforts in the achievement of climate neutrality by sequestering carbon, enhancing biodiversity, and mitigating the urban heat island effect. Additionally, these spaces offer substantial social and educational benefits, including improving mental health, reducing stress, and fostering a sense of community among students and staff. The study also mentions the challenges of maintaining these green spaces which are funding constraints and socio-cultural barriers and emphasizes that with strong institutional support, stakeholder engagement, and integration into academic programs, these challenges can be overcome. The findings suggest that the successful integration and management of green spaces not only advance climate neutrality goals but also enhance the overall quality of life on campus, making them a vital part of any university's sustainability strategy. The insights gained from this case study provide a valuable framework for other universities aimed at achieving similar goals by demonstrating that green spaces are a practical and effective tool for promoting environmental stewardship and sustainability in higher education. However, future research can focus on longitudinal studies to ascertain the long-term impacts of green spaces on climate neutrality and campus sustainability. There should be comparative studies across different geographic and cultural contexts to identify universal and context-specific best practices. And finally, future studies should investigate innovative funding and maintenance models to ensure the sustainability of green spaces.

References

Abercrombie, N., Gatrell, A. C., & Thomas, C. J. (1998). The university campus environment. *Higher Education Research & Development, 17*(2), 133–148.

Adekunle, O., & Basorun, J. (2016). The evolving roles of landscaping in campus space management: Ekiti state university, ado-Ekiti, Nigeria in focus. *Journal of Environmental Protection, 7*, 1380–1388. https://doi.org/10.4236/jep.2016.710119

Areola, A. A. (2022). Environmental justice and green spaces in Ibadan Metropolis, Nigeria: Implications on sustainable development in urban construction. *Environmental Sciences Proceedings, 15*(1), 57.

Aronson, M. F. J., et al. (2017). Biodiversity in the city: Key challenges for urban green space management. *Frontiers in Ecology and the Environment, 15*(4), 189–196.

Beatley, T. (2011). *Biophilic cities: Integrating nature into urban design and planning*. Island Press.

Belle, B. M., & Deserti, A. (2024). Urban greening plans: A potential device towards a sustainable and co-produced future. *Sustainability, 16*(12), 5033.

Braun, V., & Clarke, V. (2006). Using thematic analysis in psychology. *Qualitative Research in Psychology, 3*(2), 77–101.

Capaldi, C. A., Dopko, R. L., & Zelenski, J. M. (2014). The relationship between nature connectedness and happiness: A meta-analysis. *Frontiers in Psychology, 5*, 976. https://doi.org/10.3389/fpsyg.2014.00976

Capaldi, C. A., Passmore, H.-A., Nisbet, E. K., Zelenski, J. M., & Dopko, R. L. (2015). Flourishing in nature: A review of the benefits of connecting with nature and its application as a wellbeing intervention. *International Journal of Wellbeing, 5*(4), 1–16.

Chawla, L. (2015). Benefits of nature contact for children. *Journal of Planning Literature, 30*(4), 433–452.

Choi, H. (2022). Future cities. In *The Routledge handbook of sustainable cities and landscapes in the Pacific rim* (1st ed., pp. 11–21). Routledge.

City of Copenhagen. (2012). *Copenhagen climate adaptation plan*. Retrieved from https://kk.sites.itera.dk/apps/kk_pub2/index.asp?mode=detalje&id=942

City of Copenhagen. (2015). *Urban nature in Copenhagen 2015–2025*. City of Copenhagen.

Creswell, J. W., & Poth, C. N. (2016). *Qualitative inquiry and research design: Choosing among five approaches*. Sage Publications.

Dhaliwal, S. (2017). Types of vegetable gardens. In *Handbook of vegetable crop* (pp. 18–33). Kalyani Publishers.

Elmqvist, T., Zipperer, W. C., & Güneralp, B. (2016). Urbanisation, habitat loss and biodiversity decline: Solution pathways to break the cycle. In K. C. Seto, W. D. Solecki, & C. A. Griffith (Eds.), *The Routledge handbook of urbanization and global environmental change* (pp. 139–151). Routledge.

European Commission. (2019). *The European green deal*. Retrieved from https://ec.europa.eu/green-deal

European Commission. (2020). *The European green deal*. Retrieved from https://ec.europa.eu/clima/policies/eu-climate-action/law_en

Farahani, L. M., & Maller, C. (2018). Perceptions and preferences of urban green spaces: A literature review and case study of Rawalpindi and Islamabad. *Journal of Urban Design, 23*(1), 1–18.

Felsten, G. (2009). Where to take a study break on the college campus: An attention restoration theory perspective. *Journal of Environmental Psychology, 29*(1), 160–167.

Fields in Trust. (2018). *Revaluing parks and green spaces: Measuring their contribution to individual wellbeing*. Fields in Trust.

Govindarajulu, D. (2014). Urban green space planning for climate adaptation in Indian cities. *Urban Climate, 10*, 35–41.

Greater London Authority. (2018). *London environment strategy*. GLA.

Gupta, K., Kumar, P., Pathan, S. K., et al. (2012). Urban neighborhood green index—A measure of green spaces in urban areas. *Landscape and Urban Planning, 105*, 325–335.

Haq, S. M. A. (2011). Urban green spaces and an integrative approach to sustainable environment. *Journal of Environmental Protection, 2*(5), 601–608.

Hartig, T., Mitchell, R., de Vries, S., & Frumkin, H. (2014). Nature and health. *Annual Review of Public Health, 35*, 207–228.

Herath, P., & Bai, X. (2024). Benefits and co-benefits of urban green infrastructure for sustainable cities: Six current and emerging themes. *Sustainability Science, 19*(4), 1039–1063. https://doi.org/10.1007/s11625-024-01475-9

Hernández, J. G. V., & Wielgołaska, J. Z. (2021). Urban green infrastructure as a tool for controlling the resilience of urban sprawl. *Environment, Development and Sustainability, 23*, 1335–1354.

Hipp, J. A., Gulwadi, G. B., Alves, S., & Sequeira, S. (2016). The influence of the campus environment on students' perceived quality of life. *Landscape and Urban Planning, 148*, 102–114.

Hsieh, Y. L., & Yeh, S. C. (2024). The trends of major issues connecting climate change and the sustainable development goals. *Discover Sustainability, 5*, 31. https://doi.org/10.1007/s43621-024-00183-9

Ignatieva, M., & Hedblom, M. (2018). An alternative urban green space management strategy. *Journal of Environmental Planning and Management, 61*(11), 1976–1995.

Kowarik, I. (2011). Novel urban ecosystems, biodiversity, and conservation. *Environmental Pollution, 159*(8–9), 1974–1983.

Kuo, F. E. (2015). How might contact with nature promote human health? Promising mechanisms and a possible central pathway. *Frontiers in Psychology, 6*, 1093.

Li, J., & Nassauer, J. I. (2020). Cues to care: A systematic analytical review. *Landscape and Urban Planning, 201*, 103821. https://doi.org/10.1016/j.landurbplan.2020.103821

Marselle, M. R., Lindley, S. J., Cook, P. A., & Bonn, A. (2021). Biodiversity and health in the urban environment. *Current Environmental Health Reports, 8*(1), 146–156. https://doi.org/10.1007/s40572-021-00288-6

McFarland, A. L., Waliczek, T. M., & Zajicek, J. M. (2008). The relationship between student use of campus green spaces and perceptions of quality of life. *HortTechnology, 18*(2), 232–238.

Miles, M. B., Huberman, A. M., & Saldaña, J. (2014). *Qualitative data analysis: A methods sourcebook*. Sage Publications.

Milesi, C., Running, S. W., Elvidge, C. D., Dietz, J. B., Tuttle, B. T., & Nemani, R. R. (2005). Mapping and modeling the biogeochemical cycling of turf grasses in the United States. *Environmental Management, 36*(3), 426–438. https://doi.org/10.1007/s00267-004-0316-2

Mostafavi, M., & Doherty, G. (2010). *Ecological urbanism*. Lars Müller Publishers.

Paris Agreement. (2015). *United Nations Framework Convention on Climate Change (UNFCCC)*. Retrieved from https://unfccc.int/process-and-meetings/the-paris-agreement/the-paris-agreement

Pataki, D. E., Carreiro, M. M., Cherrier, J., Grulke, N. E., Jennings, V., Pincetl, S., Pouyat, R. V., Whitlow, T. H., & Zipperer, W. C. (2011). Coupling biogeochemical cycles in urban environments: Ecosystem services, green solutions, and misconceptions. *Frontiers in Ecology and the Environment, 9*(1), 27–36. https://doi.org/10.1890/090220

Patton, M. Q. (2015). *Qualitative research & evaluation methods: Integrating theory and practice*. Sage Publications.

Public Health England (PHE). (2020). *Improving access to green spaces: A new review for 2020*. Public Health England.

Rall, E., Bieling, C., Zytynska, S., & Haase, D. (2017). Exploring city-wide patterns of cultural ecosystem service perceptions and use. *Ecological Indicators, 77*, 80–95. https://doi.org/10.1016/j.ecolind.2017.02.00

Roseland, M. (2012). *Toward sustainable communities: Solutions for citizens and their governments*. New Society Publishers.

Seyfang, G. (2010). Community action for sustainable housing: Building a low-carbon future. *Energy Policy, 38*(12), 7624–7633.

Shanahan, D. F., Bush, R., Gaston, K. J., Lin, B. B., Dean, J., Barber, E., & Fuller, R. A. (2016). Health benefits from nature experiences depend on dose. *Scientific Reports, 6*, 28551. https://doi.org/10.1038/srep28551

Swanwick, C., Dunnett, N., & Woolley, H. (2003). Nature, role and value of green space in towns and cities: An overview. *Built Environment, 29*(2), 94–106.

Tan, P. Y., Wang, J., & Sia, A. (2013). Perspectives on five decades of the urban greening of Singapore. *Cities, 32*, 24–32.

Tudorie, M., Vallés-Planells, M., Gielen, E., Arroyo, P., & Galiana, F. (2020). Campus green spaces in university sustainability plans. *Journal of Cleaner Production, 261*, 121079.

U.S. Environmental Protection Agency (EPA). (2021). *Greenhouse gas emissions: Understanding global warming potentials*. Retrieved from https://www.epa.gov/ghgemissions/understanding-global-warming-potentials

UN. (2015). *Transforming our world: The 2030 agenda for sustainable development*. United Nations.

UNFCCC. (2015). *Paris agreement*. United Nations Framework Convention on Climate Change. Retrieved from https://unfccc.int/process-and-meetings/the-paris-agreement/the-paris-agreement

Zhang, Y., Pan, C.-L., & Liao, H.-T. (2021). Carbon neutrality policies and technologies: A scientometric analysis of social science disciplines. *Frontiers in Environmental Science, 9*, 761736. https://doi.org/10.3389/fenvs.2021.761736

Chapter 15
Climate Finance: Valuing Nature, Action, Global Movements, and Local Framings

Mukesh Babu Gupta, Anil Kumar, Anup Kumar Roy, and Neha Nupoor

Abstract Notably, climate funding will be essential for supporting the sector in responding to and transitioning from its current path as much of economic activity shifts towards a low-carbon economy, increasing resilience. This chapter contributes to the literature on climate funding and is guided by four main focal points: assessing nature, funding for action, seeing global scales and being local. The paper explores the theoretical basis of economic appraisal of nature and asserts that market mechanisms often overlook ecosystems, as well biodiversity. It promotes a holistic appraisal of the ecology, society, and culture in nature. In this context, the essay discusses how climate finance acts as an enabler to drive action and argues that financial flows must converge with climate objectives besides calling for new financing mechanisms capable of mobilizing investment in sustainable development. State policy, international cooperation, and climate finance agendas are influenced by the Paris Agreement and UN Sustainable Development Goals. The study concludes with a theoretical synthesis that bridges the world attention on global imperatives and how they can inform real-world context, locally. It posits a more holistic, inclusive approach to climate fund management that upholds planet and people prosperity.

Keywords Climate finance · Climate change · Global movements · Biodiversity · Sustainability

M. B. Gupta
Faculty of Commerce, Sarala Birla University, Ranchi, Jharkhand, India

A. Kumar
School of Business, University of Petroleum and Energy Studies, Dehradun, Uttarakhand, India

A. K. Roy
Department of Commerce, Guru Ghasidas Vishwavidyalaya, Bilaspur, Chhattisgarh, India

N. Nupoor (✉)
Faculty of Management, Sarala Birla University, Ranchi, Jharkhand, India

R. Singh and D. Crowther (eds.), *Climate Neutrality and Sustainable Ecosystems*, Approaches to Global Sustainability, Markets, and Governance,
https://doi.org/10.1007/978-981-96-7181-6_15

15.1 Introduction

Climate financing has emerged as an important way of fighting climate change but it requires more than just money in order to be effective. In climate funding there is need to value the environment, promote pragmatic approaches, understand the trends of the world, and recognize the views of the region. Due to the need for successful financial interventions to be relatable with local settings, needs and knowledge systems is why debates often emphasize on responding from a climate financing perspective based on locally held views, Joshi et al. (2024). Through a review of the literature, this work demonstrates how climate money supports sustainable development by analysing the complex interrelationships between global policies or programmes and local contexts. In view of this, the research showed that nature appreciation is beyond economic assessments and needs ecological and social fundamentals for sustainability. The study also suggests that while global movements affect overall policy, climate funding activities often win by learning from local contexts, which consider such social, economic, and ecological realities of the world (Nwokolo et al., 2023a). As it relates, understanding these connected matters expand the knowledge of how climate money may be most effectively spent to combat climate change and advance the more general goal of a green and fair Earth. The financing of climate action is emerging as the critical lifeline of the international financial structure. It looks for the reduction of climate change and the fostering of a low-carbon economy (Boissinot et al., 2016). But it is not just a small segment of the financial industry; its value relies greatly in the progression it undergoes. Richardson (2014) goes further to say that all financial activities will at some point involve climate problems and standards. The developed countries plan to provide $100 billion/year by the year 2020 with a view of aiding the poor countries to mitigate climate change. It is argued that funding is very important for local adaption initiative and many other national climate change programmes (Bloch, 2011). With regard to climate funding, multilateral development banks are depended on for the long-term, high-risk activities that help to protect the environment. However, the connection of financial systems and climate risk management and the reinforcement of international cooperation are still the issues (Richardson, 2014). Climate finance is a very important tool towards addressing the challenges of climate change in the world. Green bonds and climate insurance may help bring an answer to the financing gap and help development sustainable solutions in the private sector (Boissinot et al., 2016). These strategies help in enhancing investment and financial freedom. It ensures initiatives of such acts are for the improvement of the environment and the economy. Since governments and organizations have come to appreciate these strategies, co-investment is required to amplify the impact of such change efforts and bring about transformative change (Bhattacharyay, 2021). It means that cooperation can foster new financing solutions that use both public and private money to build greater communities in the worldwide. The involvement of the local communities will guarantee project ownership and devotion since projects are aligned to their specific needs (Steffen & Michaelowa, 2022). Data analytics may also enhance the understanding of the performance of such efforts, so changes may be made quickly where necessary to meet community needs (Mohamadi, 2021). By intentional use and partnership with technology, we are able to identify and design possibilities that are able to evolve. Our solutions address current problems intervening in people's lives and they create the basis for future change and power (Alonso & Marqués, 2019). This makes individual initiatives and collective one as well and provides a portable system of support so that communities can address the needs as they transform over the time. In this way, we can learn how to rise up from failures and create new works and thus, give people the chance to decide their destinies. This will in turn cultivate a culture of ongoing innovation and cooperation towards success as Meltzer (2016) explains. Education and skill development programmes could further this foundation putting in place measures that enable people to harness the talents and help in the society.

15.2 Valuing Nature

Appreciation in climate financing does not revolve around the donation of funds towards the combat of climate change. Climate funding is required to better understand natural products and systems. Some of the worst impacts of climate change are on developing countries because the source of their financial growth is in biodiversity resources and they have little capacity to adjust (Echeverry, 2016). These sources of finance enable these countries to fight climate change by enhancing sustainability as well as social cohesiveness (Jakob et al., 2015). Currently, climate financing focuses on the sequestration of greenhouse gases in developed countries instead of funding adaptation programmes to the high climate risk regions in developing nations. Such a strategy has impacted the country with poor climate change adaptation and infrastructure 2105 (Michaelowa & Namhata, 2022; Pickering et al., 2015).

15.2.1 Climate Finance Raising and Use

Climate change cannot be tackled by money alone. This page looks at climate finance; specifically, climate ideas, climate realities, climate schemes, and climate stories. The evidence proves that climate funding serves the cause of sustainable development as it helps to unveil interdependency between words and deeds at the international and domestic levels. This paper finds that the concept of sustainability is closely associated with ecological and social aspects which are not captured by economic evaluations of environmental recreation. Particularly, analysis presented here shows indeed global movements have an impact on policy. This way, it can be seen that focusing on local socioeconomic and environmental factors, enhances the climate financing programmes. These related problems are explored to explain how climate finance might more effectively manage the climate change, sustainability, and equity.

The findings suggest a financially integrated architecture fitting local stimulus and worldwide demands for environmental climate funding for the benefit of the society. To carry out the understated evaluations on ecosystems and natural resources require climate funding. While the developing countries depend on biodiversity to transform their economy and therefore are highly vulnerable to the effects of climate change since they cannot adapt (Echeverry, 2016). Climate finance enhances these nations environment and society (Jakob et al., 2015). Climate money today funds developed countries' greenhouse gas emission cuts, not developing countries' climate vulnerability. That has led to poor climate change adaptation and infrastructure (Michaelowa & Namhata, 2022; Pickering et al., 2015). Poor countries have currently an open budget in terms of biodiversity that they need to fund in order to reverse the trend in biodiversity loss and natural resource depletion.

Raising and Effective Use of Climate Finance

It was, also notes that there is need to increase climate money for adaptation in the poor countries and argues for the use of debt-swap for nature and green bonds for dealing with critical challenges being posed by climate change. It can thus be argued that through engaging with many stakeholders, it is possible to improve upon these financial systems and support the growth of vulnerable areas through sharing information and technology (Das, 2022). More so, there is a call for proper monitoring and evaluation of the effectiveness of these financial tools to enhance the objectives of ecological conservation and climate change mitigation on needy developing nations (Fujikura, 2022). Before climate funding decisions are made, indigenous peoples and local communities should be involved in the decisions that are made about how the funds are used, in order to ascertain the focus of projects to address climate change and reflect the culture of people most affected (Fujikura, 2022; Okoyeuzu & Ukpere, 2022). Therefore, the gender-responsive climate finance initiatives are possible to be assistance for women and possible to support for gender equal within the sustainable development. Perhaps, increasing the knowledge base and communication supported by capacity-building programmes may also help local communities in creating their climate adaptation and mitigation plans and ensure continuity and sustainability (Nwokolo et al., 2023b). It is possible for climate financing initiatives to increase biodiversity conservation and climate resilience in poor nations via the integration of scientific advancements and novel technologies: this may be achieved through collaborations with universities and other research based institutions.

15.2.2 Natural Resource Valuation and Climate Finance

Sustaining the environment for climate finance purpose requires that the natural resources and ecosystems as well be valued. Valuation helps economists to study natural capital and make investments that is good for humans since it gives monetary values to nature and the environment. As pointed out by Williams (2017), this valuation affords a holistic and tenable, systematic finance strategy. It does so using metrics that assess the inherent size and state of nature, plus accounts that record the costs of maintaining and maintaining it. It is through the policies that the policymakers may help in the evaluation of the socioeconomic costs and benefits of environmental protection and the sustainable development of climate finance for environmental management through quantifying the ecosystem service values as pointed out by Ponera (2013). There is also a need of addressing issues related to the fast accumulation of funds in support of climate and nature initiatives (Sharma & Kaushik, 2015). This is so because degradation of environment and under-provision of natural capital are costly to the economy. Akhter et al. (2013) proposed that integrating natures based solution in the financial decision might possibly enhance climate change resilience and overall sustainability. Chitengi and Sakapaji. (2022) opine that climate solutions that are environment-linked financial instruments such as green bonds and sustainable-linked loans could spur private sector participation in climate change adaptation and enhanced nature conservation. This may lead to creating new sources of financing for PPF and boosting of public private partnerships.

15.3 Action

Climate finance sources are mostly dispersed and they are grouped into a wide range of institutions. Structures such as Multilateral Development Banks also known as MDBs, and Green Finance such as green bonds and international climate finance from the developed countries (Rossati & Rossati, 2013). In addition, these sources are imperative in financing climate change mitigation activities since they offer funding for initiatives to mitigate carbon emissions as well as low-carbon development initiatives sustainable development (Richardson, 2014). Such MDBs as World Bank and the European are among the most common examples of such organizations. Global Investment Bank, have played colossal roles in the propagation of climate finance in the world's regions and industries, especially considering adaptation and the comprehensive agendas of implementation of mitigation measures (Zhu et al., 2013). Some of the advanced financial tools such as the green bonds have even grown in size and market rapidly, take-up, becoming a more than a trillion-dollar industry and occupying a significant share (Khan et al., 2024). Debt markets across the world, so as to support renewable energy investments in physical capital stock other climate change project (Banga, 2019). Climate finance from the developed countries for developing countries from different international sources helps the developing countries transform into ready and sustainable sober carbon free economies. The developed nations have been able to adopt gigantic social policies in the developing nations, funded by carbon credits are now due to be halted and scrapped investment in renewable energy projects and climate change measures that could have been financed by carbon credits, economically unsustainable, by playing a very significant role worldwide in combating climate change.

Further, availability of technology and expertise from these partners has ensured that local people played an active role in the development decision-making processes and succeeded its application towards the lasting effectiveness of climate action initiatives (Varadarajan & Chitra, 2022). Moreover, the partnership GPS is collaborative because it has made the emerging of new partnerships easy systems that facilitate the continuous provision of support and sharing of experiences and other forms of benchmarks with the stakeholders. A cause and effect relationship that goes beyond the handling of individual projects to cause a chain effect of altering how climate finance is conceptualized and practised in the global level (Steffen & Michaelowa, 2022). Therefore, these partnerships apart from promoting the equipping of communities and institutions with sustainable and innovative productive inputs have helped facilitate the implementation of the intended positive changes, climate solutions but have also encouraged organizational development in ideas as well as information dissemination, considered crucial in facing up with the complex that climate change issues present (Zhu et al., 2013).

For this reason, these partnerships have not only improved the relative success of climate action initiatives but have also set the ground for even better and more long-lasting solutions to be adopted at the global level (Sathaye et al., 2006). Moreover, the multi-stakeholder engagements between governments, business, and non-profit making organizations within these partnerships have indeed resulted to development of new technology and policies. That has minimized on the emission of carbon and encouraged on the conservation of the environment worldwide. Also, stewardship involve the provision of benchmarks and information so that the knowledge can be passed with partners that has stimulated innovation and the pace of deployment of environment-friendly technologies in different sectors and therefore results in the process of changing towards a more sustainable future (Sathaye et al., 2006). Mobilizing climate finance for supporting climate action at global level has been described by the following strategies:

Creativity and variety in the approaches to managing the de-risking process.

Debentures or Bonds are some of the financial instruments. Multilateral Development Banks (MDBs) should promote private capital mobilization via specific de-risking approaches designed to encourage investment in climate projects.

Survey also showed that the actual stocktake was (Sathaye et al., 2006), both green bonds and Islamic Sukuk, as well as other conventional bonds, have become evident as critical instruments for financing climate change mitigation projects, including the use of such approaches as the blended.

Finance and public-Private Partnership and improving their efficiency (Cheng, 2017).

Finally, in the least developed country, where aid for trade can catalyse climate finance in regard to improving the feasibility of projects and attracting private financing for projects through co-financing and capacity building.

These strategies summarize the idea of group cohesion, as a fundamental approach to performing well. Private sector is necessary to guarantee that the funding gets to some of the needy societies effectively (Sathaye et al., 2006). They can bring the development of specific bilevel cooperation with potential solutions, concerning national educational systems and that can contribute to create proper adjustments.

Sustainability options that meet local demand as they additionally play a part in achieving other climate interventions on a global scale, in the long run. Improvement of a stronger and a more sustainable society that will benefit everyone. Moreover, integrating local knowledge and professionalism of these collaborations may improve more effectiveness of intervention assuring as they are good for the culture, and the environment as well, according to Newman and Perl (2014). Private sector investments, particularly green bonds, enable sustainable investment decision-making and increase transparency around environmental, social, and governance (ESG) factors. Public funding was previously the focus. Recently pledged US $1.7 billion at CoP26 emphasizes the need for inclusive climate financing that protects Indigenous and local community women and girls' rights and funding. Nations should develop comprehensive financial policies to handle climate threats, while poor economies need international aid. Carbon finance and energy help low-carbon economic investments, therefore helping climate action.

15.3.1 Technology in Climate Finance

Thus, at the same time with increasing the transparency and differentiation of sustainability-related information, initiatives to promote sustainable finance might also amplify market challenges, transparency through applying the concept of the blockchain and AI in financing transferring to implement nature-based solutions in an effective and efficient way, that is taking minimum time to allow (Sullivan, 2012). It is possible for clients and other stakeholders, so as to track and evaluate performance in the real time using these technologies (Clapp & Sillmann, 2019). This enables assessment of nature-based solutions and the effective utilization of changes that are initiated based on research analysis. According to Nesshöver et al. (2017) ability of passing knowledge from one generation to the other, therefore, preserving and reactivating classical epistemology of interaction between man and the natural environment.

15.3.2 Indicators for Measuring the Success of Climate Finance Initiatives

Management GGDP is the measure of economic development under the sustainable development and climate, or 'War on Climate Change', is good for climate financing programmes measurement with mitigation. The effectiveness of climate finance is also quantified in relation to environmental taxes, investments into renewable energy, and climate finance allocations (Morin-Labatut & Akhtar, 1992). National climate finance institutions (NCFIs) may assist to determine the effectiveness of climate finance. The information based on NCFI Indicators on Legislative Frameworks, money and money-movement and management, openness and accountability may assess the country characteristic, follow-up climate financing activities (Natarajan, 2002). These metrics look at such factors as economic returns and to assess the relevance of financial policies for climate goals to mobilize, manage, and allocate resources. Such gender-sensitive climate financing frameworks may also capture the needs of women and girls and risks, the levels of gender, and favourable aspects for climate change, such as community resilience. People's variability of opinions and ideas can let NCFIs develop innovative climate financing solutions for a more extensive range of concerns and prospects (Rabeil et al., 2014).

15.4 Worldwide Mobilization and Funding for Climate Change

Climate financing supports the development of renewable energy projects as well as climate change adaptation projects. Of course, one cannot talk about the contribution of Swiss organizations to the struggle against climate change on a global level. This comprises of both public and private funds with the CTB implication mainly from the Clean Technology Fund, Green Climate Fund, and the World Bank Group (Richardson, 2014). Migration Management therefore requires increased funding support to support climate migrants improve their life and could change themselves in response to the circumstances and the host population of individuals (Ye, 2014). Global warming and the necessity of the importance of the non-renewable energy sources are emphasized in the international level and the analysis, issue of climate, and the implication of political will. Therefore, climate migrants should be protected and accommodated to ensure that they have a better future to new communities with dignity and well-being through firm and compound policies (Gopalakrishnan, 2018). In order to address climate migration's complex problems, governments, international organizations, and local communities therefore must collaborate (Warren, 2020).

Early warning system and climate change adaptation-oriented studies can possibly decrease the psychological impacts of climate change migration to affected and threatened groups and the modelling of climate change resilient communities in vulnerable regions (Yan, 2023).

Policies that protects rights of climate finance and education and training may also help. Industry structures are also important, at least, as Ledeneva and Lukyanets (2022) discussed little earlier. In order to deal with climate migration's intricate, it is international relations that individuals, governments, organizations, and communities have to address concerns, Singh et al. (2025).

Climate migration and infrastructure development with focus on sustainable infrastructure and renewable energy in regions prone to the effects of climate change may offer employment opportunities and better standard of living for the locals and immigrants, thus resulting in a better-standardized life, inclusive future that will also prove to be sustainable (Barua, 2020). Education on green practices and green existence may assist reduce carbon impact of human society and live sustainably for instance coving climate change (Islam & Khan, 2020).

The large-scale solutions to climate change hence require individual measures, legislations, actions, and even multilateral cooperation. Governments, non-governmental solutions, and plans but the engineering profession in organizations and the corporate sector have the key role to play in developing and implementing innovative climate, modify change mitigation policies, and ensure the future sustainability (Pinkse & Kolk, 2012). Climate change concerns require the mobilization of stakeholders to deploy different kinds of knowledge and resources. Investing in state of the art technologies such as carbon capture and storage might go a long way in healing, realize the aim of achieving carbon neutrality and zero-carbon energy systems (Nwokolo et al., 2023c). Creating a worldwide culture of environmental responsibility and resilience needs public awareness. It is more important to watch and cultivate a culture of environmental responsibility and improve the resilience of the people, as it needs the awareness of volunteers, and individual initiatives to reduce impacts on the environment and to adapt environmentally friendly practices (Souza et al., 2017). Setting up of international linkages and agreements making to arrive at obligatory green house the achievement of the set gas emission reduction goals is central to climate change intervention globally (Widerberg & Pattberg, 2015). Policies that promote the adoption of environment-friendly practices among the firms include financing in renewable sources of energy may assist in achieving the acceleration of the low-carbon economy as well as a sustainable future of the country and globally was indicated to be unknown by Le, Nguyen et al. in a forecast for the year 2024. CCS and renewable energy storage solutions may help increase efficiency and effectiveness of sustainable energy systems. The impact of individual contribution towards environment is very important (Rose, 2014).

Investing in green infrastructure and enforcing sustainable practices can also help businesses become more environmentally friendly, creating economic growth and reducing carbon emissions (Liberalesso et al., 2020). Engaging local communities and stakeholders in decision-making helps promote culturally sensitive and socially inclusive sustainable programmes that succeed and benefit the environment (Castro-Arce & Vanclay, 2020).

15.4.1 Implications for Research and Policy

The different local framings of climate financing and their policy and practical consequences deserve more study. Key research areas:

- Case studies of effective local climate financing initiatives.
- Comparative examination of localization strategies.
- Assessing the effect of local climate funding on vulnerable populations.
- Policy suggestions to improve local climate financing access and effectiveness.

15.4.2 Analysis of Present Patterns and Trends in the Allocation of Financial Resources for Addressing Global Climate Issues

Some significant changes affecting climate action financing. The main patterns and trends are listed below:

> Due to increased climate risk awareness and urgency, global climate financing flows have been rising. The Paris Agreement requires developed nations to raise $100 billion yearly for climate funding. The private sector has invested more in climate financing than the governmental sector (Kumar et al., 2024). This shows increased awareness of climate-friendly investments' economic potential. Reforestation, sustainable agriculture, and ecosystem restoration are becoming popular cost-effective resilience and adaptation measures. Climate financing remains unequally distributed across regions. Despite their climate change vulnerability, developing nations, especially in Africa and tiny island governments, get less funding (Nwokolo et al., 2023d). Asia and Latin America get most climate financing for mitigation programmes. Governments and companies are issuing green bonds to fund climate initiatives at an exponential rate. Sustainable bonds and loans that relate financial conditions to climate and sustainability results are also popular. Climate financing transparency and accountability are being prioritized.

15.4.3 The Function of Developed and Developing Nations

Industrialization in developed nations is the main cause of greenhouse gas emissions. This historical duty supports their climate action funding for poorer nations. Developed nations pay multilateral institutions like the GCF, GEF, and World Bank and provide bilateral contributions to climate financing. These grants are essential for developing country climate mitigation and adaptation efforts. Many industrialized nations send part of their Official Development Assistance (ODA) to climate-related programmes to connect development with climate goals.

The developed nations also raise private funds for climate financing by promoting advantageous policies, guarantees, and blended financing methods that lower private investment risk. Technology leaders in developed countries may share climate-friendly innovations and knowledge with poor nations to help them act. Through their leadership in international talks and governance bodies like the UNFCCC, developed nations shape global climate policy. Climate financing helps developing nations plan, execute, and monitor climate policies by building institutional and technical capacity.

15.5 Local Climate Finance Framings: Interpretation and Application

Climate finance is the funds used to address climate change issue including the fund used in the fight against this problem. Chafer (2006) rightly points that it is crucial to global climate change policies, but it is implemented and understood differently at regional levels. This article is on how communities and regions observe and spend climate money.

15.5.1 Interpretation

Climate funding is what local framings entail as far as how communities and regions perceive and act on climate funding is concerned. These framings are affected by several factors. It is clear that community priorities depend on the climate change vulnerability. Other regions may focus on achieving hurricane-proof structures, and rural communities will be concerned about salty water intrusion. Districts may focus on the efficient use of water and the crops' durability. Cultural ideas and perception and use of climate finance depends on the following socioeconomic factors. Traditional knowledge and practices impacting on climate finance project acceptability and integration.

15.5.2 Regional Approaches to Climate Finance

Local framings are how communities and regions perceive and execute climate funding. Several things affect these framings. Community priorities vary by climate change vulnerability. Coastal towns may prioritize sea-level rise fortifications, whereas agricultural districts may emphasize water management and crop resilience. Climate financing addresses needs and obtains support by engaging local stakeholders including community leaders, NGOs, and the commercial sector. Local communities prioritize agriculture productivity and water resource management programmes. Small Pacific island states, facing grave dangers from rising sea levels, frame climate funding around relocation and infrastructure initiatives. These countries want quick and considerable international financial aid. They struggle to get funding to needy populations locally. Planning local initiatives around national and international climate plans. Ensure climate financing initiatives provide results by monitoring their efficacy. Meanwhile, climate financing may enable local communities to manage their climate resilience measures. Local frames may inspire new, context-specific answers. Climate financing may improve community well-being by supporting development objectives.

15.5.3 Climate Finance Localization

This requires knowing how to redirect global climate financing to local people and initiatives. Accessibility, affordability, and local concerns are included. Traditional climate adaptation and mitigation techniques are in many local communities. These methods can be scaled up with climate money. Social fairness and gender equality should guide climate funding. Local framings emphasize inclusive decision-making and benefit-sharing. Climate Justice highlights wealthy nations' historical responsibility for climate change and the necessity for recovery funding for disadvantaged people. Local framings help communities to recognize climate issues and propose locally driven solutions. Climate resilience needs strong financial systems like insurance and risk transfer. Climate financing may benefit rural and underprivileged populations by creating green employment and sustainable lives, Bhushan and Singh (2024).

15.5.4 Challenges and Opportunities

Localizing climate funding is becoming more important, but problems remain. Lack of understanding, competence, and eligibility prevents many local communities and small businesses from accessing climate funding. Their data access is restricted. Data on local climate effects and financial demands is lacking, hindering resource allocation. Complex bureaucracy and poor governance maybe some of the reasons that may hider climate money to reach the intended local communities. Nevertheless, there are benefits of localizing climate funding as highlighted below. One is Innovative Financial Instruments.

15.6 Findings

This theoretical paper stresses that climate finance process and mechanism should be recognized as much more layered and varied than traditional economics could depict, embrace the economic value of nature, the policy shift to and implementation of climate actions, the interrelatedness of global movements, and the significance of local framings (Kumar et al., 2024). Biodiversity and ecosystem expenses thus need to be valued in terms of more than just economic framework components but also in ecological, social, and cultural terms to reveal the synergistic relationships between nature and people.

When looking at climate finance as an enabler of action, it is therefore important to consider the urgent need for development of innovative and sustainable financial instruments which will effectively support the global climate goals and developmental objectives.

Therefore, in a conclusion, this paper supports a comprehensive approach to climate finance that takes into account the global needs and local context as well as economy and ecology and social concerns. This should be pursued not only in a climate change perspective, but more importantly in an equity and justice perspective as we endeavour towards a low-carbon future for the future generations.

15.7 Conclusion

Climate financing may better reduce and adapt to climate change by identifying and addressing local communities' particular needs, cultural settings, and governance systems. Local climate financing programmes need more stakeholder participation, transparent governance, and thorough monitoring. Asia and the Pacific is one of the most sensitive areas to climate change, and its consequences are predicted to worsen. The area is a major greenhouse gas (GHG) emitter, and unless immediate action is made, this tendency will worsen. Most agree that these adaptation and mitigation issues will need a large increase in national and international public and private financial resources. The paper summarizes the most important trends, difficulties, and possibilities for funding localized climate change solutions in the area, using examples from many nations. Developed and poor nations have complementary but challenging roles in climate financing. Developing nations employ financial and technological resources to enhance resilience and minimize climate consequences, while developed countries lead in supplying them. Strengthening these responsibilities, boosting access to funding, and encouraging stakeholder engagement are crucial to global climate finance success.

References

Abdullah, S. (2014). Local community resilience in the context of global climate change: A case from Maluku Indonesia. *Sociology and Anthropology, 2*(7), 309–316. https://doi.org/10.13189/SA.2014.020708

Aberman, N.-L., Ali, S., Behrman, J. A., Bryan, E., Davis, P., Donnelly, A., Gathaara, V., Koné, D., Nganga, T., Ngugi, J., Okoba, B., & Roncoli, C. (2015). Climate change adaptation assets and group-based approaches: Gendered perceptions from Bangladesh, Ethiopia, Mali, and Kenya. *SSRN Electronic Journal*. https://doi.org/10.2139/SSRN.2564556

Akhter, S., Raihan, F., Sohel, M. S. I., Syed, A., Das, S. K., & Alamgir, M. (2013). Coping with climate change by using indigenous knowledge of ethnic communities from in and around Lawachara National Park of Bangladesh. *Journal of Forest and Environmental Science*. https://doi.org/10.7747/JFS.2013.29.3.181

Alonso, A., & Marqués, J. M. (2019). Financial innovation for a sustainable economy. *SSRN Electronic Journal*. https://doi.org/10.2139/SSRN.3471742

Banga, J. (2019). The green bond market: A potential source of climate finance for developing countries. *Journal of Sustainable Finance & Investment*. https://doi.org/10.1080/20430795.2018.1498617

Barua, S. (2020). Financing sustainable development goals: A review of challenges and mitigation strategies. *Business Strategy & Development, 3*(3), 277–293.

Belay, E. (2014). The socio cultural aspects of building climate resilience in the face of changing environment among agricultural communities in the Blue Nile Basin.

Bhattacharyay, B. N. (2021). Managing climate-related financial risk: Prospects and challenges. In *New frontiers in conflict management and peace economics: With a focus on human security*. Emerald Publishing. https://doi.org/10.1108/S1572-832320210000029004

Bhushan, M., & Singh, R. (2024). Talent management facilitates net-zero transition through employee green behavior. In *Net zero economy, corporate social responsibility and sustainable value creation: Exploring strategies, drivers, and challenges* (pp. 117–129). Springer.

Bloch, D. (2011). Financing the climate economy. *SSRN Electronic Journal*. https://doi.org/10.2139/SSRN.1883581

Boissinot, J., Huber, D., & Lame, G. (2016). Finance and climate: The transition to a low-carbon and climate-resilient economy from a financial sector perspective. *OECD Journal: Financial Market Trends*. https://doi.org/10.1787/FMT-2015-5JRRZ76D5TD5

Campos, A. (2015). Adapting and improving resilience to climate change in communities (Moravian community as a pilot), by creating new capabilities based on the implementation of a new water culture; protection and management of natural resources. *International Archives of the Photogrammetry, Remote Sensing and Spatial Information Sciences*. https://doi.org/10.5194/ISPRSARCHIVES-XL-7-W3-1453-2015

Castro-Arce, K., & Vanclay, F. (2020). Transformative social innovation for sustainable rural development: An analytical framework to assist community-based initiatives. *Journal of Rural Studies, 74*, 45–54.

Chenet, H., Hilke, A., & Duan, W. (2017). Finance sector alignment with international climate goals reviewing options and obstacles. *SSRN Electronic Journal*. https://doi.org/10.2139/SSRN.3322324

Cheng, F.-T. (2017). Public-private partnerships to address global environmental problem: Sino-US cooperation on climate change and energy. *Journalism and Mass Communication*. https://doi.org/10.17265/2160-6579/2017.03.004

Chitengi, S., & Sakapaji. (2022). Integrating local and indigenous ecological knowledge (IEK) systems into climate adaptation policy for resilience building, and sustainability in agriculture. *International Journal of Sustainable Development Research*. https://doi.org/10.11648/j.ijsdr.20220801.12

Chu, E., Anguelovski, I., & Carmin, J. (2016). Inclusive approaches to urban climate adaptation planning and implementation in the global south. *Climate Policy*. https://doi.org/10.1080/14693062.2015.1019822

Clapp, C., & Sillmann, J. (2019). Facilitating climate-smart investments. *One Earth, 1*(1), 57–61. https://doi.org/10.1016/J.ONEEAR.2019.08.009

Cordonier Segger, M.-C., & Phillips, F.-K. (2015). Indigenous traditional knowledge for sustainable development: The biodiversity convention and plant treaty regimes. *Journal of Forest Research*. https://doi.org/10.1007/S10310-015-0498-X

Das, M. (2022). Allocation of climate funds and the adaptation infrastructure gap. https://doi.org/10.30560/sdr.v4n1p42

Doughty, C. A. (2016). Building climate change resilience through local cooperation: a Peruvian Andes case study. https://doi.org/10.1007/S10113-015-0882-2

Echeverry, C. A. (2016). El sistema contable ante el cambio climático.

Fujikura, R. (2022). Financing in climate change Adaptation. https://doi.org/10.1007/978-981-19-2924-3_2

Gopalakrishnan, R. (2018). International financial mechanism in climate change mitigation and adaptation.

Green, D., & Raygorodetsky, G. (2010). Indigenous knowledge of a changing climate. https://doi.org/10.1007/S10584-010-9804-Y

Islam, M. R., & Khan, N. A. (2020). Threats, vulnerability, resilience and displacement among the climate change and natural disaster-affected people in South-East Asia: An overview. In *Climate change mitigation and sustainable development* (pp. 111–138). Routledge.

Jakob, M., Steckel, J. C., Flachsland, C., & Baumstark, L. (2015). *Climate finance for developing country mitigation: Blessing or curse?* Routledge. https://doi.org/10.1080/17565529.2014.934768

Joshi, A., Singh, R., Kumar, V., & Rani, S. (2024). Net zero strategy: Exploring opportunities, confronting challenges for developing countries. In *Transition towards a sustainable future: Net zero policies and environmental sustainability* (pp. 233–252). Springer.

Kelling, S. (2018). *Technology developments for biodiversity monitoring and conservation*. Elsevier. https://doi.org/10.3897/BISS.2.25833

Kellner, K., & Bosch, O. J. H. (2003). The value of local and indigenous knowledge for the development of information systems for conservation management. In *Conserving biodiversity in arid regions*. Springer. https://doi.org/10.1007/978-1-4615-0375-0_8

Khan, S., Singh, R., Alnahas, J., Abbate, S., & Centobelli, P. (2024). Navigating the smart circular economy: A framework for manufacturing firms. *Journal of Cleaner Production, 480*, 144007.

Kumar, K., Singh, R., Sharma, A., & Lehri, B. P. S. (2024). Climate change adaptation strategies for achieving net-zero economy. In *Transition towards a sustainable future: Net zero policies and environmental sustainability* (pp. 163–183). Springer.

Le, P., et al. (2024). Navigating the green path: The role of climate incentives in shaping eco-innovation and firm policies. In *Climate governance and corporate eco-innovation: A framework for sustainable companies* (pp. 65–96). Springer.

Ledeneva, V. Y., & Lukyanets, A. S. (2022). The role of finance in managing climate migration and preventing climate travel. https://doi.org/10.26653/2076-4650-2022-5-6-01

Liberalesso, T., et al. (2020). Green infrastructure and public policies: An international review of green roofs and green walls incentives. *Land Use Policy, 96*, 104693.

Makondo, C. C., & Thomas, D. S. G. (2018). Climate change adaptation: Linking indigenous knowledge with western science for effective adaptation. https://doi.org/10.1016/J.ENVSCI.2018.06.014

Meltzer, J. P. (2016). Financing low carbon, climate resilient infrastructure: The role of climate finance and green financial systems. *SSRN Electronic Journal*. https://doi.org/10.2139/SSRN.2841918

Michaelowa, K., & Namhata, C. (2022). Climate finance as development aid. In *Handbook of international climate finance* (pp. 62–82). Edward Elgar Publishing.

Mohamadi, F. (2021). Climate finance at a glance. https://doi.org/10.1007/978-3-030-68740-3_16

Moller, H., Berkes, F., Lyver, P., & Kislalioglu, M. (2004). Combining science and traditional ecological knowledge: Monitoring populations for co-management. https://doi.org/10.5751/ES-00675-090302

Mori, M. (2012). Sustainability issues in finance. https://doi.org/10.3280/RISS2012-001003

Morin-Labatut, G., & Akhtar, S. (1992). Traditional environmental knowledge: A resource to manage and share.

Natarajan, B. (2002). Conservation of biological diversity and indigenous traditional knowledge as exemplified by the cases of Asia and Africa. https://doi.org/10.4314/INDILINGA.V1I1.26324

Nesshöver, C., Assmuth, T., Irvine, K. N., Rusch, G. M., Waylen, K. A., Delbaere, B., Haase, D., Haase, D., Jones-Walters, L., Keune, H., Kovács, E., Krauze, K., Külvik, M., Rey, F., van Dijk, J. J., Vistad, O. I., Wilkinson, M., & Wittmer, H. (2017). The science, policy and practice of nature-based solutions: An interdisciplinary perspective. https://doi.org/10.1016/J.SCITOTENV.2016.11.106

Newman, J., & Perl, A. (2014). Partners in clime: Public-private partnerships and British Columbia's capacity to pursue climate policy objectives. https://doi.org/10.1111/CAPA.12051

Nwokolo, S. C., Singh, R., Khan, S., & Kumar, A. (2025). Threats to the rapidity of sustainability transitions posed by technological changes. In *Technological pathways for Africa's net-zero economy* (pp. 39–75). Elsevier. https://doi.org/10.1016/b978-0-443-31486-5.00002-1

Nwokolo, S. C., Singh, R., Khan, S., Kumar, A., & Luthra, S. (2023a). Africa's awakening to climate action. In *Africa's path to net-zero: Exploring scenarios for a sustainable energy transition* (pp. 299–310). Springer.

Nwokolo, S. C., Singh, R., Khan, S., Kumar, A., & Luthra, S. (2023b). Global investment and development in Africa. In *Africa's path to net-zero: Exploring scenarios for a sustainable energy transition* (pp. 15–58). Springer.

Nwokolo, S. C., Singh, R., Khan, S., Kumar, A., & Luthra, S. (2023c). Impacts of climate change in Africa. In *Africa's path to net-zero: Exploring scenarios for a sustainable energy transition* (pp. 237–262). Springer.

Nwokolo, S. C., Singh, R., Khan, S., Kumar, A., & Luthra, S. (2023d). Remedies to the challenges of renewable energy deployment in Africa. In *Africa's path to net-zero: Exploring scenarios for a sustainable energy transition* (pp. 59–74). Springer.

Okoyeuzu, C. R., & Ukpere, W. I. (2022). Dynamics of biodiversity loss and financial system stability nexus in developing countries. https://doi.org/10.21511/ee.13(1).2022.07

Parajuli, D. R., & Das, T. (2013). Indigenous knowledge and biodiversity: Interconnectedness for sustainable development.

Pickering, J., Skovgaard, J., Kim, S., Roberts, J. T., Rossati, D., Stadelmann, M., & Reich, H. (2015). Acting on climate finance pledges: Inter-agency dynamics and relationships with aid in contributor states. *World Development, 68*, 149–162.

Pinkse, J., & Kolk, A. (2012). Addressing the climate change—Sustainable development nexus: The role of multistakeholder partnerships. *Business & Society, 51*(1), 176–210.

Plante, S., Vasseur, L., & da Cunha, C. (2015). Engaging local communities for climate change adaptation: A case study in Quebec, Canada. https://doi.org/10.1016/B978-0-12-802748-6.00018-8

Ponera, S. K. (2013). Interaction between indigenous and external knowledge in influencing adaptation to climate change impacts among communities of north pare mountains, Tanzania.

Rabeil, T., Newby, J., & Harouna, A. (2014). Innovation, traditional knowledge and awareness lead to good practices for avoiding biodiversity loss in the Sahara. https://doi.org/10.1080/14888386.2014.936038

Richardson, B. J. (2014). The evolving marketscape of climate finance. https://doi.org/10.1163/18786561-00402008

Rojas Blanco, A. V. (2006). Local initiatives and adaptation to climate change. https://doi.org/10.1111/J.1467-9523.2006.00311.X

Rose, K. (2014). Adopting industrial organizational psychology for eco sustainability. *Procedia Environmental Sciences, 20*, 533–542.

Rossati, D., & Rossati, D. (2013). Inter-institutional dynamics of global climate finance: Complementarity and competition in the emerging practices of coordination. https://doi.org/10.2139/SSRN.2401309

Sathaye, J., Price, L., Kumar, S., du Can, S., Warfield, C., & Padmanabhan, S. (2006). Partnerships for clean development and climate: Business and technology cooperation benefits.

Sharma, M., & Kaushik, R. P. (2015). Plant genetic resources and indigenous traditional knowledge conservation toward resilience to climate change. https://doi.org/10.1007/978-981-10-0060-7_11

Singh, R., Gupta, A., & Kandpal, V. (2025). ESG excellence in the digital world: Circular economy e-business solutions. In *Fintech for ESG and the circular economy* (pp. 193–204). Wiley.

Souza, A. A. A., et al. (2017). Resilience for sustainability as an eco-capability. *International Journal of Climate Change Strategies and Management, 9*(5), 581–599.

Steffen, B., & Michaelowa, A. (2022). Policies for climate finance: Status and research needs. https://doi.org/10.1371/journal.pclm.0000083

Sullivan, S. (2012). Financialisation, biodiversity conservation and equity: Some currents and concerns.

Tiwari, P., Kushwaha, A., & Sharma, Y. K. (2021). Climate resilience with traditional Wisdom in Asia. https://doi.org/10.9734/AJESS/2021/V14I230352

Toxopeus, H., & Polzin, F. (2021). Reviewing financing barriers and strategies for urban nature-based solutions. https://doi.org/10.1016/J.JENVMAN.2021.112371

Turner, N. J., & Spalding, P. (2013). We might go back to this; drawing on the past to meet the future in northwestern north American indigenous communities. https://doi.org/10.5751/ES-05981-180429

Valderrama Rincón, N. (2023). Innovative ways to mobilise private sector capital in climate change adaptation investments in developing countries—Mechanisms and forward-looking vision from practitioners' standpoint. https://doi.org/10.1007/978-3-031-28728-2_16

Varadarajan, D. B., & Chitra, M. (2022). Global climate financing of multilateral development banks - An analysis. https://doi.org/10.34293/economics.v10i2.4747

Warren, P. (2020). Blind spots in climate finance for innovation. https://doi.org/10.1016/J.ACCRE.2020.05.001

Wendling, L., Huovila, A., Zu Castell-Rüdenhausen, M., Hukkalainen, M., & Airaksinen, M. (2018). Benchmarking nature-based solution and smart city assessment schemes against the sustainable development goal indicator framework. https://doi.org/10.3389/FENVS.2018.00069

Widerberg, O., & Pattberg, P. (2015). International cooperative initiatives in global climate governance: Raising the ambition level or delegitimizing the UNFCCC? *Global Policy, 6*(1), 45–56.

Williams, M. (2017). Climate finance: Why does it matter for women? https://doi.org/10.1057/978-1-137-46101-8_12

Wong, S. (2016). Can climate finance contribute to gender equity in developing countries. https://doi.org/10.1002/JID.3212

Yan, Q. (2023). A discussion on the role of financial institutions in fighting climate change. https://doi.org/10.54097/jid.v3i1.8413

Ye, Y. (2014). Climate finance, Africa and China's role. https://doi.org/10.7552/0-1-120

Zhu, Q., Wu, J., & Wang, Z. (2013). Modeling of climate financing based on MRICES. https://doi.org/10.5846/STXB201203310450

Chapter 16
Mitigating Ice Cap Melt: Implications of Climate Neutrality Concepts for Climate Policy and Practice

Ipseeta Satpathy, Ankita Nayak, and Vishal Jain

Abstract Sea ice, the buildup of which only occurred naturally, has enormous dangers for the different global ecosystems, existing sea levels, and human cities. The strategy now being used to minimize these effects is attaining climate neutrality, which refers to net-zero greenhouse gas emissions. They focus on analyzing climate neutrality concepts and their application to climate policy and practice, with the view of slowing melted ice caps. They contain an overview of major policy initiatives, essentially reducing carbon emissions, and the part played by international cooperation. This includes topics concerning energy system transition and the part that sound policies, equity, and technology development have to play in addressing climate goals. As indicated in the paper, it is required to stabilize the climate policies to address the scientific targets for reversing the worst rates of ice cap melting.

Keywords Ice cap melt · Climate neutrality · Net-zero emissions · Climate policy · Sustainable energy · Carbon reduction

16.1 Introduction

Climate neutrality is a noble international desire in that it seeks to minimize the impact of greenhouse gas in adding to carbon emissions. These are elimination of fossil fuel usage, use of sustainable energy, changed carbon dioxide rules through

I. Satpathy
School of Management, KIIT University, Bhubaneswar, India
e-mail: ipseeta@ksom.ac.in

A. Nayak (✉)
Trident Academy of Technology, Bhubaneswar, India

V. Jain
Department of Computer Science and Engineering, School of Engineering and Technology, Sharada University, Greater Noida, India
e-mail: vishal.jain@sharda.ac.in

R. Singh and D. Crowther (eds.), *Climate Neutrality and Sustainable Ecosystems*, Approaches to Global Sustainability, Markets, and Governance,
https://doi.org/10.1007/978-981-96-7181-6_16

afforestation and application of innovation technology. The significance of climate change mitigation is based on the ability to either minimize or delay the effects of climate change, regulate climatic patterns on the Earth and spare other systems, both natural and human, from deterioration (Wolf et al., 2021). Climate neutrality therefore is a difficult procedure that requires such elements as policies, technologies, and transforming societies. It includes the source identification, emission control, and cooperation between the countries. The climate targets equal commitments made by countries and corporations promising to reduce greenhouse gas emissions use green technology and fund research for the environment such as the Paris Accord (Capros et al., 2019). Greenhouse gas stock determines ice melt and sea-level rise in specialty segments of ice cap preservation and thus climate neutrality is highly essential across the value chain. Sustainable management of climate enhances temperature stabilization with reduction of chemicals affecting Ice and Pollutants. Hence, can be considered advocated to the principles of sustainable development, poverty, health, and environment. Climate neutrality increases the quality of the world and makes it immune to the climate becoming worse. It is the unexpected driving force of the modern world to change the paradigm in tackling climate change (Wang et al., 2021). Climate neutrality is helpful in the protection of ice caps as they are sensitive to global temperatures because of greenhouse gases. Higher temperatures means that the existing ice caps melt more profoundly and cause higher seawater levels and changes in the streams of ocean. The realization of climate neutrality decreases the rate at which ice caps are melting and the sea-level rises thus mitigating climate change (Fu et al., 2022). Climate neutrality is achieved through decreasing CO_2 emissions for electricity generation, transport fuel consumption, and industrial activities, enhancing the technologies for carbon capture and storage, increasing renewable energy, energy efficiency, and sustainable land management. It includes using carbon capture and storage technologies further, and direct air capture of any emissions and stopping ice cap melting. The following specific measures are also required; Darkening of white ice surface (Gizer et al., 2022). Climate neutrality seeks to lower black carbon, which is a greenhouse gas resulting from the burning of biofuels and fossils, thus erasing the icy mantle's ability to reflect light and melt. The process includes international partnership and coordinating with climate conventions, such as the Paris accord, supporting groups worldwide in preservation of ice cap, and improving strategies as well as plans (Zhang et al., 2021). Climate neutrality in ice cap is useful in regulating the socio-economic impact of the melting ice caps so as to minimize on the dangers on shore dwellers and aquatic animals. Thus, it protects ice cap resources, reducing the extent of their displacement, as well as of financial risks, for concerned groups. As it concerns the availability of freshwater, ice caps are rather important and their thinning hinders it. Admixing climate neutrality in ice cap preservation enhanced the environment thus enhancing mankind (Ekemezie & Digitemie, 2024). Reducing climate neutrality concepts is important when it comes to the conservation of ice caps due to their impact on climate change. They provide a multiple-level solution to control the melting of the ice cap, fixation of climate variables, and legal requirements for global climate change. Here, being neutral is a way to take action which will help to save ice caps and climate for the future (Yoro et al., 2020).

16.2 From Glacier to Ice Sheet: An Evaluation of the Nature and Process of Ice Formation

It creates through the deposition of snow in the polar and high altitude regions where temperatures are low to admit presence of snow. Snow also accumulates on the ground and forms layers on top of another. Ice cap formation is a process where snow layers get pressure from above, new snow in a weak structure letting layers collapse and turn into firn which is snow mix sintered between it and the glacial ice (Yancheshme et al., 2020). Ice caps form due to the accumulation of snow and the transition from firn to glacial ice. They are ice masses less than 50,000 km^2 in area, with the difference being made to ice sheets. Ice cap formation is dependent on local or regional climate, topography, and atmospheric moisture. In the polar region, low temperatures cause ice buildup. Ice flow is dynamic, with deeper ice exerting more pressure at the accumulation center. Movement of ice within an ice cap is determined by factors like temperature, land lay, accumulation rate, and loss rate (Bell & Seroussi, 2020). Ice caps play a crucial role in controlling Earth's climatic conditions through their high albedo impact, which helps cool the Earth's surface and reduce global warming impacts. They also hold a significant amount of freshwater, which can affect sea levels and ocean circulation. The melting of ice caps can affect coastal communities and ecosystems, while fresh meltwater from ice caps alters water salinity and density, preventing ocean currents. The disappearance of ice caps also impacts regional climate (Malla et al., 2022). The interaction between ice caps and atmospheric conditions affects precipitation patterns and temperature distribution. Melting ice can lead to changes in albedo, enhancing sunlight absorption and heat absorption. Climate change can be established by ice cap mass and spatial coverage. Monitoring ice cap activity can help understand the effects of global warming on ice caps and future behavior. Ice cap changes also influence atmospheric circulation, pressure distribution, wind patterns, and weather systems (Steffen et al., 2020). The melting rate of ice caps is rising worldwide and this can be labeled phase four of climate change mostly in polar and high altitude regions. Sond missions such as CryoSat-2 and GRACE portray an increase in the rate of melting and it is expressed that Greenland Ice Cap melts at the rate of 286 billion metric tons annually for the period between 1992 to 2018. This is as a result of increased surface melting, reduction in the thickness of the ice sheet, and berg-calving (Ciracì et al., 2020). The phenomenon has led to the melting of the known West Antarctic Ice Sheet whose rate is now at 252 billion metric tons per year through the thinning of existing floating ice shelves. This has extended ice flow from areas deep within the Antarctica continent and this has made sea levels rise all over the world. Other glaciers of the world including Himalayan, Andean, and Alpine glaciers are also retreating rapidly imposing threat to development of water supplies and also affecting other ecosystems. Unfortunately more action is required to break this trend (Golledge et al., 2019). This mostly occurs due to climatic change whereby temperatures of the entire globe rise because of emissions of gases such as carbon dioxide, methane, and other greenhouses. The Arctic is experiencing a higher increase in temperatures as discussed in Arctic Amplification where the temperature increases are over two times the general global average or temperature increase and this has made ice to melt at a faster pace and at the same time the ability of ice to reflect sunlight has been compromised as more heat is absorbed leading to ice cap melting (Rignot, 2021). In this region the ice cap is melting more due to factors such as reduced amount of precipitation, change in climate and the amount of snow. This weakens the capability of ice cap to regenerate itself and the formation of ice at a faster pace. Other factors which influence ice include the increase in the ocean temperature, leading to a decrease in the thickness of the ice shelves and the increase in flow of ice into the ocean which causes change in sea level (Bevis et al., 2019). Soot, dust, and algae accumulation on the ice surfaces bring its albedo down hence enhancing its capacity to trapping heat from the sun. This leads to faster melting deepening the color and the rate of melting thus the need for a bowl with thinner walls. Global warming is therefore facilitated by such feedback mechanisms as methane emitted by permafrost. Also, ice cap is known to melt and the water collected may find its way into the oceans resulting in further heat (Golledge et al., 2019).

16.3 Understanding and Assessing Climate Neutrality: Conception and Units of Assessments

Climate neutrality is a global goal of climate mitigation aiming for equal emissions and removals of greenhouse gases, reducing an economy's overall carbon footprint. This involves minimizing pollutant use and emissions, transitioning to renewable energy sources, and changing consumption patterns. Emission reductions are prioritized to address climate change impact, while remaining emissions can be compensated through carbon capture and storage, reforestation, or renewable energy investments (Salvia et al., 2021). Climate neutrality aims to eliminate greenhouse gas emissions, reducing climate change impacts like intense weather and rising ocean water levels. It supports sustainable development, provides a cleaner environment, and promotes economic growth with environmentally friendly technologies. It requires self-commitment to climate change treaties and policies, emission reporting legitimacy, and global partnership (Huang et al., 2022). Climate neutrality involves implementing emission cuts, carbon offsetting options, and effective policies to reduce greenhouse gas emissions. Strategies include avoiding emissions, implementing energy conservation measures, adopting electric cars, and promoting efficient agriculture and waste management practices. Solar, wind, and hydro power emit fewer pollutants than fossil fuels, and promoting sustainable transportation and recycling are also crucial (Capros et al., 2019). Carbon emissions can be minimized by use of carbon capture and storage technologies, reforestation, afforestation and in Renewable Energy Projects. There is also the use of capture of methane that results from the decomposition of organic wastes in landfills and livestock. Governments need to use policies to motivate change such as carbon taxes as well as the cap and trade mechanism enabling the provision of low-emitting technologies. Singularity features such as storage technologies and intelligent grids enable emissions decline to occur more effectively. In the light of such examples, it can be highlighted that campaigns initiated at the corporate and community level can help to achieve climate neutrality or even reduce the emission level for guaranteeing the future of the world by meeting the needs of the community (Zhao et al., 2022). The procedures of estimating greenhouse emissions and attaining the state of climate neutrality are based on diverse factors and indices. These entails emissions from owned or controlled sources, emissions from the purchase of electricity, steam, heating, cooling, and finally emissions, comprising of the entire value chain. The evaluation indicators will comprise the percentage of reducing total emissions, reduction of emissions tons per offset, and the effectiveness of the emission reduction technology (Chen et al., 2022). Carbon audit tools and carbon calculators are used for determining carbon footprints for individual activities or activities in a process. The credibility, storage capacity, and durability of carbon offset projects serve as critical that defines them. Accreditation and third-party verification help to avoid wrong emission and offset numbers while transparency is vital for organizations to be able to measure climate change impacts and risks appropriately (Zhang et al., 2022). In climate neutrality, it is imperative to obtain the balance of greenhouse gases and to eliminate the same. The new goals of the Danish capital include CO_2-neutral Copenhagen by 2025: targets in the fields of energy supply and energy conservation have been defined and the climate ambitions of Copenhagen Municipality have been set on a new course. Microsoft plans not just to be carbon neutral by 2030 and avoid emitting any carbon in the first place by reducing energy intensity, severing ties with fossil fuels, and investing in carbon capture technology. As such, both companies are also reporting their progress and in the process becoming examples for others to follow (Hansen & Agger, 2022). The first program is FCPF, a program funded by the World Bank which focuses mainly on the emission reduction through policies on protection of forests and sustainable land management. Aim at working with countries in maintaining carbon stock and forests: it encourages climate neuters. A not-for-profit Climate Neutral Group provides carbon audit, carbon management, or emission reduction strategy, as well as qualify to become a carbon neutral entity (Sacherer et al., 2022).

16.4 Understanding the Effects of Climate Neutrality on Ice Cap Melt: Evidence and Insight

Climate neutrality reduces emission of greenhouse gases in a way that assists in influencing the ice caps with rates of melting. This serves to limit scores of the greenhouse gases such as CO_2 and methane which escalate the atmospheric GHG's which support global temperature boost. Seasonal ice caps predominantly in the Arctic and Antarctic region are highly sensitive to temperature variations and accelerate the rate of melting of ice (Devilliers et al., 2021). Climate neutrality was proposed to halt global temperature increase to cause melting in the polar region through reduction in CO_2 emission and pressure in the ice caps. It assists in limiting ice sheets really and the part of their melting amounts to ocean waters. Climate neutrality also refers to afforestation and the preservation of native carbon stores and avoiding climate change or the decay of ice caps. It supports green power technologies, reducing air emissions and conserving ice structures, thus no contribution toward the general warming of global temperatures (Osman et al., 2021). According to the climate policy document 'The Framework for Arctic Cooperation', the Arctic Council stresses climate neutrality as a tool to address the environmental changes in the Arctic region. Apart from funding the procurement of products that curb emissions, the council has financed various projects and programs for cutting emissions throughout the states of the region with specific benefits of ice cap protection and less ice thaw (Winther et al., 2022). To mitigate warming effects on the Arctic ice cap, 'cleaner energy sources shall be adopted, energy must be conserved, and carbon neutralization projects promoted by the council'. The actual program is called 'Greening the Arctic', and it deals with the efficient and sustainable development of the Arctic taking measures against using fossil fuels and emission of greenhouse gases that thaw the ice. Further, grant support for climate-neutral technologies is available (Arruda & Johannsdottir, 2023). The Greenland Climate Research Centre (GCRC) has focused on climate neutrality projects to save the ice cap in Greenland. They have mapped the extent of ice sheets, managed stabilized ice caps, and adopted climate neutrality policies with global other bodies such as the United Nations. This brings awareness on ice cap melting (Mattmüller et al., 2022). The world's climate agreement, under the umbrella of the United Nations Framework Convention on Climate Change, is seeking for climate neutrality and a global temperature increase below 1.5 °C above pre-industrial levels. Commitments to reduce carbon dioxide emissions are made while developed nations with large portions of their territories covered by ice caps commit themselves to provide both, financial and technological contributions to other nations for their process of climate neutrality and climate change adaptation (Mor et al., 2023). Climate neutrality is on the principle that the amount of carbon produced or within the atmosphere equal the amount of carbon absorbed hence affecting temperatures or rather the climate. This also plays an important role in sustaining the climatic system of the earth in that it moderates the emission of greenhouse gases which lead to global warming. Due to this the emissions of other greenhouse gases are cut due to the efforts of achieving zero net emissions to limit the effects of climate change and the rise in global temperatures. This in turn reduces temperatures fluctuations and small seasons of heat and crafted excessive and wild weather disturbances (Chen et al., 2022).To climate neutrality, the earth's climate also changes the circulation of air around the globe, other than controlling temperature. Patterns like airflow and the jet stream for instance are as a result of contrast in temperature around the whole world. Climate neutrality negates any chance of the continuation of the warming of the polar stratosphere and stops the reduction of polar icy masses thus safeguarding gradients. There is also an induced stabilization and potentially an increase in forecasts instead of the frequent change in regional climate and, therefore, a reduction in shocking, intensive storms and precipitations. It is possible to speak about climate neutrality as a notion that can be defined by the interfering power and which influences oceanic conditions. About changes in currents and the occurrence of even more unstable climate conditions, global mean temperatures especially those that are capable of a slower rate of rise in SST are particularly useful. In support of this, climate neutrality if it seeks to mitigate climate change can also help in stabilizing sea surface temperatures that in turn help in regulating the transfer of heat in the oceans thus helping to stabilize the weather systems (Yang et al., 2022). The roles of climate neutrality are an important influence on the climate of the relevant regions and enhancing of the strength of the ice cap. In fact, global temperature has a direct impact on regional climate systems, and using positive changes that would turn us into a climate-neutral nation we can protect these systems from the consequences of global warming. The serious climate-neutral effects include the continuity of stable gradients in the climate forms of the region. The climate on the regional levels is taken from the differences in temperature between the polar and the equator and mechanisms that are responsible for the air movements such as the jet stream. In other words, when one reduces amounts of greenhouse gases, temperatures regulate and provide Gradual gradients. It also assists in regulating up some regional climate systems to alter hasty changes in the temperatures affecting local weather. In the areas of location of ice caps, whose state largely depends on the climate, climate neutrality by itself affects the stability of the ice caps, especially in the polar countries. They discharging remarkable parts in local climate and global mean sea level in general (Yang et al., 2022). During global warming the ice caps undergo a faster change of temperature which leads to sea-level rise and even change in climate regimes. In this way, by achieving such climate neutrality it can be guaranteed that there is no overheating of polar regions to cause increased melting of ice caps and therefore the vulnerability of such areas. This preservation is important for the climatic conditions within the region since ice caps have a part to play in reflection of some part of solar radiation hence affecting temperature. For example, in the Arctic region where it was discovered that there is enhanced ice cap melt, efforts toward climate neutrality can do something minimal in slowing down the rate of the occurrence. People's actions that seek to reduce the emission of greenhouse gases reduce the rate of global warming and consequently reduction of the thickness of Arctic seas. Traveling that has become popular in the last one or two decades requires stabilization to prevent further changes to the climatic system of the north. Likewise, it can save the ice sheet stability in the Antarctic area that acts as the largest regulator the world sea level. This climate change is fatal to the Antarctic ice sheet and its impacts including rising temperatures will definitely lead to the melting of the ice and consequently, decadal sea-level rise. Climate neutrality slows the rate of warming as to preserve the ice solidification of Antarctica and reduce the further addition to the sea level (Holland et al., 2019).

16.5 The Intersection of Climate Policy and Ice Cap Conservation: Measuring Effectiveness and Outcomes

Modern climate politics allowing climate neutrality and ice shield preservation are complex and comprise of international treaties, national legislation, and local programs. All these policies are meant to cut greenhouse emissions, guard the stability of global temperatures, and preserve icy areas of the Earth. The Paris Agreement is an essential international climate policy, designed for preventing global average temperature rise to above 2 °C, and ideally 1.5 °C. The agreement requires countries to define and realize their NDCs that contain goals for the greenhouse gas emission cut. As the following analysis will show, the Agreement's success in preserving the ice caps rises and falls with the success in averting global temperature increases. This means that by curbing global warming, we slow down the pace at which ice melts and this in turn means that polar ice caps are secured. It also calls on countries to improve their national climate policies including adaptation and climate risk reduction; special aspects are polar ice cap protection and sea-level rise (Huovila et al., 2022). Almost all the countries of the world have adopted their policies about the Paris Agreement but according to their situations. For instance, the European Union recently a green deal that seeks to see the EU area achieve net zero by 2050 this policy framework includes renewable energy and energy efficiency policy as well as sectors policy. The EU also uses funding for the improvements of carbon storage and, therefore, reduction of climatic effects in polar areas. In the United States, a new federal climate policy is the Inflation Reduction Act which covers renewable investments, energy efficiency, as well as emissions targets on the way to climate neutrality. The act also funds climate adaptation measures that have effects that are incidental to ice cap preservation owing to the overall effects of climate change. Other levels of government also have a significant responsibility for climate change mitigation and management; these include county and municipal levels as well. Some of these urban policies are targets such as carbon neutrality by the year 2025 that have already been set for Copenhagen, through measures such as installing more renewable energy, improving transport, and enhancing energy efficiency. Such acts strengthen local as well as global climate neutrality, for the knock-on impact is enhanced ice cap stability in a given region. Measures which include the Clean Air Act in the United States and the Emission Trading System (ETS) in the EU are legal instruments that provide the guidelines that need to be followed in the reduction of greenhouse gases from industries and businesses. These regulations are useful to reduce the emissions from large facilities such as generators of energy, industries, a sector, which is greatly contributing to increase the global warming rate along with ice melting rate (Bryndum-Buchholz et al., 2022). Altogether, there are several flaws and drawbacks related to climate policies that still reduce their efficiency in controlling the process of ice cap melt and attaining climate neutrality. Lack of adequate aspiration and effort are clearly key issues; many of even the present day pledges including the NDCs under the Paris Agreement lack the sort of zeal needed to limit warming to 1.5 °C. The climate targets and actual reductions issue is worrying as the absence of or low targets means that GHG emissions are still possible, which influences the ice cap melt and climate change. Delays in implementation and enforcement add to the mix to fashion the functionality of climate policies. Laws may be vague and may not be implemented to the letter, and sometimes there may be a lack of supervision to enforce the laws. It is common to have a difference between what policies aim at achieving, and what transpires on the ground; industries have blanket ways of avoiding or working around policies (Roelfsema et al., 2020). Climate initiatives face limitations due to inadequate funding, resource constraints, and insufficient coordination between policy domains. These issues, particularly in developing nations, lead to minimal implementation of climate measures. Climate policies often focus on different bubbles, resulting in fragmentation and missed opportunities for extensive climate action. There are short-term challenges and if these are adopted, little contribution to climate change will be made: such include focusing on immediate fiscal gains or political tenure. There is a lack of data and monitor and this complicates the situation even more (Allen et al., 2018). Climate change policies have the right approaches toward evaluations and assessments, but have low data availability in the developing regions. Political and economic impediments including opposition from fossil fuel interest and economic issues are among the barriers to policy implementation. Addressing these gaps requires raising climate ambition, enhancing implementation, improving financing, and better integrating policies. This can lead to better ice cap melt management and climate neutrality (Dechezleprêtre et al., 2022).

16.6 Blending Information Technology and Practice in the Quest for Climate Neutrality

The director discusses the advancement of reflective coating materials on ice, which increase albedo, or sunlight reflection, on the Earth's surface. These coatings absorb less solar energy, slowing ice melting. They are typically white or low-colored and have been piloted in areas like the Arctic, where ice melt is a significant issue (Vega-Bosch et al., 2022). Another outstanding technology advancement is the carbon capture and storage technology (CCS) that goes to the basic source of global warming, that is the release of CO_2 in the atmosphere. CCS entails the collection of carbon dioxide emissions from sources like power plants and industrial processes then transferring and injecting it to storage facilities for a long time. This technology does not only minimize the amount of greenhouse gases in the atmosphere but lessens the pace at which global warming occurs, which in one way or another leads to the preservation of the ice cap. CCS is gradually being improved in efficiency and cost, through creating better ways and materials that enhance the capture and reduce the storage disadvantages. Aside from these enumerated types of technologies, there is current research and development to improve technological approaches to the preservation of ice caps (Sherman, 2019). Investment in R&D takes place in improving the effectiveness of present technologies, discovering new materials for the reflectivity of coats, and constraining innovative CCS methods. For instance, current studies are focusing on the testing of nanomaterials and new generation polymers in the making of reflective coatings to increase the efficiency and longevity of the technology. In the field of CCS, research activities are underway that involve the use of technologies such as direct air capture (DAC) to directly capture CO2 from the atmosphere cut greenhouse gas levels and, thus, promote ice cap preservation. Other strategies currently in consideration include the use of tools such as siring stratospheric aerosol injectors and ocean fertilizers with a view of altering global temperatures and the stability of ice caps. SAI refers to the process of dispersing reflective particles in the upper stratosphere so as to deflect the sunlight and keep the same away from the Earth, while ocean fertilization involves promoting the growth of phytoplankton which in turn reduces the levels of CO_2 and other green gases in the atmosphere. While these methods are still in the experimental stages they are possible approaches to the ice cap melt and global warming problems (Zhu et al., 2022). Another task of R&D in halting ice cap melt is the development of better knowledge about ice cap ecosystems and their relationship to climate parameters. Research programs explore the processes related to the formation and existence of ice caps, and their declining and dissolving processes give important information about the effects of the temperature increase, changes in atmosphere circulation, and ocean currents on ice caps. That is significant knowledge, which is required to elaborate the specific initiatives and technologies in order to counteract particular issues connected with ice melt. For instance, research on mechanisms explaining the ice-albedo feedback in climate change is useful to inform the formulation of albedo coatings that increase ice reflectance and decelerate melt. Another key area in R&D is innovation in the technology (Lockley et al., 2020). Scholars are always involved in the effort to design and enhance techniques that will reduce the effects of ice melt. For example, growth in the reflective coatings industry dissimilarities in the search for different types of material and methods to enhance performance as well as sturdiness of coatings available. CCS studies seek to enhance the processes and look for ways of cutting costs in a bid to make them practical for commercial use, thus cutting costs. Also, ongoing research and development of geoengineering processes is concerned with geoengineering techniques that could in one way or directly affect the global temperatures and ice cap stability including the s.a.i (stratospheric aerosol injection) and ocean fertilization (Kang et al., 2019). Field investigations and monitoring programs are also important in R&D. A good number of researchers of Arctic and Antarctic areas undertake extensive fieldwork to gather information concerning the state of ice caps, their melt rates, as well as the alterations in the environment. This data is used to verify the models, monitor the trends, and evaluate the effectiveness of applied measures of mitigation. Phenological monitoring includes satellite surveillance and airborne investigations that supply direct and fresh information on ice cap shifts, which allow recognition of the short-term consequences of climate change and assessing the efficiency of remedial actions. Another factor is involved in the research and development of ice cap preservation and one of them is collaborative research. Multilateral arrangements including IPCC or the Arctic Council comprised of scientists, policymakers, and interest participants to organize the knowledge, efforts, and solutions. Partnership in research means that more information gets out to the public on the strategies being employed toward achieving internationally endorsed climatic objectives. Policy and economic research are also involved in R&D by explaining the economic consequences of ice cap preservation technologies and policies. Research on the economics of different types of mitigation measures, economic values of ice caps, and the economic losses that might stem from their melting impacts on the communities and the different industries also aid in policymaking and resource mobilization to enhance the effectiveness of climate change policies (Lindenschmidt et al., 2022).

16.7 Climate-Related Socio-Economic Issues and Possible Interventions for Climate Neutrality and Ice Cap Preservation

The impacts of melting of ice caps as seen are far reaching and have a wide ramification on indigenous people and their economies more especially in the regions of the Arctic and Antarctica. The same people include Inuit populations in the Arctic region and the Sámi of Northern Europe; for these people loss in ice means loss as a people, that is loss of how they existed. These are communities, which depend on ice, sea coast, and snow for hunting, fishing, and mobility. Some indicated that because of the thinning and the retreat of sea ice, the movements of marine animals like seals and whales, which are important to them are affected. Whereas before, ice was created during the cold spring and persisted into the autumn, hunters, and fishers were affected, so there was a shortage of food supply, forcing people to buy foods from stores, which are too expensive for most people. The social aspect is also impacted as more knowledge and skills connected to ice-encompassing cultural activities are eradicated (Rasul & Molden, 2019). The cognitive/cultural justification for the violation of culture and social continuity in these enclosed communities is that these populations are predominantly physically engaged with the environment; thus, transformations in the structure of the ice disrupt conventional practices that have been previously maintained by the communities. In addition, ice becomes thinner and its presence affects the security of the travel ways as the thinner ice becomes more insecure to cross or travel across thereby enhancing the likelihood of incidental hazards and the sealing off of access to hunting areas. At a macro-level, there are issues of economic effect in local economies of regions that are mainly influenced by ice cap ecosystems. Melting of ice can affect businesses involving ice for instance in tourism by reducing the attractions that involve glaciers and ice fishing. In Greenland Iceland and many other places, the melting of the ice caps impacts on the tourism industry which is very vital to the affected countries (Steiner et al., 2021). Also, diminished ice cover affects the generation of electricity through flowing water, especially in places where glaciers contribute water to river used to generate electricity through hydropower plants. Melting ice and receding glaciers also change the flow of rivers and decrease their stability in turn, which is bad news for hydroelectric power generation and local energy provision. It also causes sea level to rise and this lead to coastal erosion hence losing vital pieces of land in the process. This makes them prone to flooding and losses of infrastructure for the coastal dwellers. The adaptation to these changes becomes an added expense which burdens local budgets and resources and results to economic difficulties and probable displacement of residents. All in all, the disastrous consequences of the ice melting for the indigenous people and the local economy can be observed in the changes of subsistence practices, cultural identity, and tourism as well as in infrastructural impacts. Mitigating these impacts entails sound approaches that factor the environmental and socio-economic variables (Bogdanova et al., 2021). The role in helping communities to adapt can be played out in the following ways: 'means to support the communities structurally must also accommodate the requirement of the ice caps melt'. Financial and logistics support have to be rendered to the communities to enable them respond to the dynamic conditions. This includes capital investment in public works such as those for flood and coastal defense, including sea walls and flood barriers. For indigenous communities, such aid might also imply subsidizing traditional practices and technologies that are peculiar to new conditions, for instance, better equipment for hunting and fishing. On management practices, management practices that are adaptive must be put in practice. These practices include consultation with the communities that are most affected to come up with and institute unique techniques for managing the communities affected. This could involve partnering with indigenous knowledge owners to be able to develop solutions that are contextually appropriate and sustainable while acknowledging and incorporating contemporary developments in technology (Carey et al., 2021). The use of indigenous knowledge in development of climate resilience packages would improve on the strengths of the adaption strategies. Promoting the intervention and prevention type of programs such as community-based resilience programs will assist the community in preparing for the effects of ice cap melt. Such programs could include education and capacity building for the local leaders and the community members on climate risk management involving preparedness on disaster response, and more to do with the protection and preservation of natural resources. Local capacity and resilience involve the ability of the people of a certain region to deal more effectively with burdens some of which are likely to be as a result of environmental alteration. Continued backing of research and educational endeavors that target the identification and management of the requirements of these communities is crucial (Jeff Birchall & Bonnett, 2019). Investigative projects can give crucial information of the effects of the melt in ice caps and how effective measures can be adopted. The educational campaigns may play the role of making people informed of climate change and its impacts to enable them to come up with measures and be involved with climate change activities. Strengthening policy as well as advocacy initiatives at community, country, and global levels is can contribute greatly toward the creation of the right environment and that can support the implementation of policies that meet the needs of vulnerable people and their communities. This entails lobbying for policy systems whose economic help and defense of endangered groups and development is sustainable. Working with political authorities and with the relevant organizations at the international level might contribute to attracting funding and resources for adaptation as well as facilitating the representation of the interests of the affected population at the decision-making level (Larsen et al., 2021).

16.8 Conclusion

The ongoing melting down of polar caps is a very vivid example of the need to have sound climate policies that espouse climate neutrality. This study underlines that taking on climate neutrality as the path of change makes it possible not only to arrest the continuing decline in the ice cap but also to stop climate change processes on the planet. But, climate neutrality per se cannot be realized without reinventing technology, deep-cutting emission reductions, and holistic policy changes. Policymakers, industries, and societies have to now embrace the process of decarbonization that features the shift from fossil fuel-centric energy systems to renewable energy sources, improvement of our energy utilization efficiency, and carbon capture and storage. Moreover, climate policies should also be progressive, guaranteeing the weak gargantuan harm, and bearing in mind that costs of climate change conversion are evenly afforded. The importance arising from the climate practice is huge. Decision-makers should not be concerned with short-term objectives of economic gain, they should understand that the cost of not implementing policies that focus on climate change is far worse than the cost of funding such policies. Global politics of climate change is therefore important as climate change occurs at the international level and would therefore require international collaboration in the reduction of greenhouse gases for the sustainability of adaptation measures. Furthermore, the idea of 'climate neutrality' should not only be considered within a framework of carbon offsetting but also encompass ecological and social parameters to guarantee that protecting ice caps also benefits the rest of the planet. Although the problem of slowing down the ice cap melt is rather enormous, adopting climate neutrality as one of the core values in climate policy remains a doable and essential task. With the implementation of policies that promote the achievement of a climate-neutral world, much can be achieved in conserving the polar ice caps, protecting vulnerable ecosystems, and development measures that can ensure that future generations will enjoy a sustainable environment.

References

Abbatt, J. P., Leaitch, W. R., Aliabadi, A. A., Bertram, A. K., Blanchet, J.-P., Boivin-Rioux, A., Bozem, H., Burkart, J., Chang, R. Y., Charette, J., Chaubey, J. P., Christensen, R. J., Cirisan, A., Collins, D. B., Croft, B., Dionne, J., Evans, G. J., Fletcher, C. G., Galí, M., & Yakobi-Hancock, J. D. (2019). Overview paper: New insights into aerosol and climate in the Arctic. *Atmospheric Chemistry and Physics, 19*(4), 2527–2560. https://doi.org/10.5194/acp-19-2527-2019

Allen, C., Metternicht, G., & Wiedmann, T. (2018). Prioritising SDG targets: Assessing baselines, gaps and interlinkages. *Sustainability Science, 14*(2), 421–438. https://doi.org/10.1007/s11625-018-0596-8

Arruda, G., & Johannsdottir, L. (2023). Regulatory environment on climate finance, EU climate diplomacy and policy, arctic investment protocol. In *Climate change adaptation and green finance* (pp. 23–41). Routledge. https://doi.org/10.4324/9781003312178-2

Azimi Yancheshme, A., Momen, G., & Jafari Aminabadi, R. (2020). Mechanisms of ice formation and propagation on superhydrophobic surfaces: A review. *Advances in Colloid and Interface Science, 279*, 102155. https://doi.org/10.1016/j.cis.2020.102155

Bell, R. E., & Seroussi, H. (2020). History, mass loss, structure, and dynamic behavior of the Antarctic ice sheet. *Science, 367*(6484), 1321–1325. https://doi.org/10.1126/science.aaz5489

Bevis, M., Harig, C., Khan, S. A., Brown, A., Simons, F. J., Willis, M., Fettweis, X., van den Broeke, M. R., Madsen, F. B., Kendrick, E., Caccamise, D. J., van Dam, T., Knudsen, P., & Nylen, T. (2019). Accelerating changes in ice mass within Greenland, and the ice sheet's sensitivity to atmospheric forcing. *Proceedings of the National Academy of Sciences, 116*(6), 1934–1939. https://doi.org/10.1073/pnas.1806562116

Bogdanova, E., Andronov, S., Soromotin, A., Detter, G., Sizov, O., Hossain, K., Raheem, D., & Lobanov, A. (2021). The impact of climate change on the food (in)security of the Siberian indigenous peoples in the arctic: Environmental and health risks. *Sustainability, 13*(5), 2561. https://doi.org/10.3390/su13052561

Bryndum-Buchholz, A., Boerder, K., Stanley, R. R. E., Hurley, I., Boyce, D. G., Dunmall, K. M., Hunter, K. L., Lotze, H. K., Shackell, N. L., Worm, B., & Tittensor, D. P. (2022). A climate-resilient marine conservation network for Canada. *FACETS, 7*, 571–590. https://doi.org/10.1139/facets-2021-0122

Capros, P., Zazias, G., Evangelopoulou, S., Kannavou, M., Fotiou, T., Siskos, P., De Vita, A., & Sakellaris, K. (2019). Energy-system modelling of the EU strategy towards climate-neutrality. *Energy Policy, 134*, 110960. https://doi.org/10.1016/j.enpol.2019.110960

Carey, M., McDowell, G., Huggel, C., Marshall, B., Moulton, H., Portocarrero, C., Provant, Z., Reynolds, J. M., & Vicuña, L. (2021). A socio-cryospheric systems approach to glacier hazards, glacier runoff variability, and climate change. In *Snow and ice-related hazards, risks, and disasters* (pp. 215–257). Elsevier. https://doi.org/10.1016/b978-0-12-817129-5.00018-4

Chen, L., Msigwa, G., Yang, M., Osman, A. I., Fawzy, S., Rooney, D. W., & Yap, P.-S. (2022). Strategies to achieve a carbon neutral society: A review. *Environmental Chemistry Letters, 20*(4), 2277–2310. https://doi.org/10.1007/s10311-022-01435-8

Ciracì, E., Velicogna, I., & Swenson, S. (2020). Continuity of the mass loss of the world's glaciers and ice caps from the grace and grace follow-on missions. *Geophysical Research Letters, 47*(9), 86926. https://doi.org/10.1029/2019gl086926

Dechezleprêtre, A., Fabre, A., Kruse, T., Planterose, B., Chico, A. S., & Stantcheva, S. (2022). *Fighting climate change: International attitudes toward climate policies (No. w30265)*. National Bureau of Economic Research.

Devilliers, M., Swingedouw, D., Mignot, J., Deshayes, J., Garric, G., & Ayache, M. (2021). A realistic Greenland ice sheet and surrounding glaciers and ice caps melting in a coupled climate model. *Climate Dynamics, 57*(9-10), 2467–2489. https://doi.org/10.1007/s00382-021-05816-7

Ekemezie, I. O., & Digitemie, W. N. (2024). Carbon capture and utilization (CCU): A review of emerging applications and challenges. *Engineering Science Technology Journal, 5*(3), 949–961. https://doi.org/10.51594/estj.v5i3.949

Fu, J., Li, P., Lin, Y., Du, H., Liu, H., Zhu, W., & Ren, H. (2022). Fight for carbon neutrality with state-of-the-art negative carbon emission technologies. *Eco-Environment Health, 1*(4), 259–279. https://doi.org/10.1016/j.eehl.2022.11.005

Gizer, S. G., Polat, O., Ram, M. K., & Sahiner, N. (2022). Recent developments in CO_2 capture, utilization, related materials, and challenges. *International Journal of Energy Research, 46*(12), 16241–16263. https://doi.org/10.1002/er.8347

Golledge, N. R., Keller, E. D., Gomez, N., Naughten, K. A., Bernales, J., Trusel, L. D., & Edwards, T. L. (2019). Global environmental consequences of twenty-first-century ice-sheet melt. *Nature, 566*(7742), 65–72. https://doi.org/10.1038/s41586-019-0889-9

Hansen, K. B., & Agger, A. (2022). Copenhagen CO2 neutrality in 2025? A polycentric analysis of urban climate governance in Copenhagen 2006–2020. *Environmental Policy and Governance, 33*(3), 288–300. https://doi.org/10.1002/eet.2030

Huang, G., Xu, Z., Qu, X., Cao, J., Long, S., Yang, K., Hou, H., Wang, Y., & Ma, X. (2022). Critical climate issues toward carbon neutrality targets. *Fundamental Research, 2*(3), 396–400. https://doi.org/10.1016/j.fmre.2022.02.011

Huovila, A., Siikavirta, H., Antuña Rozado, C., Rökman, J., Tuominen, P., Paiho, S., Hedman, Å., & Ylén, P. (2022). Carbon-neutral cities: Critical review of theory and practice. *Journal of Cleaner Production, 341*, 130912. https://doi.org/10.1016/j.jclepro.2022.130912

Jeff Birchall, S., & Bonnett, N. (2019). Thinning Sea ice and thawing permafrost: Climate change adaptation planning in Nome, Alaska. *Environmental Hazards, 19*(2), 152–170. https://doi.org/10.1080/17477891.2019.1637331

Kang, S., Zhang, Q., Qian, Y., Ji, Z., Li, C., Cong, Z., Zhang, Y., Guo, J., Du, W., Huang, J., You, Q., Panday, A. K., Rupakheti, M., Chen, D., Gustafsson, Ö., Thiemens, M. H., & Qin, D. (2019). Linking atmospheric pollution to cryospheric change in the third pole region: Current progress and future prospects. *National Science Review, 6*(4), 796–809. https://doi.org/10.1093/nsr/nwz031

Larsen, J. N., Schweitzer, P., Abass, K., Doloisio, N., Gartler, S., Ingeman-Nielsen, T., Ingimundarson, J. H., Jungsberg, L., Meyer, A., Rautio, A., Scheer, J., Timlin, U., Vanderlinden, J.-P., & Vullierme, M. (2021). Thawing permafrost in Arctic coastal communities: A framework for studying risks from climate change. *Sustainability, 13*(5), 2651. https://doi.org/10.3390/su13052651

Lindenschmidt, K.-E., Alfredsen, K., Carstensen, D., Choryński, A., Gustafsson, D., Halicki, M., Hentschel, B., Karjalainen, N., Kögel, M., Kolerski, T., Kornaś-Dynia, M., Kubicki, M., Kundzewicz, Z. W., Lauschke, C., Malinger, A., Marszelewski, W., Möldner, F., Näslund-Landenmark, B., Niedzielski, T., & Zdralewicz, M. (2022). Assessing and mitigating ice-jam flood hazards and risks: A European perspective. *Water, 15*(1), 76. https://doi.org/10.3390/w15010076

Lockley, A., Wolovick, M., Keefer, B., Gladstone, R., Zhao, L.-Y., & Moore, J. C. (2020). Glacier geoengineering to address sea-level rise: A geotechnical approach. *Advances in Climate Change Research, 11*(4), 401–414. https://doi.org/10.1016/j.accre.2020.11.008

Malla, F. A., Mushtaq, A., Bandh, S. A., Qayoom, I., Hoang, A. T., & Shahid-e-Murtaza. (2022). Understanding climate change: Scientific opinion and public perspective. *Climate Change.* https://doi.org/10.1007/978-3-030-86290-9_1

Mattmüller, R. M., Thomisch, K., Van Opzeeland, I., Laidre, K. L., & Simon, M. (2022). Passive acoustic monitoring reveals year-round marine mammal community composition off Tasiilaq, Southeast Greenland. *The Journal of the Acoustical Society of America, 151*(2), 1380–1392. https://doi.org/10.1121/10.0009429

Mor, S., Aneja, R., Madan, S., & Ghimire, M. (2023). Kyoto protocol and Paris agreement: Transition from bindings to pledges—A review. *Millennial Asia*. https://doi.org/10.1177/09763996221141546

Osman, M. B., Smith, B. E., Trusel, L. D., Das, S. B., McConnell, J. R., Chellman, N., Arienzo, M., & Sodemann, H. (2021). Abrupt common era hydroclimate shifts drive West Greenland ice cap change. *Nature Geoscience, 14*(10), 756–761. https://doi.org/10.1038/s41561-021-00818-w

Rasul, G., & Molden, D. (2019). The global social and economic consequences of mountain cryospheric change. *Frontiers in Environmental Science, 7*, 91. https://doi.org/10.3389/fenvs.2019.00091

Rignot, E. (2021). Sea level rise from melting glaciers and ice sheets caused by climate warming above pre-industrial levels. *Uspekhi Fizicheskih Nauk*. https://doi.org/10.3367/ufnr.2021.11.039106

Roelfsema, M., van Soest, H. L., Harmsen, M., van Vuuren, D. P., Bertram, C., den Elzen, M., Höhne, N., Iacobuta, G., Krey, V., Kriegler, E., Luderer, G., Riahi, K., Ueckerdt, F., Després, J., Drouet, L., Emmerling, J., Frank, S., Fricko, O., Gidden, M., & Vishwanathan, S. S. (2020). Taking stock of national climate policies to evaluate implementation of the Paris agreement. *Nature Communications, 11*(1), 2096. https://doi.org/10.1038/s41467-020-15414-6

Sacherer, A.-K., Hoch, S., Dalfiume, S., & Kassaye, R. (2022). Financing forest conservation and restoration through climate policy instruments: Lessons from the CDM and REDD+. In *Enhanced direct access: The first decade* (pp. 293–317). Elgar. https://doi.org/10.4337/9781784715656.00020

Salvia, M., Reckien, D., Pietrapertosa, F., Eckersley, P., Spyridaki, N.-A., Krook-Riekkola, A., Olazabal, M., De Gregorio Hurtado, S., Simoes, S. G., Geneletti, D., Viguié, V., Fokaides, P. A., Ioannou, B. I., Flamos, A., Csete, M. S., Buzasi, A., Orru, H., de Boer, C., Foley, A., & Heidrich, O. (2021). Will climate mitigation ambitions lead to carbon neutrality? An analysis of the local-level plans of 327 cities in the EU. *Renewable and Sustainable Energy Reviews, 135*, 110253. https://doi.org/10.1016/j.rser.2020.110253

Sherman, R. (2019). Carbon dioxide snow cleaning applications. In *Developments in surface contamination and cleaning: Applications of cleaning techniques* (pp. 97–115). Elsevier. https://doi.org/10.1016/b978-0-12-815577-6.00003-7

Steffen, W., Richardson, K., Rockström, J., Schellnhuber, H. J., Dube, O. P., Dutreuil, S., Lenton, T. M., & Lubchenco, J. (2020). The emergence and evolution of earth system science. *Nature Reviews Earth Environment, 1*(1), 54–63. https://doi.org/10.1038/s43017-019-0005-6

Steiner, N. S., Bowman, J., Campbell, K., Chierici, M., Eronen-Rasimus, E., Falardeau, M., Flores, H., Fransson, A., Herr, H., Insley, S. J., Kauko, H. M., Lannuzel, D., Loseto, L., Lynnes, A., Majewski, A., Meiners, K. M., Miller, L. A., Michel, L. N., Moreau, S., & Wongpan, P. (2021). Climate change impacts on sea-ice ecosystems and associated ecosystem services. *Elementa: Science of the Anthropocene, 9*(1), 7. https://doi.org/10.1525/elementa.2021.00007

Vega-Bosch, A., Santamarina-Campos, V., Colomina-Subiela, A., & Carabal-Montagud, M.-Á. (2022). Cryogenics as an advanced method of cleaning cultural heritage: Challenges and solutions. *Sustainability, 14*(3), 1052. https://doi.org/10.3390/su14031052

Wang, F., Harindintwali, J. D., Yuan, Z., Wang, M., Wang, F., Li, S., Yin, Z., Huang, L., Fu, Y., Li, L., Chang, S. X., Zhang, L., Rinklebe, J., Yuan, Z., Zhu, Q., Xiang, L., Tsang, D. C. W., Xu, L., Jiang, X., & Chen, J. M. (2021). Technologies and perspectives for achieving carbon neutrality. *The Innovation, 2*(4), 100180. https://doi.org/10.1016/j.xinn.2021.100180

Winther, J.-G., Hinzman, L., & Holmén, K. (2022). Arctic climate change—Perspectives on international scientific cooperation. In *Global development in the arctic* (pp. 149–164). Routledge. https://doi.org/10.4324/9781003246015-12

Wolf, S., Teitge, J., Mielke, J., Schütze, F., & Jaeger, C. (2021). The European green deal—More than climate neutrality. *Intereconomics, 56*(2), 99–107. https://doi.org/10.1007/s10272-021-0963-z

Yang, D., Wang, W., Gueymard, C. A., Hong, T., Kleissl, J., Huang, J., Perez, M. J., Perez, R., Bright, J. M., Xia, X., van der Meer, D., & Peters, I. M. (2022). A review of solar forecasting, its dependence on atmospheric sciences and implications for grid integration: Towards carbon neutrality. *Renewable and Sustainable Energy Reviews, 161*, 112348. https://doi.org/10.1016/j.rser.2022.112348

Zhang, J., Sewell, C. D., Huang, H., & Lin, Z. (2021). Closing the anthropogenic chemical carbon cycle toward a sustainable future via CO2 valorization. *Advanced Energy Materials, 11*(47), 2102767. https://doi.org/10.1002/aenm.202102767

Zhang, Z., Hu, G., Mu, X., & Kong, L. (2022). From low carbon to carbon neutrality: A bibliometric analysis of the status, evolution and development trend. *Journal of Environmental Management, 322*, 116087. https://doi.org/10.1016/j.jenvman.2022.116087

Zhao, X., Ma, X., Chen, B., Shang, Y., & Song, M. (2022). Challenges toward carbon neutrality in China: Strategies and countermeasures. *Resources, Conservation and Recycling, 176*, 105959. https://doi.org/10.1016/j.resconrec.2021.105959

Zhu, Q., Chua, M. H., Ong, P. J., Cheng Lee, J. J., Le Osmund Chin, K., Wang, S., Kai, D., Ji, R., Kong, J., Dong, Z., Xu, J., & Loh, X. J. (2022). Recent advances in nanotechnology-based functional coatings for the built environment. *Materials Today Advances, 15*, 100270. https://doi.org/10.1016/j.mtadv.2022.100270

Chapter 17
Sustainable Growth Strategies: Integrating Climate Neutrality into BRICS National Development Plans

Srikanth Potharla, Hiranya Dissanayake, and Otilia Manta

Abstract This chapter discusses how conventional national development strategies and climate neutrality action plans can be incorporated in BRICS countries: Brazil, Russia, India, China, and South Africa. Considering the fact that all the countries in BRICS make a major contribution to the emissions, the concept of achieving climate neutrality is relevant for sustainable development and economic growth. With reference to the topic, the chapter underscores the need for policy comprehensiveness, encouragement of technology advancement and sound governance systems. They analyze the importance of institutional credibility and the challenges of internal constraints addressed through international collaboration. In addition, the chapter outlines main barriers to climate policies, including political volatility and regulatory risks. This paper has made an argument that through use of green technologies as well as through dealing with inequal income the BRICS countries have option to solve two problems: economic growth and environmental impacts. Last section provides recommendations for policymakers to enhance institutionalization of renewable energy infrastructure and recognize the long-term development advantages of decarbonization and structure green growth for sustainable development.

S. Potharla
ICFAI Business School (IBS), IFHE, Hyderabad, Telangana, India

H. Dissanayake (✉)
Department of Accountancy, Faculty of Business Studies and Finance, Wayamba University of Sri Lanka, Kuliyapitiya, Sri Lanka
e-mail: hiranya@wyb.ac.lk

O. Manta
Romanian Academy, Victor Slavescu Centre for Financial and Monetary Research, Bucharest, Romania

Romanian American University, Bucharest, Romania

O. Manta
e-mail: otilia.manta@rgic.ro

R. Singh and D. Crowther (eds.), *Climate Neutrality and Sustainable Ecosystems*, Approaches to Global Sustainability, Markets, and Governance,
https://doi.org/10.1007/978-981-96-7181-6_17

Keywords Climate neutrality · National plans · BRICS nations · Green technologies · Sustainable development

17.1 Introduction

The impacts of climate change, including rising temperatures and shifting weather patterns, are exacerbating droughts and accelerating desertification, which in turn contribute to widespread food shortages and growing water scarcity (Nwokolo et al., 2023c). Achieving climate neutrality is critical for BRICS nations, given their significant contribution to global emissions and rapid economic growth. As emerging economies, BRICS countries (Brazil, Russia, India, China, and South Africa) face the challenge of integrating climate neutrality goals into national plans without compromising development. This integration is essential not only for reducing emissions but also for ensuring sustainable economic growth.

With heavy reliance on fossil fuels, BRICS nations must adopt a multipronged policy approach that diversifies economies and fosters green technologies to reduce dependence on unsustainable practices (Gillani, 2022). Technological innovation, particularly in renewable energy, can help balance growth with sustainability, though its impact varies across countries (Zhang, 2021). The Environmental Kuznets Curve (EKC) hypothesis provides insight into how economic growth initially worsens environmental degradation but improves as wealth increases (Ahmad et al., 2020).

Institutional integrity and governance are also critical. Strong governance frameworks help reduce corruption and enforce climate policies effectively (Gillani, 2022). International cooperation through technology transfer and climate financing is vital for BRICS nations to meet their climate goals (Halsnæs & Shukla, 2008).

Integrating climate neutrality into national plans allows BRICS countries to achieve environmental sustainability and long-term economic resilience.

Brazil's climate action performance, highlighting its "Insufficient" rating in achieving climate neutrality (Climate Action Tracker, 2024). Historically, Brazil's emissions peaked around 2005, exceeding 2000 $MtCO_2e$/year, largely driven by deforestation and land-use changes. Post-2005, emissions significantly declined due to efforts in reducing deforestation, dropping below 1000 $MtCO_2e$/year by 2010. However, emissions remained steady between 2010 and 2020, with no significant improvements.

Brazil's current climate policies align with an "Insufficient" rating, consistent with a global warming trajectory by 3 °C. The country's Nationally Determined Contribution (NDC) target for 2030 is rated as "Almost Sufficient," aligning with a sub-2 °C world, but it still falls short of the 1.5 °C Paris Agreement goal. The image also emphasizes the critical role of the land-use and forestry sector, which has historically driven Brazil's emissions.

To meet the 1.5 °C pathway, Brazil needs to significantly strengthen its policies and reduce emissions, particularly in the land-use sector. The gap between current actions and the 1.5 °C compatible pathway highlights the need for more robust climate action to meet global temperature goals.

The overall climate action rating for the Russian Federation as Critically Insufficient. While Russia's total emissions were above 1990 levels before the breakup of the Soviet Union, they have been declining since. After early 2000s CO_2 emissions have stabilize at around 2000 $MtCO_2$ equivalent per annum globally.

There is a section designed to reflect on Russia's current policies and actions, and all have been labeled under a classification of "Highly Insufficient" signifying that today's Russia supports a warming path of more than 4 °C (Climate Action Tracker, 2024). Its 2030 NDC target is also considered Highly Insufficient and signifies that the country's target as of the current status still remains far from reaching the target of the Paris Agreement at 1.5 °C.

Huge emission gap between Russia current and planned policies and the needed level to achieve the 1.5 °C compatible pathway (Climate Action Tracker, 2024). This is because of its characterization of land-use and forestry as a sink rather than a mechanism to mitigate overall emissions path in the country. Based on fairness, Russian's NDC target and its policies are seriously inadequate and, therefore, call for upscaling of climate action. The final evaluation of the analyzed papers is that despite Russia is a large country, its climate policies are not adjusted to the international climate requirements and are significantly weaker.

Currently India's overall climate action is highly insufficient (Climate Action Tracker, 2024). Carbon dioxide emission without LULUCF has steadily been on the rise in India since 1990 and has slightly surpassed 4000 $MtCO_2$/year in the year 2020. India's current policies and actions are most compatible with the pathway that is labeled as "Highly insufficient" corresponding to global warming of more than 4 °C.

India has an unconditional INDC target for 2030, which is the same as having no target since it indicates that country's climate commitments are far from 1.5 °C target required in order to meet global climate goals as rated "Highly insufficient" so does India's conditional INDC targets for 2030 (Climate Action Tracker, 2024). The "Unconditional NDC target" score is as low as "Insufficient" suggesting that it represents a pathway consistent with 3 °C warming.

The land-use and forestry area of India has presented comparatively a more or less constant trend of contribution without much direct effect on total emissions. According to the 1.5 °C compatible pathway, India requires higher level policies as well as enhanced actions to slash emissions. This comparison of the existing policies as well compare it to NDC targets and the required emission reduction further accentuates the need for India to step up their climate action and shift to cleaner resources for generating power.

China's climate action rated as "Highly Insufficient" (Climate Action Tracker, 2024). China's emissions, excluding LULUCF, have risen significantly since 1990, reaching over 14,000 $MtCO_2e$/year by 2020. This sharp rise reflects China's rapid industrial growth. Despite recent efforts to curb emissions, China's current policies and actions are aligned with a trajectory consistent with a warming of over 4 °C.

China's Nationally Determined Contribution (NDC) target for 2030 is also rated "Highly Insufficient," which implies that the country's current commitments are not enough to limit global warming to below 4 °C. The gap between China's policies and the 1.5 °C pathway is significant, with the graph showing that China's emissions remain well above the level needed to be compatible with a 1.5 °C world.

The land-use and forestry sector has had a minimal impact on China's overall emissions, remaining steady throughout the period. To meet the 1.5 °C target, China must implement more aggressive emission reduction measures, especially in the industrial and energy sectors. The image highlights the need for stronger action to bring China's emissions aligned with global climate goals.

South Africa's climate action as "Insufficient," based on modeled domestic pathways and fair share analysis (Climate Action Tracker, 2024). Historically, South Africa's emissions have risen steadily, peaking around 2008 at nearly 600 $MtCO_2e$/year. Since then, emissions have slightly decreased but remain far above the required levels to meet climate goals. Current policies and actions align with a trajectory that leads to a warming of less than 3 °C, which is rated as "Insufficient."

The Nationally Determined Contribution (NDC) target for 2030 is slightly better, rated as "Almost Sufficient," meaning it aligns with a warming of less than 2 °C. However, this still falls short of the 1.5 °C goal. The graph also shows South Africa's projected emissions under planned policies remain higher than the 1.5 °C compatible pathway.

The land-use and forestry sector has had a minimal impact on total emissions, remaining stable throughout. The image highlights that South Africa needs stronger actions and more aggressive emission reduction measures to align with the 1.5 °C target. The current gap between policies, planned actions, and the 1.5 °C pathway indicates that additional efforts are essential to achieve climate neutrality.

17.2 The Need for Integrating Climate Neutrality into National Plans

Due to the escalating effects of climate change, efforts should be made, and in the strongest terms possible, principally from countries that emit the most. Brazil, Russia, India, China, and South Africa, collectively called BRICS, are home to nearly half the global population and contribute roughly a third of the world's GDP. A clear observation is that as the size of industrial sectors increases, their contribution to greenhouse gas emissions also increases. This means that climate neutrality action plans must be incorporated into the national development frameworks of these countries to which they already belong (Joshi et al., 2024).

In the first place, what can be called climate neutrality is critical to sustaining economically stable conditions in BRICS countries. Various climatic threats like disasters and resource scarcity have become prevalent and discriminately impact emerging markets. Thus, the declaration of climate neutrality for the BRICS countries can help minimize the economic losses caused by environmental deterioration of the social-economic infrastructure. Second, aligning climate policies with fiscal and monetary strategies ensures that important sectors such as agriculture, energy, and manufacturing are more resilient on behalf of the BRICS economies (Gillani, 2022).

Second, BRICS countries can set political courses that impact their citizens and the global community. They are now new economies with great emission responsibilities, which makes them timely to show positive change in the course. When these countries succeed in integrating climate neutrality into the development framework, they have the potential to shape global markets and climate regimes, thereby expediting change throughout the entire world (Halsnæs & Shukla, 2008). This transition will be smoother and more practicable when there is international cooperation through technology transfers and funding.

In addition, just as the economic rationale for integration is still based on tradition, the new breed of global competitors will be those that will embark on models of sustainable economic growth. This means that those countries that invest early in green technologies and renewable energy will stand to gain more economic value in the long run. Thus, promoting climate neutrality in their national strategies will enable BRICS countries to attract foreign investments to sustainable industries and participate in new global markets in renewable energy and green technologies (Zhang, 2021). Such a transition adds value to investments to help fuel economic growth while lessening the reliance on uncertain and fluctuating fossil fuel markets.

The other factor is the Environmental Kuznets Curve hypothesis, arguing that pollution will rise with economic development but decrease once a country reaches a level of development to invest in advanced technology (Ahmad et al., 2020). This thinking framework asserts that BRICS nations, which are presently very much into industrialization, can reduce pollution easily as a country progresses economically. However, this transition is possible only through integration at the policy level. Lack of climate neutrality in their development plans keeps BRICS countries in the vicious cycle of continued economic growth, which might never allow them to shift to the positively sustainable economy of the Ecological Kuznets Curve track.

Finally, institutional integrity and governance reforms are at the center. For climate neutrality goals to be achieved, environmental laws must be backed by strong institutions with minimal corruption and complacency (Gillani, 2022). Although there can be effective efforts towards promoting sustainable practices, stronger governance can hamper the process by giving space for practices contrary to national or international climate commitments. Thus, strengthening the institutions is needed so that climate action plans can be conceptualized and efficiently operationalized.

Integrating climate neutrality into the national plans of BRICS countries is imperative for several reasons: preserving economic sustainability in the long term, increasing companies' cross-border competitiveness, combating climate change consistent with global objectives, and guaranteeing the compatibility of institutions with sustainable development. This integration is of paramount importance for the now BRICS countries and the world facing climate change challenges and the transition to sustainable development.

17.3 Government Efforts on the Path Towards Net Zero

Government policies in the BRICS countries are pivotal in achieving climate neutrality while progressing the agreement. Among the largest world emitters of greenhouse gases, these countries have incorporated different measures to decrease emissions while facilitating development. Among the major objectives of such activities is forming the legal basis for formulating and achieving goals for reducing carbon emissions. China has set carbon neutrality by 2060, and India has committed to net-zero emissions by 2070. These commitments constitute the core of national transformation pathways from fossil energy to renewable energy and enhanced energy efficiency (Chandler et al., 2002). Be mindful that policies of that nature respond to the need to coordinate national activities with the global climate agenda. Additional measures also combine the intentions of the government to advance economic growth with climate neutrality through sponsoring green technologies. Renewable energy investments such as solar and wind power in many countries such as China and India have boosted economic growth and reduced dependency on fossil fuels. This shows that clean energy can drive economic growth without elevating carbon emissions (Zhang, 2021). In addition, the government has supported research and development to make it possible to advance technology in energy storage and smart grids that help create economic value and solutions to environmental problems (Su et al., 2021).

According to the Environmental Kuznets Curve (EKC) hypothesis, emerging economies of BRICS can reverse the degradation of the environment through sinking horns in cleaner technologies as the economies advance (Ahmad et al., 2020). This is evident in the framework of Brazil and South Africa, where strict environmental measures are used to check emission practices while enhancing industrial output. However, the effectiveness of such measures is always associated with institutional credibility and management. Corruption is a likely reason for some climate policies not working as intended, and strong implementation of environmental laws is necessary for climate policies to work appropriately (Gillani, 2022).

The countries in the BRIC group also require international cooperation as technology and climate funds support the implementation of advanced technologies by the governments and their compliance with climate targets (Halsnæs & Shukla, 2008). This paper underscores that government-led endeavors in the BRICS nations will shape climate neutrality in the future by encouraging the adoption of clean technologies, endorsing innovation, and reconciling economic development with sustainability. Thus, sustainable and sound government and international cooperation will be essential if these initiatives are to yield the intended outcomes.

17.4 Challenges in Integration

Political risks, institutional aspects, and governance challenges challenge sustainable development in BRICS countries. Energy production, economic development, and increased standard of living create tensions with emission decrease objectives. Applying conventional fossil fuel-based growth models poses challenges to shift to sustainable practices without compromising the pace of growth (Zhang, 2021). It remains imperative; it is seen that countries with comparatively low levels of political risks, such as China, can better steer climate change policies. Since the government frequently changes, South Africa cannot implement sound environmental policies (Gillani, 2022). Several studies indicate that the level of institutional capacity directly impacts the implementation of sustainability activities. Robust approaches are necessary, backed by international and national legislation that possibly covers industrial emissions. However, the institutional effectiveness is not the same in all the countries that are part of the BRICS community. Corporation maladministration and embezzlement usually act against climate policies, allowing negative environmental impacts to continue (Ahmad et al., 2020). Unclean operations corrupt public sectors and enable industries to use unsustainable resources, thus increasing their impact on the environment (Singh et al., 2025). For instance, Brazil and South Africa suffer from high corruption index, which results in poor climate measures.

It is, therefore, clear that political instability also constrains the achievability of climate change interventions. A stable environment allows for planning in the long run and effective implementation of regulations. China has been an ideal example. At the same time, uncertainty in countries such as Brazil hinders FDI in green technologies and creates an inconsistent climate action plan. Erratic political policies caused by volatile political administration deter needed advancements in using renewable energy and other sustainable development projects. They have to be eradicated through sound governance mechanisms and institutional change. Establishing autonomous institutions suited to participate in formulating and supervising climate change policies is essential. Enhancing institutional accountability and transparency will also reduce corruption and promote sustainable practices. In sum, the systemic obstacles identified in this paper, alongside the political and institutional factors expected to underpin climate neutrality in BRICS nations, underline the need to support coherent and efficient climate action plans.

17.5 Integration Management: The Action Plan

Advancing sound legal frameworks and strengthening institutional integrity are critical for improving environmental governance in BRICS nations. Effective legal structures, such as New Zealand's Resource Management Act (RMA), demonstrate how decentralizing ecological management allows local governments to enforce policies tailored to regional needs. Adopting similar frameworks can help BRICS countries align climate neutrality targets with national development priorities. Corruption remains a significant barrier to institutional effectiveness, undermining climate policies and sustainable resource management. Anti-corruption measures and legal transparency are essential to enhancing compliance and combating environmentally degrading practices (Singh et al., 2025). For instance, laws promoting renewable energy, reducing emissions, and managing natural resources sustainably can serve as a foundation for integrating climate neutrality into national plans.

To coordinate climate neutrality efforts, BRICS nations need robust governance structures. Establishing autonomous agencies to oversee environmental compliance and tie national efforts to international climate policies fosters accountability and facilitates technology transfer (Halsnæs & Shukla, 2008). However, institutional weaknesses, particularly in countries like Brazil and South Africa, hinder policy enforcement and exacerbate environmental degradation. The political independence of institutions responsible for enforcing climate policies is crucial to ensure their effectiveness. Building institutional capacity requires targeted strategies, including enhancing ecological regulators, implementing IT solutions for environmental monitoring, and making environmental information accessible to the public (Gillani, 2022). Such measures improve governance and build public trust, which is vital for the long-term success of climate policies.

Integrating effective policies and addressing institutional gaps are key to achieving climate neutrality in BRICS nations. These countries can establish a transparent and accountable system to manage environmental sustainability and align with global climate goals by focusing on anti-corruption measures, judicial transparency, and governance reforms.

17.6 Role of Energy Policies and Management

Proper legal mechanisms and resilience of institutions are potentially significant for enhancing BRICS environmental democracy, 2017. New Zealand's Resource Management Act (RMA) is a good example of how the devolution of ecological management empowers local governments to enforce policies relevant to their regions (Sharp, 2002). It is suggested that similar frameworks can also be appropriate for BRICS countries to address a link between climate neutrality and development goals. From the previous heading, corruption impedes institutional efficiency, achieving climate goals, and sustainable resource governance. On the same note, it was argued that measures against corruption and legalization are indispensable to improving compliance and suppressing environmentally detrimental actions (Singh et al., 2025). For instance, policies supporting renewable energy generation, combating pollution, and rationally using and conserving natural resources on which climate neutrality can be built already exist on paper.

Regarding climate neutrality, the BRICS nations require proper governance systems. Creating independent agencies to coordinate environmental performance and integrate national activities into international climate change contributes to greater accountability and technology diffusion (Halsnæs & Shukla, 2008). However, institutional constraints, especially in developing member countries such as Brazil and South Africa, limit policy implementation and worsen environmental degradation (Carlsson & Lazdinis, 2004). Therefore, political independence must be in institutions assigned to implement climate policies. Environmental management calls for developing strategies to improve ecological regulators' institutional capacity, deploying information technology solutions for environmental control and data sharing, and ensuring fulsome public access to essential environmental records (Gillani, 2022). They enhance good governance and engender public support, a crucial element missing in climate change policies.

Ultimately, the question of brass-tack-type climate policies and institutional change becomes fundamental to reaching climate neutrality for BRICS countries. These countries can build a sharp and responsible system for ecological stability and compliance with worldwide climate initiatives by paying vital focus on anti-corruption actions, a clear judiciary, and control reforms.

17.7 Technology Management and Environmental Technology

The study uncovers the value of technology management in realizing climate neutrality and economic growth for BRICS nations. The nations that embrace such policies can simultaneously lower emissions through investment in renewable energy, energy efficiency, and low-emissions technologies, hence supporting sustainable development (Zhang, 2021). The availability of energy storage systems and electric vehicles and fair use of green energy technologies like solar and wind power make the BRICS growing economy less detrimental to the environment and have more negative impacts. China and India have emerged as the world's largest markets in the clean energy sector, primarily given the huge investments in systems such as solar, wind, and hydropower systems.

Consequently, these BRICS nations will have to incorporate these pressing innovations into their existing economic structures to unleash the full potential of green technologies. It can be subsidies, tax credits, R & D support, and policies encouraging renewables and environmentally friendly products (Gillani, 2022). For instance, the Porter Hypothesis claims that environmental regulations can increase sales growth and improve technological advancement in industries aiming to adopt cleaner technologies, thereby increasing the competitiveness of industries at the global level. Policies put in place by governing bodies, such as smart energy systems and EV charging infrastructure, facilitate low-carbon advancement and security of energy supply (Su et al., 2021).

Other funding strategies, such as green bonds and international climate funds, also facilitate funding for sustainable technologies. BRICS countries can offer domestic policies that are in congruence with global climate change targets to source for green projects' funding and engage the world economy that is opening up for green economy investment (Halsnæs & Shukla, 2008). Moreover, the PPPs can add value to the research and development of literature, which makes the competitive economies of BRICS in green technology markets.

As is clear from this discussion, there is no reason why adopting green technologies cannot be done in a manner that does not bring impoverishment in the BRICS nations but leads to carved-out low-carbon economies. In technological integration into policies, fiscal, and infrastructure framework, these countries are poised to realize sustainable economic growth and stem climate change effects.

17.8 Governance Structures and Political Stability

17.8.1 Enhancing Governance for Integration

Enhanced governance is important if BRICS countries are to achieve sustainable economic growth and development without harming the environment. Management also points to environmental policies that must be formulated, set, and complied with (Singh et al., 2025). Specifically, great institutional quality, integrity in the rule of law, a clear and transparent set of rules and regulations, and powerful accountability systems help the BRICS countries go together with development and climate strategically. In this sense, institutions can be empowered to closely assess the effects on the environment, authorize or prohibit certain activities in industries and companies, and achieve climate neutrality (Gillani, 2022).

Another necessary step for raising the level of governance is the formation of autonomous non-executive committees emphasizing the environment. Such a body must be non-political in implementing set rules and regulations and ensuring compliance with the legal frameworks between the public and private entities. For instance, China has provided institutions coordinating environmental laws with the economic systems to enforce sustainability regulations to industries without limiting the economy (Su et al., 2021). On the other hand, if a country has weak institutions, it cannot implement environmental laws; hence, too much environmental exploitation occurs.

Political stability has the most influence on the enforcement of environmental laws. Since ecological planning and policy deployment take time to achieve the intended goals, it is easier in countries with stable political settings, such as China. Stability allows the government to work on the sustainable goal and guarantee an equal reaction to the environmental regulation transgressions. On the other hand, political instability, as evidenced by Brazil and, to some extent, South Africa, means that environmental policies are challenging to implement since the change in government means a policy change. Such inconsistency results in flip-flop policies, declining institutional capacity, and investor confidence, which hinders climate neutrality.

17.8.2 Addressing Regulatory Uncertainty

Fluctuations in regulation, which stem from political volatility, affect FDI in environmental technologies across the BRICS nations. Due to these reasons, investors don't risk investing in green technologies in countries with unstable environmental policies or often change due to fluctuating political situations. Ambiguity in the regulatory system destabilizes the operating conditions in the business environment, discouraging long-term investments in developing environmentally friendly and sustainable systems.

For example, political instability has repeatedly brought drastic changes to environmental policies in countries such as South Africa and Brazil, forcing investors to wonder about the stability of investing in green technologies. Such uncertainty distorts the prospects for raising the capital needed to roll out scaled-up environment initiatives like renewable energy generation plants and smart grid infrastructure in BRICS nations. To counter this, the members of the BRICS group must now focus on policy stability, which means no sudden changes due to unfair political sentiment.

However, such an absence of legal regulation may slow the implementation of environmentally friendly innovations. For instance, firms might be reluctant to invest in fresh technologies or commit to better environmental standards if they expect future governments to undo them. Thus, Policy certainty is crucial to encouraging mass dynamization mechanisms in which event businesses are assured that deploying clean technologies will generate long-term returns (Gillani, 2022).

The other solution to decreasing regulatory risk is enhancing the institutional environments. This, in turn, should make BRICS' climate policies more stable, despite political changes, if only state institutions are more robust. This consists of developing legal requirements promoting the deployment of clean technologies and clear and stable market rules that companies must follow.

The stability of political leadership and responsive and accountable government institutions are important prerequisites for BRICS countries to adopt sound environmental policies within growth tracks. FDI will be solved by seeking to reduce regulatory uncertainty in the environmental field, creating better institutions and clear policies useful in ecological goals.

17.9 Policy Approaches and Economic Models

17.9.1 Multipronged Policy Approaches

Overall, the democratic, authoritative BRICS nations have a great problem in the case of sustainable development with the prevailing long-established economic structures that prefer short-term gains. This means that to surmount these challenges, a comprehensive policy solution is possible, simultaneously solving environmental, social, and financial issues. These policies include policy measures, economic instruments and technological measures, international cooperation measures, and a lack of measures to support the strategy of economic development soon with the environmental protection objectives in the long run (Nwokolo et al., 2023a).

F7r instance, improving environmental standards that encourage emission-cutting measures and compensating industries for embracing cleaner techniques to assist in the economy's growth rather than sharpening ecological degradation. As one of the strongest suppliers of renewable energy, China has provided an example of using a set of measures where state legal regulation has been complemented by significant investments in solar and wind technologies that contribute to developing the green economy (Su et al., 2021). They keep industries in the process of changing towards a more sustainable model as they continue to expand.

Besides, the combinative technology approach involves public-private partnerships to enable technological development. Political rulers can support research and development in environmentally friendly technologies with private companies to equip different enterprises with tools that would allow for the latter to decrease their direct negative impact on the environment while staying financially profitable. Such cooperation can also bring technology swaps and best practices, thus helping the BRICS countries to progress faster in transforming their economies into cleaner ones (Halsnæs & Shukla, 2008).

The second important factor in complex intervention is embedding international climate agreements in national policy. Considering their binding national climate policies, which outline commitments to international climate regulation, climate support can obtain global climate financing and technical assistance to solve domestic climate problems (Gillani, 2022). The rise of global collaboration is important as we seek to address the more insurmountable challenges that stem from the current strains of economic models that earned high profits at the cost of a sustainable environment.

17.9.2 Addressing Income Inequality

The enemy of income disparity must be fought to realize the broad environmental objectives espoused by BRICS countries. Social injustice is bound to prompt environmental exploitation since individuals with low standards of living cut down trees and overuse the resources to live. Income disparities are firmly set in BRICS countries, where some of the population has very low income, such as Brazil and South Africa. It also indicates that these nations could lessen the pressure on natural resources, consider inequity, and advance the utilization of more viable environmental frameworks (Gillani, 2022).

Income inequality also affects the effectiveness of environmental governance. When wealth is concentrated among a small elite, it can lead to unequal access to resources and opportunities for sustainable development. Those of lower income are somehow locked out of any decision-making process regarding the environment, which leaves poor policies in the hands of the classy few. This is important because environmental policies are more likely to be fair, and certain societal sectors will experience the benefits of sustainability measures.

Also, tackling income distribution can enable countries to establish more sustainable economies that are ready for a change to the green economy. Education, Health Care, and Jobs are areas that governments can spend money on since they ensure people are provided for, and they leave environmentally unfriendly jobs for other better practices. Such a transition is useful in fighting poverty, copying jobs in renewable energy, recycling, efficient waste disposal, and organic farming Jadhav and Katti (2012).

To achieve a sustainable governance agenda, income inequality in BRICS countries must be managed. Understanding these nations' social and environmental goals and the current growing economic gaps that initiate unsustainability, with new appropriate policy anomalies, the BRICS nations can help create a more sustainable world and a more equitable BRIC society.

17.10 Conclusion

17.10.1 Key Insights

Emerging economies within the BRICS grouping (Brazil, Russia, India, China, and South Africa) find themselves at a crossroads as mainstreaming climate neutrality and associated interventions into conventional national development plans crumbles into economic viability on one hand and ecological sustainability on the other. These problems look like the specific obstacles peculiar to each country: the post-ministerial domination of fossil fuels and the different levels of institutional development. However, all the strategies aim at the firm's sustainable growth while avoiding rising carbon emissions. The authors believe that the long-term policy package using measures that promote green technologies, enhance institutional integrity, and dovetail economic growth policies with sustainable development policies has been crucial for achieving economic growth in the short term while addressing sustainable development goals (Gillani, 2022). Besides, internal barriers and the need to align national policies with international climate change goals can be overcome with adequate international cooperation and innovation in given policies.

When formulated into climate neutrality, the following strategies offer BRICS nations the chance to adopt robust economic growth without significantly impacting the environment (Kumar et al., 2024). Efficient use of clean energy through the development of renewable energy has been observed to foster economic growth in countries such as China & India, where policies towards technological innovation and sure-shot regulatory frameworks have already been implemented, which has exemplified the notion of leading economic growth & reducing emissions hand in hand.

17.10.2 Recommendations

To further integrate climate neutrality into national plans, BRICS policymakers should:

1. Strengthen institutional frameworks: It is important for the institutions implementing environmental policies to be strong, independent, and free from corruption. It is recommended that governments should develop the capacity, avail training, and establish anti-corruption mechanisms (Gillani, 2022).
2. Promote public-private partnerships: Therefore, governments should promote partnerships between governmental authorities and companies to enhance the speed of green technology implementation. It is useful to point out that these partnerships may support the development and deployment of renewable energy technologies (Nwokolo et al., 2023d; Zhang, 2021).
3. Foster international cooperation: If BRICS countries link their national climate targets to international treaties, they can access climate funding and technology sharing. This can assist in closing the divide between domestic developmental demands and international environmental obligations (Halsnæs & Shukla, 2008; Nwokolo et al., 2023b).
4. Address income inequality: hence, it can be understood that the reduction of income inequality could mean that pressure on natural resources could be reduced, as well as the inclusiveness of environmental policies. Governments should develop green jobs, education, and healthcare for the poor.
5. Invest in renewable energy infrastructure: Research, development, and extension support to renewable energy technologies, including wind, solar, and hydropower, are crucial. The governments should encourage industries to move from conventional power sources, mainly fossil energy sources (Su et al., 2021).

17.10.3 Future Outlook

There is great potential for sustainable development in the advanced integration of climate neutrality action plans into the national development strategies of the BRICS countries. Since these countries target the reduction of carbon emissions and promote green growth, they will be part of leading economies in the shift to a low-carbon economy. The benefits are twofold: efficiency in economic and ecological predictability. An inclusive green economy conserves natural wealth, creates green occupations, enhances energy security, and minimizes the dangers of climate change and the probability of devastating events (Gillani, 2022).

In the long run, successful integration will also enhance the global reputation of BRICS countries as they are ready to shoulder international leadership in climate negotiations. In addition, when these countries change into sustainable ones, they will act as examples for other developing countries to know that sustainable development for developing economies can and should be achieved without deteriorating the earth, its environment, and its natural resources.

In conclusion, it must be mentioned that, in principle, BRICS nations have the tools and opportunities to incorporate climate neutrality into conventional strategies. These countries should thus encourage the adoption of a strategic approach combining different means within a sustainable and inclusive development framework, hence facilitating the creation of positive equities for their people and the global community.

References

Ahmad, M., Jiang, P., Majeed, A., Umar, M., Khan, Z., & Muhammad, S. (2020). The dynamic impact of natural resources, technological innovations and economic growth on ecological footprint: An advanced panel data estimation. *Resources Policy, 69*, 101817. https://doi.org/10.1016/J.RESOURPOL.2020.101817

Chandler, W., Secrest, T. J., Logan, J., Schaeffer, R., Szklo, A. S., Schuler, M. E., Dadi, Z., Kejun, Z., Yuezhong, Z., Huaqing, X., Shukla, P. R., Tudela, F., Davidson, O., Mwakasonda, S., Spalding-Fecher, R., Winkler, H., Mukheibir, P., & Alpan-Atamer, S. (2002). *Climate change mitigation in developing countries, Brazil, China, India, Mexico, South Africa, and Turkey*. Pew Center on Global Climate Change.

Climate Action Tracker. (2024). *Trackling global climate action: Since 2009*. Retrieved from https://climateactiontracker.org/

Gillani, D. (2022). Climate change and political letdown: Understanding environmental degradation through the prisoner's dilemma. *Journal of Political Studies, 29*, 44. Retrieved from https://heinonline.org/HOL/Page?handle=hein.journals/jlo29&id=122&div=&collection=

Halsnæs, K., & Shukla, P. (2008). Sustainable development as a framework for developing country participation in international climate change policies. *Mitigation and Adaptation Strategies for Global Change, 13*(2), 105–130. https://doi.org/10.1007/S11027-006-9079-9/METRICS

Joshi, A., Singh, R., Kumar, V., & Rani, S. (2024). *Net zero strategy: Exploring opportunities, confronting challenges for developing countries* (pp. 233–252). Springer. https://doi.org/10.1007/978-981-97-5756-5_11

Kumar, K., Singh, R., Sharma, A., & Lehri, B. P. S. (2024). *Climate change adaptation strategies for achieving net-zero economy* (pp. 163–183). Springer. https://doi.org/10.1007/978-981-97-5756-5_8

Nwokolo, S. C., Singh, R., Khan, S., Kumar, A., & Luthra, S. (2023a). Africa's awakening to climate action. *CSR, Sustainability, Ethics and Governance, 1850*, 299–310. https://doi.org/10.1007/978-3-031-44514-9_9

Nwokolo, S. C., Singh, R., Khan, S., Kumar, A., & Luthra, S. (2023b). Global investment and development in Africa. *CSR, Sustainability, Ethics and Governance, 1850*, 15–58. https://doi.org/10.1007/978-3-031-44514-9_2

Nwokolo, S. C., Singh, R., Khan, S., Kumar, A., & Luthra, S. (2023c). Impacts of climate change in Africa. *CSR, Sustainability, Ethics and Governance, 1850*, 237–262. https://doi.org/10.1007/978-3-031-44514-9_7

Nwokolo, S. C., Singh, R., Khan, S., Kumar, A., & Luthra, S. (2023d). Remedies to the challenges of renewable energy deployment in Africa. *CSR, Sustainability, Ethics and Governance, 1850*, 59–74. https://doi.org/10.1007/978-3-031-44514-9_3

Singh, R., Gupta, A., & Kandpal, V. (2025). ESG excellence in the digital world. In *Fintech for ESG and the circular economy* (pp. 193–204). Wiley. https://doi.org/10.1002/9781394234035.CH10

Su, C. W., Xie, Y., Shahab, S., Faisal, C. M. N., Hafeez, M., & Qamri, G. M. (2021). Towards achieving sustainable development: Role of technology innovation, technology adoption and CO_2 emission for BRICS. *International Journal of Environmental Research and Public Health, 18*(1), 277. https://doi.org/10.3390/IJERPH18010277

Zhang, H. (2021). Technology innovation, economic growth and carbon emissions in the context of carbon neutrality: Evidence from BRICS. *Sustainability, 13*(20), 11138. https://doi.org/10.3390/SU132011138

The manufacturer's authorised representative in the EU is Springer Nature Customer Service Centre GmbH, Europaplatz 3, 69115 Heidelberg, Germany. If you have any concerns regarding our products, please contact ProductSafety@springernature.com

Printed and bound by CPI Group (UK) Ltd, Croydon, CR0 4YY
01/10/2026
02225915-0008